MICROBIAL ECOLOGY IN SUSTAINABLE AGROECOSYSTEMS

Advances in Agroecology

Series Editor: Clive A. Edwards

Agroecosystems in a Changing Climate, Paul C.D. Newton, R. Andrew Carran, Grant R. Edwards, and Pascal A. Niklaus

Agroecosystem Sustainability: Developing Practical Strategies, Stephen R. Gliessman

Agroforestry in Sustainable Agricultural Systems, Louise E. Buck, James P. Lassoie, and Erick C.M. Fernandes

Biodiversity in Agroecosystems, Wanda Williams Collins and Calvin O. Qualset

The Conversion to Sustainable Agriculture: Principles, Processes, and Practices, Stephen R. Gliessman and Martha Rosemeyer

Global Economic and Environmental Aspects of Biofuels, David Pimentel

Integrated Assessment of Health and Sustainability of Agroecosystems, Thomas Gitau, Margaret W. Gitau, and David Waltner-Toews

Interactions between Agroecosystems and Rural Communities, Cornelia Flora

Land Use Intensification: Effects on Agriculture, Biodiversity, and Ecological Processes, David Lindenmayer, Andrew Young, Saul Cunningham, and Linda Broadhurst

Landscape Ecology in Agroecosystems Management, Lech Ryszkowski

Microbial Ecology in Sustainable Agroecosystems, Tanya E. Cheeke, David C. Coleman, and Diana H. Wall

Multi-Scale Integrated Analysis of Agroecosystems, Mario Giampietro

Soil Ecology in Sustainable Agricultural Systems, Lijbert Brussaard and Ronald Ferrera-Cerrato

Soil Organic Matter in Sustainable Agriculture, Fred Magdoff and Ray R. Weil

Soil Tillage in Agroecosystems, Adel El Titi

Structure and Function in Agroecosystem Design and Management, Masae Shiyomi and Hiroshi Koizumi

Sustainable Agriculture and New Biotechnologies, Noureddine Benkeblia

Sustainable Agroecosystem Management: Integrating Ecology, Economics and Society, Patrick J. Bohlen and Gar House

Tropical Agroecosystems, John H. Vandermeer

MICROBIAL ECOLOGY IN SUSTAINABLE AGROECOSYSTEMS

EDITED BY
TANYA E. CHEEKE
DAVID C. COLEMAN
DIANA H. WALL

CRC Press
Taylor & Francis Group
Boca Raton London New York

CRC Press is an imprint of the
Taylor & Francis Group, an **informa** business

CRC Press
Taylor & Francis Group
6000 Broken Sound Parkway NW, Suite 300
Boca Raton, FL 33487-2742

Printed in the United States of America on acid-free paper
Version Date: 20120525

International Standard Book Number: 978-1-4398-5296-5 (Hardback)

Library of Congress Cataloging-in-Publication Data

Microbial ecology in sustainable agroecosystems / editors, Tanya E. Cheeke, David C. Coleman, and Diana H. Wall.
p. cm. -- (Advances in agroecology)
Includes bibliographical references and index.
ISBN 978-1-4398-5296-5 (alk. paper)
1. Soil microbial ecology. 2. Agricultural ecology. I. Cheeke, Tanya E. II. Coleman, David C., 1938- III. Wall, Diana H. IV. Series: Advances in agroecology.

QR111.M384 2012
631.4'6--dc23 2012002441

Visit the Taylor & Francis Web site at
http://www.taylorandfrancis.com

and the CRC Press Web site at
http://www.crcpress.com

Cover Photo Credits

Center Photo

Intercropping on a small farm on San Juan Island, WA, USA. Photo credit: C. A. Miles, Washington State University

Top Row, Left to Right

1. *Lumbricus terrestris*. Photo credit: Tami S. Ransom, Environmental Studies, Fulton School of Liberal Arts, Salisbury University
2. Oribatid mite SEM. Photo Credit: Sue M. Lindsay, Australian Museum, Sydney, Australia

Center Row, Left to Right

1. Mycophagous amoeba SEM. Photo credit: V. Gupta, CSIRO Adelaide, South Australia
2. A collembolan collected in Durham, New Hampshire, USA. Photo Credit: Amanda B. Daly, Graduate Student, University of New Hampshire, Department of Natural Resources and the Environment
3. *Agrobacterium tumefaciens* SEM. Photo Credit: Jing Xu, Doctoral Student with Clay Fuqua, Department of Biology, Indiana University Bloomington

Bottom Row, Left to Right

1. Arbuscular mycorrhizal fungi in corn roots. Photo credit: Tanya E. Cheeke, Portland State University
2. *Acrobeles complexus* SEM—bacterial feeding nematode. Photo credit: M. Mundo, K. Bozhilov, and J. Baldwin, University of California, Riverside

Contents

Foreword

This is the 18th volume to be published in the series, *Advances in Agroecology*. Books in this series have addressed a broad range of topics related to the role of agroecological inputs into sustainable agroecosystems and their effects on crop yields and productivity. Soil microbial ecology is a critical component in the maintenance of biological inputs into sustainable crop production, in minimization of the use of synthetic inorganic fertilizers and pesticides, and in transforming and recycling organic matter.

The chapters in this book are based on a 2009 Ecological Society of America symposium, "How Can Microbial Ecology Contribute to the Sustainability of Agricultural Systems." The presentations were so good that it was decided to turn them into a book. The lead authors of chapters include a number of well-established soil ecologists who have made major contributions to understanding the functioning of soil systems. These key contributors are supported by inputs from university faculty, postdoctoral researchers, and graduate students who contribute as chapter authors and coauthors.

The topics covered include overviews of the diversity and complexity of microbial activity and interactions in soil systems and the many ways in which microorganisms may be manipulated and managed to improve the functions of crop rhizospheres and thereby maximize crop yields and overall productivity. A broad range of soil microbial processes is reviewed in terms of the importance of microbial heterogeneity and how microorganisms may be manipulated to increase crop productivity and yields.

The emphasis throughout the book is on interdisciplinary and innovative research designed to increase microbiological inputs into crop production. Interactions between soil-inhabiting invertebrates and microorganisms in functional food webs are reviewed to assess how they may be managed to maintain nutrient supplies and overall crop productivity. A major emphasis is on how the availability of nitrogen, and other nutrients, may be increased through the manipulation of organic matter inputs and the encouragement of microbial activity.

The broad range of critical soil microbiological topics that are covered in the book includes inputs by soil microorganisms into wheat-farming systems; the importance of arbuscular mycorrhizal fungi in making nutrients more available to crops; and benefits and environmental problems associated with the use of crops that have been genetically modified with *Bacillus thuringiensis*. The overall theme on the role of microorganismal biodiversity in agricultural soils is brought together in discussions on soil fertility, erosion minimization, disease management, and the maximization of nutrient utilization uptake. These have together been termed "ecosystem services" by some scientists. The book should appeal to a broad range of soil scientists as a key reference source on soil microbiological inputs to agroecosystems.

Clive A. Edwards
Editor in Chief
Advances in Agroecology

Preface

Tanya E. Cheeke, David C. Coleman, and Diana H. Wall

While soil ecologists continue to be on the forefront of research on biodiversity and ecosystem function, there are few interdisciplinary studies that incorporate ecological knowledge into sustainable land management practices. Conventional, high-fossil-fuel, input-based agricultural systems can reduce soil biodiversity, alter soil community structure and nutrient cycling, and lead to greater dependence on energy-intensive practices. These practices can lead to habitat destruction, greenhouse gas production, and damage to soil ecosystems. This book brings together soil ecologists, microbial ecologists, and agroecologists working globally to demonstrate how research in soil ecology can contribute to the long-term sustainability of agricultural systems.

Many of the benefits of sustainable agricultural systems—stable yields, minimized external inputs, and reduced fossil fuel usage—rely on soil ecological theory and are key to sustaining soils worldwide. Current challenges in sustainable agricultural research could help to drive new questions in the blossoming field of soil microbial ecology. To improve linkages among soil ecologists, microbial ecologists, and agroecologists, for this volume we invited scientists whose research overlaps the subdisciplines of soil ecology, microbial ecology, and agroecology. Several of the chapters focus on incorporating soil ecological or microbial ecological theory into agricultural practice to improve agricultural productivity and sustainability. Other chapters consider challenges in sustainable agricultural research and the need for coalescing new avenues of research in agriculture and soil ecology. A concluding chapter addresses how to sustain soil diversity and ecosystem services in agriculture.

These chapters were prepared by many of the contributing panelists from a symposium at the 2009 Ecological Society of America annual meeting, "How Can Soil Microbial Ecology Contribute to the Sustainability of Agricultural Systems?" that was held in Albuquerque, New Mexico. The authors range from long-time ecological researchers to graduate students and early-career scientists, representing a wide spectrum of experience, ages, diversity, and research interests in this area. The authors have identified five key areas of research that can be combined to support and direct sustainable land management practices: agriculture, biodiversity, ecosystem services, integrated soil ecology research, and policy. These recommendations can be used to direct and influence agricultural and environmental policy and guide future research in sustainable agricultural systems management.

We hope this book will generate synergy among soil ecologists, microbial ecologists, and agroecologists and inspire new and innovative lines of research.

Acknowledgments

We thank the many friends, coauthors, and colleagues who made the symposium and book possible. For sponsorship of the symposium at the Ecological Society of America (ESA) meetings in Albuquerque, New Mexico, in August 2009, we are grateful to the following: W. J. Landesman (symposium coorganizer); ESA Soil Ecology Section (Nancy Johnson, president); ESA Microbial Ecology Section (Brendan Bohannan); and ESA Agroecology Section (Dan Pavuk, secretary).

Each chapter was reviewed by two or three external reviewers. Their comments and suggestions were helpful to the authors of the chapters and thus were vital to the success of this volume. We express our deep appreciation to Dr. Suellen Melzer-Drinnen at Colorado State University for her help with the final preparation for the book.

Research and the ideas developed by several of the authors were supported, in part, by research grants from the U.S. Department of Agriculture, the U.S. Environmental Protection Agency Science to Achieve Results Fellowship program, and the National Science Foundation. In the interest of saving space, we are omitting mention of individual grant numbers.

Last and not least, we appreciate the support of Dr. Clive Edwards and CRC Press for providing encouragement for this book project.

About the Editors

Tanya E. Cheeke is a PhD candidate and an Environmental Protection Agency Science to Achieve Results (EPA STAR) Fellow in the Department of Biology at Portland State University (PSU) and a visiting scholar at Indiana University. Her dissertation research examines symbiotic plant-fungal relationships in transgenic *Bt* maize and assesses the effects of the cultivation of *Bt* crops on the abundance and diversity of arbuscular mycorrhizal fungi in the soil ecosystem. Tanya received her bachelor of science in sustainable agriculture/environmental science from the Evergreen State College. At PSU, she has served as the graduate student representative for the Promotion and Tenure Committee and the Faculty Hiring Committee and was elected by her colleagues to serve as secretary/treasurer (2007–2008), president (2008–2010), and faculty liaison (2010–2011) of the PSU Biology Graduate Student Association. In 2009, she organized an Ecological Society of America symposium "How Can Soil Microbial Ecology Contribute to the Sustainability of Agricultural Systems?" and was an organizer for the 2010 Evolution annual meeting in Portland, Oregon. In 2011, she served on the Student Advisory Panel for the EPA STAR Conference and was an organizer and moderator for the Recruitment and Retention of Underrepresented Students session in Washington, DC. In 2009, she was awarded the PSU President's Award for Outstanding University Service for her mentoring and leadership services and was awarded Best Student Oral Presentation at the 2009 ESA annual meeting in Albuquerque, New Mexico. She has given numerous talks at scientific conferences and has received multiple grants and awards for her research and outreach activities. After earning her PhD, she hopes to become a university professor in the biological sciences. Her research interests include soil ecology, agroecology, climate change, and biotech risk assessment.

David C. Coleman has been a lifelong soil ecologist with interests in soil biodiversity and biogeochemical cycling. He conducted research at the Savannah River Ecology Laboratory of the University of Georgia (1965–1971) and the Natural Resource Ecology Laboratory, Colorado State University (1972–1985), and from 1985 has been at the University of Georgia in the Institute of Ecology and later the Odum School of Ecology. He served as distinguished research professor of ecology from 1985 onward and has been a professor emeritus since 2005.

David received the Distinguished Service award from the Soil Ecology Society in 1999 and the Distinguished Ecosystem Scientist award from the Natural Resource Ecology Laboratory in 2000. He served as co-chief editor of *Soil Biology and Biochemistry* from 1998 to 2010 and serves on the editorial boards of several other soil biology journals. He has published over 285 refereed journal articles and books and is senior author of *Fundamentals of Soil Ecology* (second edition, 2004) and the author of *Big Ecology: The Emergence of Ecosystem Science* (2010).

He received his master's and PhD degrees from the University of Oregon and spent a postdoctoral year with Professor Amyan Macfadyen at the University of Wales, Swansea, United Kingdom.

A soil ecologist and environmental scientist, **Diana H. Wall** is university distinguished professor and director of the School of Global Environmental Sustainability at Colorado State University. Diana is actively engaged in research exploring how nematode and other invertebrate diversity contribute to healthy, productive soils and thus to society and the consequences of human activities on soil globally. Her more than 20 years of research in the Antarctic Dry Valleys follows the response of soil organisms and ecosystem processes to environmental change: In Africa, she examines biodiversity in fertile and degraded soils. For her research contributions, Wall Valley, Antarctica, was designated (2005), and a new soil mite species was named (2009). She was designated as the 2010 Tansley Lecturer by the British Ecological Society. Diana holds an honorary doctorate from Utrecht University and is a fellow of the Aldo Leopold Leadership Program, the American Association for the Advancement of Science, and the Society of Nematologists. She served as president of the Ecological Society of America, the American Institute of Biological Sciences, the Intersociety Consortium for Plant Protection, the Association of Ecosystem Research Centers, and the Society of Nematologists; and chair, Council of Scientific Society Presidents. Diana received her BA and PhD from the University of Kentucky, Lexington, and is a professor of biology and a senior scientist at the Natural Resource Ecology Laboratory at Colorado State University.

Contributors

T. Martijn Bezemer
Netherlands Institute of Ecology
Wageningen, The Netherlands

Klaus Birkhofer
Department of Biology
Lund University
Lund, Sweden

Tanya E. Cheeke
Department of Biology
Portland State University
Portland, Oregon

David C. Coleman
Odum School of Ecology
University of Georgia
Athens, Georgia

Peter C. de Ruiter
Biometris Institute
Department of Land Dynamics
Wageningen University
Wageningen, The Netherlands

Richard P. Dick
School of Environment and Natural Resources
The Ohio State University
Columbus, Ohio

David D. Douds Jr.
USDA-ARS Eastern Regional Research Center
Wyndmoor, Pennsylvania

A. Stuart Grandy
Department of Natural Resources and Environment
University of New Hampshire
Durham, New Hampshire

Vadakattu V.S.R. Gupta
CSIRO Ecosystem Sciences Division
Glen Osmond, South Australia, Australia

Katarina Hedlund
Department of Biology
Lund University
Lund, Sweden

Cynthia Kallenbach
Department of Natural Resources and Environment
University of New Hampshire
Durham, New Hampshire

Terry D. Loecke
School of Natural Resources
University of Nebraska–Lincoln
Lincoln, Nebraska

John C. Moore
Natural Resource Ecology Laboratory
Colorado State University
Fort Collins, Colorado

Rita Seidel
The Rodale Institute
Kutztown, Pennsylvania

Heikki Setälä
Department of Environmental Sciences
University of Helsinki
Lahti, Finland

Carol Shennan
Department of Environmental Studies
University of California, Santa Cruz
Santa Cruz, California

Dorothy Sirrine
Department of Environmental Studies
University of California, Santa Cruz
Santa Cruz, California
Utopia Foundation
Traverse City, Michigan

Richard G. Smith
Department of Natural Resources and Environment
University of New Hampshire
Durham, New Hampshire

Sieglinde S. Snapp
Department of Crop and Soil Sciences and
W. K. Kellogg Biological Station
Michigan State University
East Lansing, Michigan

Diana H. Wall
School of Global Environmental Sustainability and
Department of Biology
Colorado State University
Fort Collins, Colorado

chapter 1

Soil ecology and agroecosystem studies

A dynamic and diverse world

David C. Coleman
University of Georgia

Vadakattu V.S.R. Gupta
CSIRO Ecosystem Sciences Division

John C. Moore
Colorado State University

Contents

1.1 Introduction

Soil organic matter (SOM) is a complex of many organic molecules of varied forms, representing one of the more important components of detritus in terrestrial ecosystems. The maintenance and rehabilitation of SOM are considered to be desirable goals in agroecosystems (Coleman et al., 1984, 1994; Bossuyt et al., 2002). SOM maintains soil structural stability and enhances water-holding capacity, soil fertility, and crop production, ensuring long-term agricultural ecosystem stability (Hassink et al., 1997; Denef et al., 2004). For these reasons, the impacts of agricultural practices on detritus in general through SOM are the focus of this chapter.

Agricultural practices (row crop, orchard, forest management, and rangeland management) that disrupt the soil profile with aggressive tillage for crop production or weed control or structural rehabilitation, remove crops and their stubble on a regular basis, or rely on inorganic fertilizers tend to promote declines in natural SOM reserves (Six et al., 2000; Coleman et al., 2004; Wang and Dalal, 2006). Ways to ameliorate the losses of SOM pools in croplands and forested lands are being explored by scientists in several countries (Elliott and Coleman, 1988; Grace et al., 1998; Six et al., 2000, 2004). These efforts were initially promoted to sustain croplands for food production but more recently are seen as mechanisms for the sequestration of C in soil due to the concern over global warming and rising levels of atmospheric CO_2 and the fact that soils represent the largest terrestrial pool of C, containing 1500 pg, twice that of the atmosphere (Schlesinger, 1996).

The importance of SOM and the processes that contribute to its formation and to soil formation has been understood for some time (Jenny, 1941). The different forms of SOM arise from the transformation of plant, microbial, and animal-derived detritus through chemical and physical processes that are mediated by weathering and biotic activity. The processes behind the transformations of detritus into SOM are central in controlling and regulating key nutrient cycles in natural, agricultural cropland, rangeland, and agroforestry systems. In this chapter, we set forth the principal mechanisms and the biotic interactions involved in the transformation of detritus into SOM. Organic detritus is produced by photosynthesizing and heterotrophic organisms and enters the soil system both above- and belowground. The physical localization of the detritus and its chemical makeup (e.g., C/N ratio and lignocellulose content) have strong influences on subsequent system behavior. In fact, soils, with their strong roles in governing biogeochemical cycles of carbon, nitrogen, and phosphorus, should be viewed as central organizing entities within terrestrial ecosystems (Coleman et al., 1998). A thermodynamic approach to ordering and dissipative processes in soil-plant systems provides a comparison of system-level anabolic and catabolic processes to achieve this synthesis (Table 1.1; Addiscott, 1995). This approach provides a

Table 1.1 Ordering and Dissipative Processes in Soil-Plant Systems Categorized as Biological or Physical

Ordering processes	Dissipative processes
Entropy decreases	Entropy increases
Biological	
Photosynthesis	Respiration
Growth	Senescence
Formation of humus	Decomposition of humus
Physical	
Water flow (profile development)	Water flow (erosion, leaching)
Flocculation	Dispersion
Aggregation	Disaggregation
Development of structure	Breakdown of structure
Larger units	Smaller units
Fewer of them	More of them
More ordered	Less ordered

Sources: Coleman et al., 1998, from Addiscott, T.M. 1995. Entropy and sustainability. *European Journal of Soil Science* 46:161–168. With permission from John Wiley and Sons.

Note: The pairs are not exact opposites.

useful theoretical framework linking biological and physical factors and mediated processes in which to consider the biogeochemical phenomena described in this chapter. The capacity for self-organization resides in the soil population. This population, made up of soil organisms, provides the entropy-maximizing characteristics that characterize soil thermodynamically (Addiscott, 2010).

Our approach incorporates the contributions of invertebrates and microbes to this process by melding and combining the earlier food web analyses of Hunt et al. (1987) and Moore et al. (2003) to provide a more mechanistic and realistic synthesis of the convergence of aboveground and belowground aspects, that is, "green and brown worlds." In the following sections, we provide an overview of both empirical and theoretical approaches to a successful integration of the green and brown worlds via the all-important detritus pathway in terrestrial ecosystems.

1.2 Studies of organic matter dynamics

The production, translocation, and decomposition of SOM in managed ecosystems, including forests and rangelands, have been studied empirically and modeled conceptually and mathematically (Jenkinson et al., 1991; Moorhead et al., 1999; Parton et al., 1987; Paustian et al., 1990, 1997). These approaches include a conjoint consideration of SOM with key biotic and abiotic processes, including C/N ratio, lignin/C ratio, and 1/percentage sand, moving SOM fractions from litter sources into different pools depending on their recalcitrance and turnover times. Employing models backed up with empirical studies has been useful for land managers who compare management treatments across large land areas (Barrett and Burke, 2000; Epstein et al., 1997; Grace et al., 2006). Models of the dynamics of SOM have been used to predict trends in SOM with reasonable accuracy. Of these, the CENTURY model (Parton et al., 1987) has proven useful. There is a specific agroecosystem version (Version 4.0), which is being incorporated into several other cropping systems models (Gijsman et al., 2002) and can effectively simulate C accumulation during soil formation (Parton et al., 1987).

The CENTURY model also offers a heuristic tool to study the key components and processes important to detritus and the formation and dynamics of SOM. CENTURY includes three important features (Figure 1.1). First, the origins, rates of production, and quantity of production of organic carbon are taken into consideration. Second, the forms of organic carbon produced and its structural quality in relation to its rate of degradation are examined. Third, the biotic and abiotic factors and processes that mediate the degradation and transformation of organic matter are included. The biotic components are based on microbes whose activities are governed in large part by soil moisture, temperature, and substrate quality. These components will require a more explicit consideration of the full suite of soil biota given that the approach lacks implicit consideration of the detrital food web interactions and habitat heterogeneity effects.

CENTURY is comprised of submodels, one that represents the inputs of organic material from plant residues and animal excreta and another that represents SOM; these are structured and parameterized for different types of ecosystems (e.g., grassland/crop, forest, and savanna). For each ecosystem type, the submodels possess different rates of primary production and partition organic matter into aboveground and belowground residues. Residue is divided into above- and belowground metabolic and structural pools that differ in quality based on lignin and N contents. The former decompose rapidly and cycle soluble and labile inputs, while the latter pools decompose more slowly and contain metabolically resistant compounds, including plant lignin. The SOM submodel simulates the flow of C and N through the above- and belowground components into multiple SOM pools with

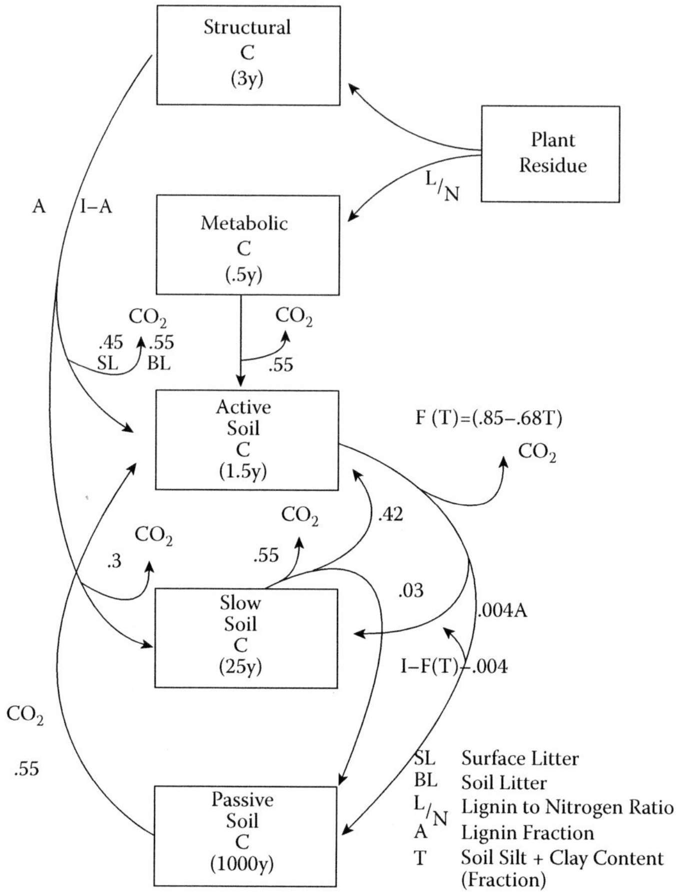

Figure 1.1 Flow diagram for the C flows in the CENTURY model (from Parton, W.J., D.S. Schimel, C.V. Cole, and D.S. Ojima. 1987. Analysis of factors controlling soil organic matter levels in Great Plains grasslands. *Soil Science Society of America Journal* 51:1173–1179. With permission). Plant residue is divided into structural and metabolic C, and soil C is considered as three pools with average turnover times of 1.5 y (active), 25 y (slow), and 1,000 y (passive).

different maximum decomposition rates (Figure 1.1). The decomposition of aboveground and belowground structural and metabolic materials is mediated by microbial activity.

The original model functioned on a monthly time step, but subsequent modifications of maximum decay rates and temperature and moisture factors have allowed the model to function on a daily time step, which may be more appropriate for agricultural, pulse-driven systems. The model partitions SOM into active, slow, and passive pools, with turnover rates of 1–5 y, tens to hundreds of years, and hundreds to thousands of years, respectively. Interactions between the compartments are influenced by soil texture, moisture, initial SOM content, and plant production dynamics.

1.2.1 *The origins of organic material*

Organic material enters soils from within and outside the ecosystem. The internal or autochthonous sources originate from primary production by plants, algae, and microbes.

The organic by-products, death, and turnover of primary producers enter the system as different forms of detritus for consumption by microbes and invertebrates. Direct consumption of primary producers by consumers leads to consumer by-products and corpses that enter the soil milieu as detritus. Outside or allochthonous inputs of organic material enter systems largely as detritus. Specific examples include algal wracks from marine habitats washing up onto the terrestrial realm, as well as wind- or water-driven materials transported from one terrestrial or aquatic habitat to another.

At this point, it is essential to note that the most significant proportion of the carbon inputs to soils coming from the plant litter (leafy and woody materials) on the soil surface and the plant root materials that enter the soils are all subject to direct consumption by invertebrates and degradation by microbes. The direct contribution by herbivores and predators is modest; for most terrestrial systems, less than 10% of total plant biomass is consumed by herbivores (Coleman et al., 2004). The remaining 90% of the aboveground and belowground plant biomass enters soils as structural and metabolic forms, with the contribution from roots dominating the SOM development. From 16% to 33% of the total carbon assimilated by plants is released directly into the soil in the form of root exudates, mucigels, and cells, contributing from 30% to 60% of the organic pool in soil (Table 1.2; Andrén et al., 1990; Heal et al., 1997). The amounts of persistent organic matter compounds in ecosystems have been the subject of intensive study by many researchers. For a synoptic review of studies making use of radiocarbon in soil systems and profiles, see the work of Trumbore (2009). Trumbore noted that there are four principal mechanisms for stabilizing soil carbon: (1) In climatic stabilization, SOM can be preserved due to freezing temperatures, low O_2 content, or high moisture content in a range of ecosystems. This type of preservation occurs in high-latitude soils and water-saturated soils in temperate and tropical environments. (2) For intrinsic recalcitrance, while all organic carbon is ultimately thermodynamically unstable, soils contain a number of compounds that may be difficult to decompose owing to their specific chemical structure. Pyrolized carbon formed in fire can be one example (Preston and Schmidt, 2006); some lipid compounds are also extremely long-lived in soil (Mikutta et al., 2006). (3) Regarding physical stabilization, there is clear evidence that the association of organic compounds with aggregates and mineral surfaces through a range of interactions (inclusion, sorption, etc.) slows their decomposition (Oades, 1984; Jastrow, 1996; Six et al., 2004; Mikutta et al., 2006). (4) For inhibition of microbial activity/inaccessibility, until recently, links between the microbial community and the amount and age distribution of SOM have received less attention than physical and

Table 1.2 Biomass Inputs into Agricultural and Agroforestry Systems (Agroecosystem, Kjettslinge, Sweden)

Experimental treatment	NPP	Net additions to soil
Barley 0[a]	484	250
Barley 120[b]	963	300
Grass ley	1464	400
Lucerne ley	1580	400

Source: Modified from Andrén, O., T. Lindberg, K. Paustian, and T. Rosswall, eds. 1990. *Ecology of Arable Land. Organisms, Carbon and Nitrogen Cycling.* Ecological Bulletins 40. Copenhagen: Munksgaard.

Note: All input means in grams of dry matter per square meter per year (gdm m^{-2} yr^{-1}). Net additions include roots and crop residues.

[a] 0 = no fertilizer.

[b] 120 = 120 kg ha^{-1} yr^{-1} nitrogen fertilizer.

chemical controls on SOM (Waldrop et al., 2000; Balser and Firestone, 2005). Breakdown of organic molecules is, for the most part, mediated by microbial enzymes, and if there are barriers to the interaction of extracellular enzymes and substrate, decomposition is slowed (Schimel and Weintraub, 2003; Allison, 2005; Ekschmitt et al., 2005). For over a century, there has been interest in the nature of recalcitrant organic compounds in SOM. For an extensive review of the controversial nature of "humic compounds" as artifacts of harsh alkaline extractants, and more realistic alternatives to them, see the work of Kleber and Johnson (2010).

1.2.2 The importance of litter quality and nutrients

Litter quality is defined in terms of the resistance of the material to decomposition. Qualitative references to quality refer to materials that are readily decomposed with short half-lives as being labile, while those that resist decomposition and possess longer half-lives are termed resistant or recalcitrant. Early indices of quality relied on the ratios of C and N or C and lignin. Under this convention, labile materials possess C:N ratios of less than 30:1, while recalcitrant materials possess C:N ratios greater than 30:1. While useful, these simple ratios have given way to more comprehensive approaches.

In both temperate and tropical ecosystems, chemically oriented indexes of litter (detritus) quality have proven to be powerful predictive tools of rates of degradation and release of nutrients. One of the more frequently employed ratios is the plant residue quality index (PRQI; Tian et al., 1993, 1995). The PRQI builds on earlier indices by including C/N, lignin, and polyphenol concentration of plant residues and is defined as

$$\text{PRQI} = [1/(a\ \text{C/N} + b\ \text{lignin} + c\ \text{polyphenols})] \times 100 \qquad (1.1)$$

where a, b, and c are coefficients of relative contribution of C/N ratio, lignin content (%), and polyphenol content (%) to plant residue quality, respectively. Varying the relative contributions of the three variables noted, such as varying the overall nutrient quality and quantity of crop residues, is quite informative. With lower values of all three variables, there was a marked mulching effect, leading to enhanced macrofaunal activities (e.g., increased termite and earthworm activity), allowing greater aeration and nutrient mineralization (Tian et al., 1995).

In tropical agricultural and agroforestry studies, a decision tree was created for testing hypotheses about the resource quality parameters that affect nitrogen release patterns and rates. The decision tree is linked to an Organic Resource Database (ORD) with detailed information on the resource quality of several hundred species of leguminous crops and agroforestry trees. This provides a systematic means of selecting organic resources for soil fertility management (Palm et al., 2001). The chemical quality of an organic material is influenced both by its carbon constituents (carbon quality) and by its nutrient content and the chemical form of the nutrients (nutrient quality). Like the PQRI (Equation 1.1), this approach includes a minimum set of resource quality characteristics for decomposition studies (Palm et al., 2001), including lignin, soluble carbon, total nitrogen and phosphorus, ash-free dry weight, and soluble phenolics if total nitrogen exceeds 1.8%. Measurement of the protein-binding capacity to assess reactive polyphenols may be useful as well in some instances. For an extensive overview of lignins and soils and major factors governing lignin formation and decomposition, see the work of Thévenot et al. (2010).

1.2.3 The importance of aggregate formation in soils

As organic matter decomposes, its fate and the influences of soil biota on the process are far from linear or additive. Some forms of organic matter contribute to the stabilization of other forms, and the collective contributions of biota to the process lead to counterintuitive results (Moore et al., 2003, 2004). Nowhere is this more evident than in the emerging understanding of the role of soil microbes and fauna in the formation and degradation of soil aggregates (Elliot et al., 1984; Elliott, 1986; Six et al., 2000; Smucker et al., 2007), the importance of consortia of microbes to the decomposition of complex organic compounds, and the relationship between changes in the flow of nutrients through food webs and the rates of mineralization (Clarholm, 1985; Ingham et al., 1985; Gupta and Germida, 1989; Beare et al., 1992; Moore et al., 2003).

Soil aggregates are composed of sand, silt, and clay particles that are bound together by SOM and inorganic compounds. Aggregates are hierarchically classified into water-stable size classes based on their solubility (Tisdall and Oades, 1982; Elliott, 1986; Six et al., 2000). The smaller size classes include a silt and clay fraction (<53 μm) and microaggregates (53–250 μm). These smaller factions combined with coarse particulate organic matter (CPOM; > 250 μm) to form small macroaggregates (250–2000 μm), which in turn all combine with organic matter in the form of free light fractions (particulate organic matter outside of aggregates [POMLF, particulate organic matter light fraction] and residues (intraaggregate particulate organic matter, or iPOM, 250–2000 μm) to form large macroaggregates (>2000 μm). The organic components originate largely from plants in the form of surface litter, structural materials, roots, and root products (sloughed cells, exudates, and mucigels). These compounds are further degraded by microbes and invertebrates, resulting in the primary binding agents that are mostly microbial in origin (Jenny and Grossenbacher, 1963).

1.2.4 The roles of soil microbes and invertebrates

The importance of plants and soil biota to soil formation and processes has been understood for some time (Tull, 1733; Darwin, 1881; Jenny, 1941). Plants introduce organic substrates into soils as products of growth, senescence, and death, while soil microbes and invertebrates utilize these organic substrates as an energy source and in the process contribute their own organic substrates through consumption, growth, and death. For microbes, the dominant constituents include microbial cell wall residues and hyphae in the case of fungi, extracellular polysaccharides, carbohydrates, and amino sugars (glucosamine, galactosamine, and muramic acid). For invertebrates, the substrates include nitrogen-rich fecal pellets, cell walls, chitinous exoskeletons, cytoplasm, and mucus secretions (Coleman et al., 2004).

Apart from the transformation and direct inputs of organic substrates, soil invertebrates indirectly affect the decomposition of organic substrates by enhancing the activities of soil microbes through grazing, comminution, and dissemination (Wallwork, 1976; Moore et al., 1988). Moderate levels of consumption of microbes by protozoa and invertebrates can stimulate further microbial growth through the principle of optimal grazing (Hilbert et al., 1981; Clarholm, 1985; Wall and Moore, 1999). Assuming that microbes exhibit logistic growth tied to available resources with a carrying capacity K, microbial growth rates are maximized at population densities of $K/2$. Higher growth rates translate to increased utilization and transformation of organic substrates and increased production of microbial-derived organic substrates. Given that invertebrates consume more nitrogen than they require for growth (i.e., invertebrates are net mineralizers of nitrogen), enhanced microbial growth through grazing leads to increased nitrogen availability for plants. When this process plays itself out within the rooting zone of plants, a positive feedback can occur—plants and the soil food web feed off one another's excess and waste. Under this scenario, plants exude excess photosynthate in the form of labile carbon substrates from their roots, which in turn are utilized by microbes and subsequently as food for microbivorous invertebrates, which release nitrogenous waste that is utilized by plants (Gupta et al., 1999; Ingham et al., 1985).

Many invertebrates feed directly on plant materials and organic substrates. The fragmentation or comminution of these materials enhances their decomposition. For plant structural materials, comminution increases the surface areas of the materials and exposes cytoplasm, thereby enabling greater access by microbes. Decomposition is further accelerated as the feeding activity often results in the translocation of nitrogen from the soil to the substrate in the form of fecal material and through fungal hyphae. Grazing by invertebrates disseminates microbes from one organic source to another as many microbes adhere to invertebrate exoskeletons and cuticles and survive passage through their digestive tracts.

The diversity of archaea and bacteria is as yet little known, but with the use of new massive parallel sequencing devices (Roche 454 and other pyrosequencing techniques; Buée et al., 2009), we are beginning to approach an understanding of the very large biodiversity of these organisms. Metagenomics encompasses the full array of microorganisms present in ecosystems at any point in time (Handelsman, 2004; Van Elsas et al., 2008). Metatranscriptomics refers to the number of microbes in a given community or ecosystem that are active, namely, respiring or synthesizing at a given time. Stable isotope probing (SIP) platforms provide additional avenues to track a specific element (C or nutrients) through microbial and faunal communities, facilitating the linkage of diversity to function. "At any moment, an estimated 10^{30} bacterial and archaeal genes are mediating essential ecological processes throughout the world" (Moran, 2009). In a number of marine microbial ecological studies, diurnal patterns have been measured. Thus, bacterioplankton communities in oligotrophic waters of the North Pacific Ocean showed greater investments in energy acquisition and metabolism (photosynthesis, oxidative phosphorylation, and C1 compound metabolism) during the day and in biosynthesis (of membranes, amino acids, and vitamins) at night, according to an analysis of RNA transcripts from these periods (Poretsky et al., 2009).

1.3 *An integrative framework*

A framework as described that links soil microbes and invertebrates directly to pedogenic processes is beginning to emerge. Soil food webs and the differences in microbial and invertebrate life-forms within them offer a starting point for connecting the components of the detrital food web to soil pedogenesis and SOM dynamics (Coleman et al., 1983, 2004). Coleman et al. (1983) identified a bacterial-based fast cycle and fungal-based slow cycle within soils. Subsequent studies revealed that these fast cycles and slow cycles were borne from the trophic interactions of detritus to bacteria and their consumers and from detritus to fungi and their consumers (Figure 1.2; Hunt et al., 1987; Moore et al., 1988; de Ruiter et al., 1996). These dominant trophic pathways, or "energy channels," are ubiquitous across ecosystem types and grounded in the basic architecture of soil food webs and their structural stability (Moore and Hunt, 1988).

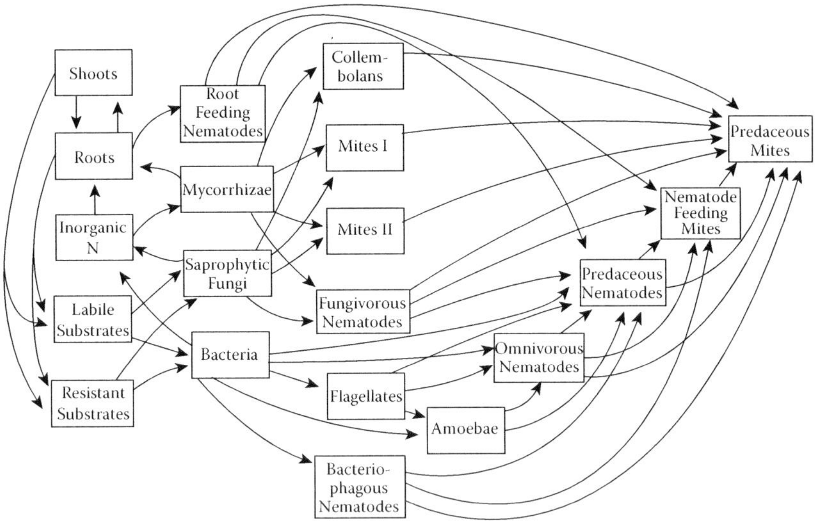

Figure 1.2 Flow diagram of aboveground and belowground detrital inputs in a shortgrass steppe ecosystem (from Hunt, H.W., D.C. Coleman, E.R. Ingham, R.E. Ingham, E.T. Elliott, J.C. Moore, S.L. Rose, C.P.P. Reid, and C.R. Morley. 1987. The detrital food web in a shortgrass prairie. *Biology and Fertility of Soils* 3:57–68. With permission from Springer). Note the separation of flows between various faunal groups, including the fungivorous Cryptostigmata, collembolans, and nematodes, and the more generalized feeders, non-cryptostigmatic mites. A majority of the nitrogen mineralization were calculated to come from the protists, feeding primarily on bacteria, and bacterial and fungal-feeding nematodes (Ingham et al., 1985).

The complexity presented in Figure 1.2 can be condensed into the dominant pathways beginning with pools of detritus or SOM that differ in quality. These pools would serve as the primary energy sources for a suite of bacteria and fungi, each of which is consumed by a host of microbial consumers and predators. Metabolic wastes and by-products that cycle back as energy sources and binding agents would be factored in much as C and N are in the current generation of models. This approach preserves the basic premise of material transformations that occur in the soil carbon models (Parton et al., 1987; Gijsman et al., 2002) and material transfers that occur in food web models (Hunt et al., 1987; de Ruiter et al., 1993) in a way that provides a common currency. A comparison (Figure 1.3) of belowground food webs from a wide range of agroecosystems is instructive. The food webs from Central Plains, Colorado, and Horseshoe Bend, Georgia, in the United States and two European ones, from Kjettslinge, Sweden, and Lovinkhoeve, the Netherlands, show varying degrees of aggregation of functional groups, depending in part on the expertise of the investigators involved. For example, in some webs, flagellates were distinguished from amoebae, and in others they were aggregated as one: protozoa. Earthworms were not found at the conventional fields of Lovinkhoeve. With much greater emphasis on aboveground and belowground interactions currently (see De Deyn and van der Putten, 2005), it is noteworthy that herbage arthropods were considered only at the Swedish site. For ease of depiction, material flows to the detrital pool, through the death rates and the excretion of waste products, were not represented in the diagrams but were taken into account in the material flow calculations and stability analyses (de Ruiter et al., 1998).

Moore et al. (2003) presented a first approximation of this approach by linking the activities of organisms within the bacterial and fungal pathways to SOM dynamics and

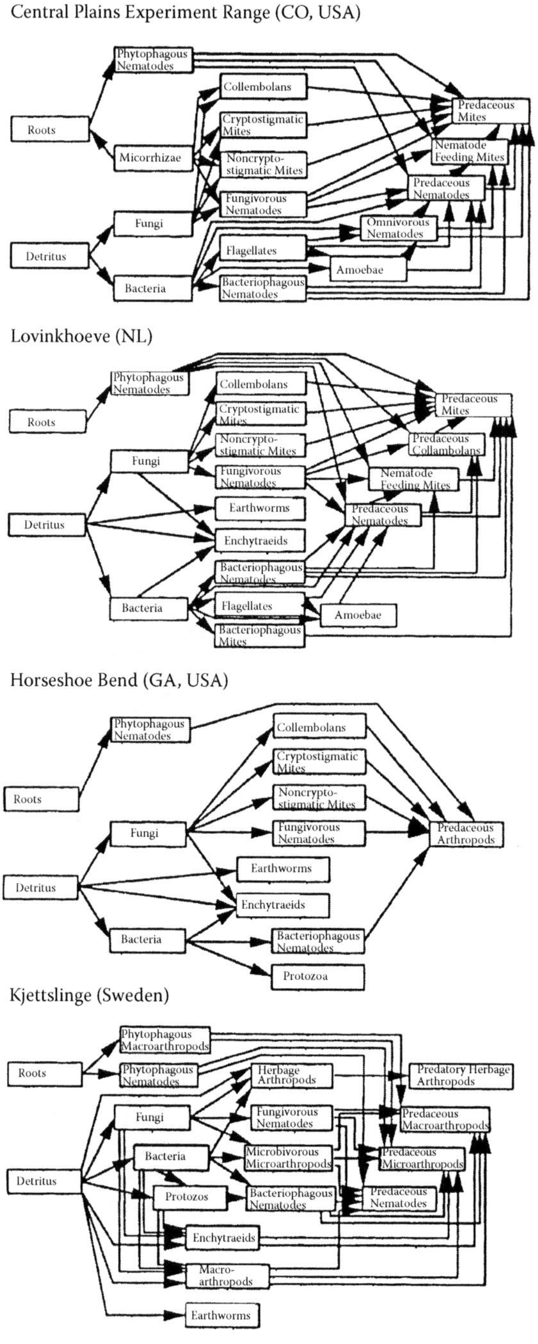

Figure 1.3 Diagrams of the belowground food webs from Central Plains (Hunt et al., 1987), Lovinkhoeve (de Ruiter et al., 1993; Brussaard et al., 1990), Horseshoe Bend (Hendrix et al., 1986), and Kjettslinge (Andrén et al., 1990). (From de Ruiter, P.C., A.-M. Neutel, and J.C. Moore. 1998. Biodiversity in soil ecosystems: the role of energy flow and community stability. *Applied Soil Ecology* 10:217–228. With permission from Elsevier.) Note that earthworms were not present in the food web from the conventional tillage management treatment at the Lovinkhoeve site.

key ecosystems processes. Here, SOM is broadly defined as nonliving organic material within soils, regardless of form or origin (e.g., detritus, dead roots, corpses of soil biota, traditional SOM). The balance in the activities of one pathway relative to the other is governed by the quality of SOM, with low C:N (<30) substrates favoring the bacterial pathway and high C:N (>30) substrates favoring the fungal pathway. This model allows for changes in the relative activities of the different pathways with natural seasonal variation in the phenology of plants and season succession in plant types, as well as abrupt or gradual changes that result from different land-use practices and increases in atmospheric CO_2. We discuss some of the last in the section that follows.

1.4 High-priority topics on detrital dynamics of concern to land managers

1.4.1 Biodiversity in agricultural and agroforestry systems

Viewing the agroecosystem holistically, it is essential to consider the many activities of the soil micro-, meso-, and macrofauna. The roles of the larger soil invertebrates (e.g., earthworms) in burrowing, soil transport, and mixing are extensive and can involve up to several tons per hectare annually of more readily assimilable organic materials (Lavelle et al., 2001). Thus, a combination of conservation tillage and enhanced residue quality pays considerable dividends for ecosystem health.

The previous history of a given agroecosystem can have strong effects on the extent of soil food web complexity. In a 1-y field station trial in Davis, California (Alfisol), including no tillage and continuous cropping, no tillage and fallow, standard tillage and continuous cropping, and standard tillage and fallow, significant enhancements in SOM storage were measured in the no tillage and continuous cropping in the top 0- to 5-cm layer, principally due to greater amounts of fungal biomass in the microbial biomass carbon. The soil food web, as denoted by the nematode diversity, was not changed in any of the treatments, probably due to the elimination of higher trophic-level nematodes during previous decades of cultivation (Minoshima et al., 2007).

A recent comparative study of decomposition rates in replicated conventional till, no-till, and old-field agricultural sites in southern Michigan, United States, found that, after one growing season, litter decomposition under conventional till was 20% greater than in old-field communities (Wickings et al., 2010). In contrast, decomposition rates in no till were not significantly different from those in either of the other two treatments. Wickings et al. (2010) suggested that agricultural intensification can increase litter decomposition rates, alter decomposer communities, and influence litter chemistry in ways that could have significant and long-term effects on SOM dynamics. In the southern Australian agricultural region, continuous cropping systems caused significant changes in the rate of decomposition of crop residues and composition of microbial and faunal communities compared to that under short-term (1- to 4-y) pasture-crop or fallow-crop rotation systems (Pankhurst et al., 1995; Gupta et al., 2008; Wang and Dalal, 2006). Such changes had a negative effect on short-term nutrient turnover and soilborne biological constraints to agricultural production, which in turn had an impact on long-term SOM dynamics and overall ecosystem health. An increased frequency of pastures in rotation generally increased soil organic C levels, whereas higher frequency of fallows caused losses in soil biota populations and amounts of labile pools of SOM (Grace et al., 1998).

The diversity of microbial (archaea, bacteria, fungi, and viruses) communities in soils has only recently been evaluated in terms of small-subunit ribosomal RNA (rRNA) genes.

Comparing desert, prairie, and rain forest sites in North and South America, Fierer et al. (2007) found minimal taxonomic overlap between sites, with the archaeal and fungal communities less even than the bacterial communities. Comparing bacterial communities in ultisols of the southern Piedmont and the southern Appalachian Mountains, Upchurch et al. (2008) noted that bacterial diversity (16S rRNA) was greater in conventional tillage (CT) and no-tillage (NT) agroecosystems at Horseshoe Bend near Athens, Georgia, versus forested sites nearby and at Coweeta LTER (Long Term Ecological Research), 160 km to the north of it. The ultimate causal factors are yet unknown but may be linked to the greater diversity of primary producers (crops and weeds) in the agroecosystem than in the forested sites. In similar sites in the southeastern Piedmont of the United States, Jangid et al. (2008) observed similar trends in agricultural ecosystems compared with nearby forested sites.

One of our agroecosystem studies in the U.S. southern coastal plain of Georgia used a series of five commercial no-tillage cotton (*Gossypium hirsutum* L.) field plots that followed a chronosequence of 4–25 y since conventional tillage. These sites are loamy to fine-loamy thermic, kaolinitic Kanhapludults and Paleaquults, with pH of 3.7–4. During the first 8 y of tillage, there was some increase in organismal abundance, but only the two older fields (8–25 y) had accumulated abundance and species richness that approached that of undisturbed sites. Total microbial biomass changed little over the years and tended to follow the marked seasonal changes in total organic matter in spring and fall seasons (Figure 1.4; Adl et al., 2006). Microbial diversity, as measured by fatty acid methyl ester (FAME) analysis, showed some shifts from bacterial to fungal dominance, with bacterial/fungal ratio highest in site 0 and lowest in sites 1 and 5 (Figure 1.5a; Simmons and Coleman, 2008). Levels of 16:1ω5c, an arbuscular mycorrhizal fungal indicator, were all low, but were lowest in the conventionally tilled field (Figure 1.5b; Simmons and Coleman, 2008). In contrast, the changes in soil faunal species richness were more pronounced across the chronosequence.

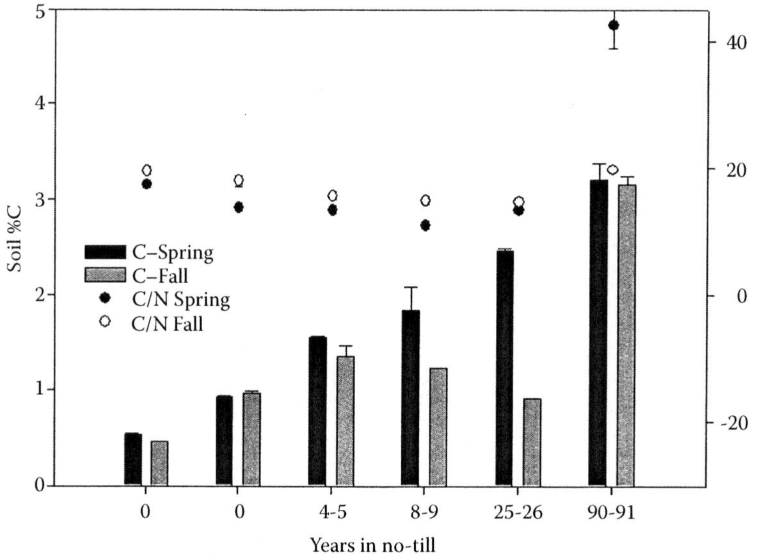

Figure 1.4 Soil organic matter in a chronosequence of cotton fields (0–25 y) and a nearby pine forest (90 y) in sandy soils in Coffee County, southern Georgia, United States. Note decrease in autumn in 4- to 25-y fields (gray bars). (From Adl, M.S., D.C. Coleman, and F. Reed. 2006. Slow recovery of biodiversity in sandy loam soils of Georgia after 25 years of no-tillage management. *Agriculture, Ecosystems and Environment* 114:323–334. With permission from Elsevier.)

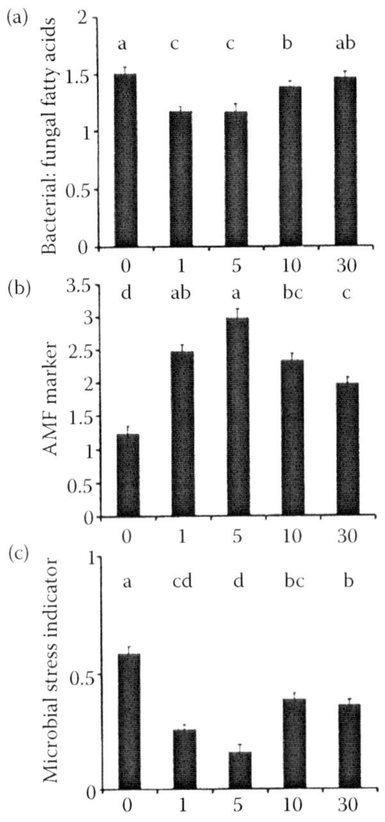

Figure 1.5 The effect of age in conservation tillage in cotton fields in southern Georgia on (a) bacterial-to-fungal ratios as measured by FAME biomarkers; (b) arbuscular mycorrhizal fungi (AMF) expressed as the relative percentage mole fraction of 16:1ω5c; and (c) an indicator of microbial stress, cy17:0: 16:1ω7c, from soils (0–5 cm depth) at five sites across a chronosequence of conservation tillage. Data shown include standard error; different letters denote significant differences at $P < 0.05$ (Tukey's). (From Simmons, B.L. and D.C. Coleman. 2008. Microbial community response to transition from conventional to conservation tillage in cotton fields. *Applied Soil Ecology* 40:518–528. With permission from Elsevier.)

Thus, there were more ciliates, flagellates, and amoebae in the older fields (Figure 1.6; Adl et al, 2006); microarthropods were more speciose but had lower total biomass in the older fields as well (Adl et al., 2006). This amplification of species diversity effects in the detritivore fauna compared to the microbial diversity was unexpected.

In our agroecosystem chronosequence study, the microbial biodiversity patterns did not conform to the proposed fungal/bacterial changes as proposed by Harris (2009; Figure 1.7). However, the faunal changes showed a greater response to the changes in management and organic matter levels. With the marked oscillations in SOM between spring and fall sampling seasons over the 2 y of this study, it is probably safest to consider these trends as on the margins of significant effects. In contrast, a comparative study of microbial diversities in conventional tillage, no-tillage, and forested ecosystems adjacent to the Horseshoe Bend Long-Term Research study in bottomlands near Athens, Georgia, microbial diversity, as measured by 16S rRNA genes, was greater in the agricultural fields than in the nearby forest (Upchurch et al., 2008). These trends seemed to be correlated with the

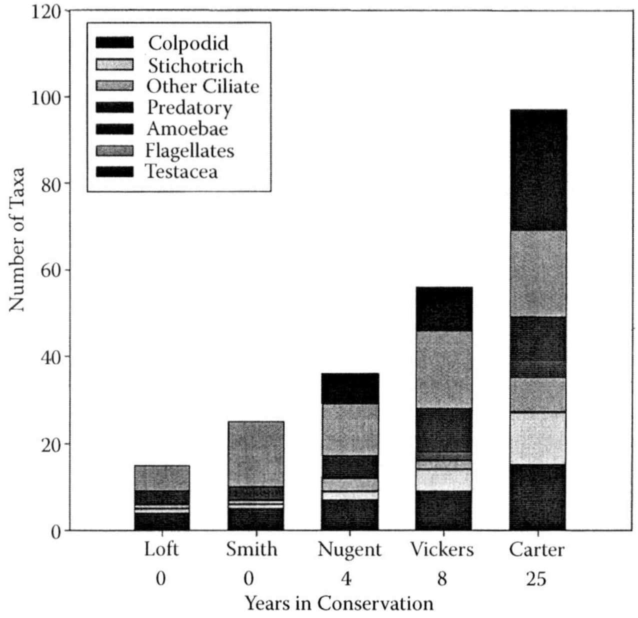

Figure 1.6 The temporal dimensions of protozoan diversity in a southern Georgia chronosequence. Note the increasing proportions of amoebae, flagellates, and testacea in the longer-term conservation tillage fields. (From Adl, M.S., D.C. Coleman, and F. Reed. 2006. Slow recovery of biodiversity in sandy loam soils of Georgia after 25 years of no-tillage management. *Agriculture, Ecosystems and Environment* 114:323–334. With permission from Elsevier.)

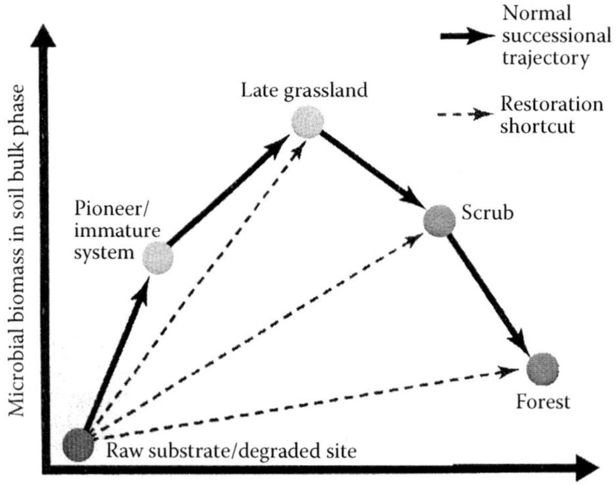

Figure 1.7 As ecosystems mature, there is a switch in dominance from bacterial to fungal biomass; total microbial biomass in bulk (nonrhizosphere) soil peaks around midsuccession because this is the period of root exudates in this phase peak. As fungal symbionts begin to dominate the rhizosphere, fewer plant exudates diffuse into the soil bulk phase. The aim of restoration to mature forest systems is to short-cut this trajectory by means of establishing fungal dominance at an early stage. (From Harris, J. 2009. Soil microbial communities and restoration ecology: facilitators or followers? *Science* 325:573–574. With permission from the American Association for the Advancement of Science.)

greater diversity of companion plants (weeds) in the agricultural plots. In a self-mulching vertisol under an irrigated cotton-based cropping system in New South Wales, Australia, Coleman et al. (2010) found that the diversity of bacteria, archaea, and eukarya (fungi, micro- and mesofauna) was more influenced by previous cropping history (i.e., inclusion of a legume crop in rotation) than short-term addition of carbon inputs (crop stubble). Proteobacteria were the most abundant phylum and β-Proteobacteria were the largest group. In contrast, Actinobacteria were the most abundant phylum in the Horseshoe Bend agroecosystem plots in the United States (Upchurch et al., 2008) and in the cereal cropping soils from the rain-fed region in South Australia (Gupta et al., 2010). Unlike bacterial communities, soil protistan and nematode communities (e.g., fungi and protozoa) were differentially influenced by cropping history and short-term manipulations. For example, stubble addition and wetting-drying treatments that influence C availability and habitat structure caused greater response in soil fungi and protozoan communities. Soil fauna such as nematodes that require stable habitable pore structure were negatively influenced by the effects of changing soil structure in response to wetting-drying events. In vertisols, with smectitic clays, biota are regularly exposed to changing habitat structure; hence, their response in terms of diversity and functionality may be strongly related to physicochemical characteristics compared to that in other soil types.

In more arid regions (e.g., large areas of the Sahel, Australia, and the arid regions of North America), the keystone roles of termites in transporting and concentrating both organic and inorganic nutrients are of similar magnitude to that of earthworms (Whitford, 2000). For more details on the roles of micro- and mesofauna in soil nutrient cycling, see the work of Coleman et al. (2004) and Coleman (2008, 2011).

1.4.2 *Changes in land use and impacts on detrital food webs*

The impacts of changes in land-use practices are of increasing interest to ecologists worldwide. Changes from agroecosystems to pasturelands and longer-term succession into forested ecosystems lead to significant changes in the detrital food web. Thus, the more frequently disturbed tillage agricultural system tends to be more bacterial dominated versus the fungal-dominated no tillage or pastureland (Wardle et al., 2004). Soil communities usually respond more slowly to changes occurring with land abandonment than in the communities aboveground (Korthals et al., 2001). Slow development rates of the belowground community may be an important factor controlling ecosystem services and the outputs (goods) provided by these restored ecosystems (Wardle et al., 2004).

Further examples of aboveground influences (e.g., herbivory) on belowground processes are becoming more evident as the result of recent experimental studies. Thus, grasshopper (*Romalea* spp.) feeding on corn leaves caused a greater exudation of labile carbon from the roots (Holland et al., 1996), which was further transmitted along the detrital food web to increased carbon uptake by microbivorous nematodes in no-tillage plots in the Horseshoe Bend agroecosystem study in the Georgia Piedmont (Fu et al., 2001; Coleman et al., 2006, 2009).

1.5 *Conclusions*

The three domains of life (archaea, bacteria, and eukarya) are well represented and active in soil systems. Much of the metabolic activity in soils is microbially mediated and faunally influenced. Because agroecosystems are the most amenable to experimental manipulation of any terrestrial ecosystems, many experiments on them have provided intriguing results.

Using conservation and no-tillage techniques, we have measured increases in biotic diversity and enhanced carbon sequestration in soils worldwide. Methods to enhance carbon sequestration at micro-, meso-, and macroscales are being pursued.

Questions remain about the primary causal factors of biodiversity in cropped, forested, and rangelands. Results from cross-ecosystem studies of microbial biodiversity have yielded counterintuitive information, with agroecosystems having greater diversity than less-disturbed ecosystems such as forests on similar soil types in the southeastern United States. Conservation tillage contributes significantly to enhanced soil faunal biodiversity when compared to conventional tillage. Interactions between SOM and soil biodiversity are worthy of further investigation.

The maintenance and rehabilitation of SOM in agroecosystems, including agroforestry systems, have drawn increasing attention from ecologists and land managers worldwide. Careful attention is being paid to proven principles of organic matter quality and its importance to the structure and functioning of the soil food web. Similarly, various types of reduced tillage practices that can influence turnover of C and nutrients added through crop residues are being recommended either to reduce losses or to improve SOM status in intensive cropping regions in Australia, Africa, South America, and India. These approaches have enabled the continuation or restoration of detrital food webs in many managed ecosystems.

Models coupled with empirical studies have proven to be useful tools in integrating organic matter quality, soil formation, and food web dynamics. Models that focus on soil formation and models that focus on the trophic interactions among soil biota both point to the importance of the quality of organic matter to the dynamics of SOM and the dynamics of soil biota. A comprehensive approach that melds the two in the manner proposed by Jenny (1941) is in the nascent stages of development.

More than 20 y ago, Elliott and Coleman (1988) urged "let the soil work for us," arguing that effective land management would require an understanding of the physical, chemical, and biological components of soil formation and detritus dynamics. As noted in this review, the prospects for land amelioration for increasing production of food and fiber are very good, with the caveat that suitable land-use policies are encouraged by private and governmental organizations. Central to the success of these efforts will be the effective management of detritus.

Some questions for the future include the following: What aspects of soil carbon dynamics, related to aggregate stability, are most strongly correlated with soil biodiversity? Can knowledge of these aspects be linked with the usage of a wide range of more marginal soils for the growth of biofuel crops or even to maintain food production to meet the demand of ever-growing world population?

References

Addiscott, T.M. 1995. Entropy and sustainability. *European Journal of Soil Science* 46:161–168.

Addiscott, T.M. 2010. Entropy, non-linearity and hierarchy in ecosystems. *Geoderma* 160:57–63.

Adl, M.S., D.C. Coleman, and F. Reed. 2006. Slow recovery of biodiversity in sandy loam soils of Georgia after 25 years of no-tillage management. *Agriculture, Ecosystems and Environment* 114:323–334.

Allison, S.D. 2005. Cheaters, diffusion and nutrients constrain decomposition by microbial enzymes in spatially structured environments. *Ecology Letters* 8:626–635.

Andrén, O., T. Lindberg, K. Paustian, and T. Rosswall, eds. 1990. *Ecology of Arable Land. Organisms, Carbon and Nitrogen Cycling*. Ecological Bulletins 40. Copenhagen.

Balser, T.C., and M.K. Firestone. 2005. Linking microbial community composition and soil processes in a California annual grassland and mixed-conifer forest. *Biogeochemistry* 73:395–415.

Barrett, J.E., and I.C. Burke. 2000. Potential nitrogen immobilization in grassland soils across a soil organic matter gradient. *Soil Biology and Biochemistry* 32:1707–1716.

Beare, M.H., R.W. Parmelee, P.F. Hendrix, W. Cheng, D.C. Coleman, and D.A. Crossley, Jr. 1992. Microbial and faunal interactions and effects on litter nitrogen and decomposition in agroecosystems. *Ecological Monographs* 62:569–591.

Bossuyt, H., J. Six, and P.F. Hendrix. 2002. Aggregate-protected carbon in no-tillage and conventional tillage agroecosystems using carbon-14 labeled plant residue. *Soil Science Society of America Journal* 66:1965–1973.

Brussaard, L., L.A. Bouwman, M. Geurs, J. Hassink, and K.B. Zwart. 1990. Biomass, composition and temporal dynamics of soil organisms of a silt loam soil under conventional and integrated management. *Netherlands Journal of Agricultural Science* 38:283–302.

Buée, M., M. Reich, C. Murat, E. Morin, R.H. Nilsson, S. Uroz, and F. Martin. 2009. 454 Pyrosequencing analyses of forest soils reveal an unexpectedly high fungal diversity. *New Phytologist* 184:449–456.

Clarholm, M. 1985. Possible roles for roots, bacteria, protozoa and fungi in supplying nitrogen to plants. In *Ecological Interactions in Soil: Plants, Microbes, and Animals,* eds. A.H. Fitter, D. Atkinson, D.J. Read, and M.B. Usher, 355–365. Blackwell, Oxford, UK.

Coleman, D.C. 2008. From peds to paradoxes: linkages between soil biota and their influences on ecological processes. *Soil Biology and Biochemistry* 40:271–289.

Coleman, D.C. 2011. Understanding soil processes: one of the last frontiers in biological and ecological research. *Australasian Plant Pathology* 40:207–214.

Coleman, D.C., C.V. Cole, and E.T. Elliott. 1984. Decomposition, organic matter turnover and nutrient dynamics in agroecosystems. In *Agricultural Ecosystems—Unifying Concepts,* eds. R. Lowrance, B.R. Stinner, and G.J. House, 83–104. Wiley/Interscience, New York.

Coleman, D.C., D.A. Crossley, Jr., and P.F. Hendrix. 2004. *Fundamentals of Soil Ecology,* 2nd ed. Elsevier Academic Press, San Diego, CA.

Coleman, D., V.V.S.R Gupta, K. Jangid, S. Wakelin, and W. Whitman. 2010. The composition and diversity of prokaryotic and eukaryotic communities from an Australian vertisol: an experimental study. Proceedings of the 19th World Congress of Soil Science, August 1–6; *Soil Solutions for a Changing World* [DVD], http://www.iuss.org. IUSS, Brisbane, Australia: 231–234.

Coleman, D.C., P.F. Hendrix, M.H. Beare, D.A. Crossley, Jr., S. Hu, and P.C.J. van Vliet. 1994. The impacts of management and biota on nutrient dynamics and soil structure in sub-tropical agroecosystems: impacts on detritus food webs. In *Soil Biota Management in Sustainable Farming System,* eds. C.E. Pankhurst, B.M. Doube, V.V.S.R. Gupta, and P.R. Grace, 133–143. CSIRO, Melbourne, Australia.

Coleman, D.C., P.F. Hendrix, and E.P. Odum. 1998. Ecosystem health: an overview. In *Soil Chemistry and Ecosystem Health,* ed. P.H. Wang, 1–20. Soil Science Society of America Special Publication No. 52, Madison, WI.

Coleman, D.C., M.D. Hunter, P.F. Hendrix, D.A. Crossley, Jr., S. Arce-Flores, B. Simmons, and K. Wickings. 2009. Long-term consequences of biological and biogeochemical changes in the Horseshoe Bend Long-Term Agroecosystem Project. In *Sustainable Agroecosystem Management. Integrating for Ecology, Economics, and Society,* eds. P.J. Bohlen and G. House, 195–209. CRC Press, Boca Raton, FL.

Coleman, D.C., M.D. Hunter, P.F. Hendrix, D.A. Crossley, Jr., B. Simmons, and K. Wickings. 2006. Long-term consequences of biochemical and biogeochemical changes in the Horseshoe Bend agroecosystem, Athens, GA. *European Journal of Soil Biology* 42:S79–S84.

Coleman, D.C., C.P.P. Reid, and C.V. Cole. 1983. Biological strategies of nutrient cycling in soil systems. *Advances in Ecological Research* 13:1–55.

Darwin, C. 1881. *The Formation of Vegetable Mould, through the Action of Worms, with Observations on Their Habits.* Murray, London.

De Deyn, G.B., and W.H. van der Putten. 2005. Linking aboveground and belowground diversity. *Trends in Ecology and Evolution* 20:625–633.

Denef, K., J. Six, R. Merckx, and K. Paustian. 2004. Carbon sequestration in microaggregates of no-tillage soils with different clay mineralogy. *Soil Science Society of America Journal* 68:1935–1944.

de Ruiter, P.C., J.C. Moore, K.B. Zwart, L.A. Bouwman, J. Hassink, J. Bloem, J.A. De Vos, J.C.Y. Marinissen, W.A.M. Didden, G. Lebbink, and L. Brussaard. 1993. Simulation of nitrogen mineralization in the below-ground food webs of two winter wheat fields. *Journal of Applied Ecology* 30:95–106.

de Ruiter, P.C., A.-M. Neutel, and J.C. Moore. 1996. Energetics and stability in belowground food webs. In *Food Webs: Integration of Patterns and Dynamics,* eds. G.A. Polis and K.O. Winemiller, 201–210. Chapman and Hall, New York.

de Ruiter, P.C., A.-M. Neutel, and J.C. Moore. 1998. Biodiversity in soil ecosystems: the role of energy flow and community stability. *Applied Soil Ecology* 10:217–228.

Ekschmitt K., M.Q. Liu, S. Vetter, O. Fox, and V. Wolters. 2005. Strategies used by soil biota to overcome soil organic matter stability—why is dead organic matter left over in the soil? *Geoderma* 128:167–176.

Elliott, E.T. 1986. Hierarchic aggregate structure and organic C, N, and P in native and cultivated grassland soils. *Soil Science Society of America Journal* 50:627–633.

Elliott, E.T., and D.C. Coleman. 1988. Let the soil work for us. In *Ecological Implications of Contemporary Agriculture*, 23–32. Ecological Bulletins 39. Copenhagen.

Elliott, E.T., K. Horton, J.C. Moore, D.C. Coleman, and C.V. Cole. 1984. Mineralization dynamics in fallow dryland wheat plots, Colorado. *Plant and Soil* 76:149–155.

Epstein, H.E., W.K. Lauenroth, and I.C. Burke. 1997. Effects of temperature and soil texture on ANPP in the US Great Plains. *Ecology* 78:2628–2631.

Fierer, N., M. Breitbart, J. Nulton, P. Salamon, C. Lozupone, R. Jones, M. Robeson, R.A. Edwards, B. Felts, S. Rayhawk, R. Knight, F. Rohwer, and R.B. Jackson. 2007. Metagenomic and small-subunit rRNA analyses reveal the genetic diversity of bacteria, archaea, fungi, and viruses in soil. *Applied and Environmental Microbiology* 73:7059–7066.

Fu, S., K.W. Kisselle, D.C. Coleman, P.F. Hendrix, and D.A. Crossley, Jr. 2001. Short-term impacts of aboveground herbivory (grasshopper) on the abundance and ^{14}C activity of soil nematodes in conventional tillage and no till agroecosystems. *Soil Biology and Biochemistry* 33:1253–1258.

Gijsman A.J., G. Hoogenboom, W.J. Parton, and P.C. Kerridge. 2002. Modifying DSSAT crop models for low-input agricultural systems using a soil organic matter-residue module from CENTURY. *Agronomy Journal* 94:462–474.

Grace, P.R., J.N. Ladd, G. Robertson, and S.H. Gage. 2006. SOCRATES—s simple model for predicting long-term changes in soil organic carbon in terrestrial ecosystems. *Soil Biology and Biochemistry* 38:1172–1176.

Grace, P.R., W.M. Post, D.C. Godwin, K.P. Bryceson, M.A. Truscott, and K.J. Hennessy. 1998. Soil carbon dynamics in relation to soil surface management and cropping systems in Australian agroecosystems. In *Management of Carbon Sequestration in Soil. Advances in Soil Science,* eds. R. Lal, J.M. Kimble, R.F. Follett, and B.A. Stewart, 175–193. CRC Press/Taylor and Francis, Boca Raton, FL.

Gupta, V.V.S.R., and J.J. Germida. 1989. Influence of bacterial-amoebal interactions on sulfur transformations in soil. *Soil Biology and Biochemistry* 21:921–930.

Gupta, V.V.S.R., C.R. Penton, R. Lardner, and J. Tiedje. 2010. Catabolic and genetic diversity of microbial communities in Australian soils are influenced by soil type and stubble management. In *Proceedings of the 19th World Congress of Soil Science; Soil Solutions for a Changing World*, August 1–6 [DVD], http://www.iuss.org. IUSS, Brisbane, Australia, 1–4.

Gupta, V.V.S.R., D.K. Roget, C.W. Davoren, R. Llewellyn, and A. Whitbread. 2008. Farming system impacts on microbial activity and soil organic matter in southern Australian Mallee Soils. In *Proceedings of 14th Agronomy Conference 2008—Global Issues*, ed. M. Unkovich, Paddock Action. http://www.regional.org.au/au/asa/2008/concurrent/crop-agronomy/5918_guptavvsr.htm, The Regional Institute, Gosford, NSW, Australia.

Gupta, V.V.S.R., P.C.J. van Vliet, L.K. Abbott, and E. Leonard. 1999. Farming system and soil biota in Western Australia. CRCSLM/CTT/2/99, p. 4, Adelaide, South Australia.

Handelsman, J. 2004. Metagenomics: application of genomics to uncultured microorganisms. *Microbiological and Molecular Biology Reviews* 68:669–684.

Harris, J. 2009. Soil microbial communities and Restoration ecology: facilitators or followers? *Science* 325:573–574.

Hassink, J., A.P. Whitmore, and J. Kubat. 1997. Size and density fractionation of soil organic matter and the physical capacity of soils to protect organic matter. *European Journal of Agronomy* 7:189–199.

Heal, O.W., J.M. Anderson, and M.J. Swift. 1997. Plant litter quality and decomposition: an historical overview. In *Driven by Nature: Plant Litter Quality and Decomposition*, eds. G. Cadisch and K.E. Giller, 3–30. CABI, Wallingford, UK.

Hendrix, P.F., R.W. Parmelee, D.A. Crossley, Jr., D.C. Coleman, E.P. Odum, and P. Groffman. 1986. Detritus food webs in conventional and no-till agroecosystems. *Bioscience* 36:374–380.

Hilbert, D.W., D.M. Swift, J.K. Detling, and M.I. Dyer. 1981. Relative growth-rates and the grazing optimization hypothesis. *Oecologia* 51:14–18.

Holland, J.N., W. Cheng, and D.A. Crossley, Jr. 1996. Herbivory-induced changes in plant carbon allocation: assessment of below-ground C fluxes using carbon-14. *Oecologia* 107:87–94.

Hunt, H.W., D.C. Coleman, E.R. Ingham, R.E. Ingham, E.T. Elliott, J.C. Moore, S.L. Rose, C.P.P. Reid, and C.R. Morley. 1987. The detrital food web in a shortgrass prairie. *Biology and Fertility of Soils* 3:57–68.

Ingham, R.E., J.A. Trofymow, E.R. Ingham, and D.C. Coleman. 1985. Interactions of bacteria, fungi, and their nematode grazers: effects on nutrient cycling and plant growth. *Ecological Monographs* 55:119–140.

Jangid, K., M.A. Williams, A.J. Franzluebbers, J.S. Sanderlin, J.H. Reeves, M.B. Jenkins, D.M. Endale, D.C. Coleman, and W.B. Whitman. 2008. Relative impacts of land-use, management intensity and fertilization upon microbial community structure in agricultural systems. *Soil Biology and Biochemistry* 40:2843–2853.

Jastrow, J.D. 1996. Soil aggregate formation and the accrual of particulate and mineral-associated organic matter. *Soil Biology and Biochemistry* 28:665–676.

Jenkinson, D.S., D.E. Adams, and A. Wild. 1991. Model estimates of CO_2 emissions from soil in response to global warming. *Nature* 351:304–306.

Jenny, H. 1941. *Factors of Soil Formation*. McGraw-Hill, New York.

Jenny, H., and K. Grossenbacher. 1963. Root-soil boundary zones as seen in the electron microscope. *Soil Science Society of America Proceedings* 27:273–277.

Kleber, M., and M.G. Johnson. 2010. Advances in understanding the molecular structure of soil organic matter: implications for interactions in the environment. *Advances in Agronomy* 106:77–142.

Korthals, G.W., P. Smilauer, C. Van Dijk, and W.H. van der Putten. 2001. Linking above- and belowground biodiversity: abundance and trophic complexity in soil as a response to experimental plant communities on abandoned arable land. *Functional Ecology* 15:506–514.

Lavelle, P., E. Barros, E. Blanchart, G. Brown, T. Desjardins, L. Mariani, and J.-P. Rossi. 2001. SOM management in the tropics: why feeding the soil macrofauna? *Nutrient Cycling in Agroecosystems* 61:53–61.

Mikutta R., M. Kleber, M.S. Torn, and R. Jahn. 2006. Stabilization of soil organic matter: Association with minerals or chemical recalcitrance? *Biogeochemistry* 77:25–56.

Minoshima, H., L.E. Jackson, T.R. Cavagnaro, S. Sánchez-Moreno, H. Ferris, S.R. Temple, S. Goyal, and J.P. Mitchell. 2007. Soil food webs and carbon dynamics in response to conservation tillage in California. *Soil Science Society of America Journal* 71:952–963.

Moore, J.C., E.L. Berlow, D.C. Coleman, P.C. de Ruiter, Q. Dong, A. Hastings, N.C. Johnson, K.S. McCann, K. Melville, P.J. Morin, K. Nadelhoffer, A.D. Rosemond, D.M. Post, J.L. Sabo, K.M. Scow, M.J. Vanni, and D.H. Wall. 2004. Detritus, trophic dynamics, and biodiversity. *Ecology Letters* 7:584–600.

Moore, J.C., and H.W. Hunt. 1988. Resource compartmentation and the stability of real ecosystems. *Nature* 333:261–263.

Moore J.C., K. McCann, H. Setälä, and P.C. de Ruiter. 2003. Top-down is bottom-up: does predation in the rhizosphere regulate aboveground dynamics? *Ecology* 84:846–857.

Moore, J.C., D.E. Walter, and H.W. Hunt. 1988. Arthropod regulation of micro- and mesobiota in belowground food webs. *Annual Reviews of Entomology* 33:419–439.

Moorhead, D.L., W.S. Currie, E.B. Rastetter, W.J. Parton, and M.E. Harmon. 1999. Climate and litter quality controls on decomposition: an analysis of modeling approaches. *Global Climate Change* 13:575–589.

Moran, M.A. 2009. Metatranscriptomics: eavesdropping on complex microbial communities. *Microbe* 4(7):1–7.

Oades, J.M. 1984. Soil organic-matter and structural stability—mechanisms and implications for management. *Plant and Soil* 76:319–337.

Palm, C.A., K.E. Giller, P.L. Mafongoya, and M.J. Swift. 2001. Management of organic matter in the tropics: translating theory into practice. *Nutrient Cycling in Agroecosystems* 61:63–75.

Pankhurst, C.E., B.G. Hawke, H.J. McDonald, C.A. Kirkby, J.C. Buckerfield, P. Michelsen, V.V.S.R. Gupta, and B.M. Doube. 1995. Evaluation of soil biological properties as potential bioindicators of soil health. *Australian Journal of Experimental Agriculture* 35:1015–1028.

Parton, W.J., D.S. Schimel, C.V. Cole, and D.S. Ojima. 1987. Analysis of factors controlling soil organic matter levels in Great Plains grasslands. *Soil Science Society of America Journal* 51:1173–1179.

Paustian, K., G.I. Ågren, and E. Bosatta. 1997. Modeling litter quality effects on decomposition and soil organic matter dynamics. In *Driven by Nature: Plant Litter Quality and Decomposition*, eds. G. Cadisch and K.E. Giller, 313–335. CABI, Wallingford, UK.

Paustian, K., L. Bergstrom, P.-E. Jansson, and H. Johnsson. 1990. Ecosystem dynamics. In *Ecology of Arable Land. Organisms, Carbon and Nitrogen Cycling*, eds. O. Andrén, T. Lindberg, K. Paustian, and T. Rosswall, 153–180. Ecological Bulletins 40. Copenhagen.

Poretsky, R.S., I. Hewson, S. Sun, A.E. Allen, J.P. Zehr, and M.A. Moran. 2009. Comparative day/night metatranscriptomics analysis of microbial communities in the North Pacific subtropical gyre. *Environmental Microbiology* 11:1358–1375.

Preston, C.M., and M.W.I. Schmidt. 2006. Black (pyrogenic) carbon: a synthesis of current knowledge and uncertainties with special consideration of boreal regions. *Biogeosciences* 3:397–420.

Schimel, J.P., and M.N. Weintraub. 2003. The implications of exoenzyme activity on microbial carbon and nitrogen limitation in soil: A theoretical model. *Soil Biology and Biochemistry* 35:549–563.

Schlesinger, W.H. 1996. *Biogeochemistry: An Analysis of Global Change*, 2nd ed. Academic Press, San Diego, CA.

Simmons, B.L., and D.C. Coleman. 2008. Microbial community response to transition from conventional to conservation tillage in cotton fields. *Applied Soil Ecology* 40:518–528.

Six, J., H. Bossuyt, S. Degryze, and K. Denef. 2004. A history of research on the link between (micro) aggregates, soil biota, and soil organic matter dynamics. *Soil and Tillage Research* 79:7–31.

Six, J., E.T. Elliott, and K. Paustian. 2000. Soil macroaggregate turnover and microaggregate formation: a mechanism for C sequestration under no-tillage agriculture. *Soil Biology and Biochemistry* 32:2099–2103.

Smucker, A.J.M., E.J. Park, J. Dorner, and R. Horn. 2007. Soil micropore development and contributions to soluble carbon transport within macroaggregates. *Vadose Zone Journal* 6:282–290.

Thévenot, M., M.-F. Dignac, and C. Rumpel. 2010. Fate of lignins in soils: a review. *Soil Biology and Biochemistry* 42:1200–1211.

Tian, G., L. Brussaard, and B.T. Kang. 1993. Biological effects of plant residues with contrasting chemical-compositions under humid tropical conditions—effects on soil fauna. *Soil Biology and Biochemistry* 25:731–737.

Tian, G., L. Brussaard, and B.T. Kang. 1995. Breakdown of plant residues with chemically contrasting compositions: effect of earthworms and millipedes. *Soil Biology and Biochemistry* 27:277–280.

Tisdall, J.M., and J.M. Oades. 1982. Organic matter and waterstable aggregates in soils. *Journal of Soil Science* 33:141–163.

Trumbore, S. 2009. Radiocarbon and soil carbon dynamics. *Annual Review of Earth and Planetary Sciences* 37:47–66.

Tull, J. 1733. The horse-hoeing husbandry: an essay on the principles of tillage and vegetation. Printed for the author, London.

Upchurch, R.A., C.-Y. Chiu, K. Everett, G. Dyszynski, D.C. Coleman, and W.B. Whitman. 2008. Differences in the composition and diversity of bacterial communities from agricultural and forest soils. *Soil Biology and Biochemistry* 40:1294–1305.

Van Elsas, J.D., A.J. Speksnijder, and L.S. van Overbeek. 2008. A procedure for the metagenomics exploration of disease-suppressive soils. *Journal of Microbiological Methods* 75:515–522.

Waldrop, M.P., T.C. Balser, and M.K. Firestone. 2000. Linking microbial community composition to function in a tropical soil. *Soil Biology and Biochemistry* 32:1837–1846.

Wall, D.H., and J.C. Moore. 1999. Interactions underground: soil biodiversity, mutualism, and ecosystem processes. *Bioscience* 49:109–117.

Wallwork, J.A. 1976. *The Distribution and Diversity of Soil Fauna*. Academic Press, London.

Wang, W.J., and R.C. Dalal. 2006. Carbon inventory for a cereal cropping system under contrasting tillage, nitrogen fertilisation and stubble management practices. *Soil and Tillage Research* 91:68–74.

Wardle, D.A., V.K. Brown, V. Behan-Pelletier, M. St. John, T. Wojtowicz, R.D. Bardgett, G.G. Brown, P. Ineson, P. Lavelle, W.H. van der Putten, J.M. Anderson, L. Brussaard, H.W. Hunt, E.A. Paul, and D.H. Wall. 2004. Vulnerability to global change of ecosystem goods and services driven by soil biota. In *Sustaining Biodiversity and Ecosystem Services in Soils and Sediments*, ed. D.H. Wall, 101–135. SCOPE Volume 64, Island Press, Washington, DC.

Whitford, W.G. 2000. Keystone arthropods as webmasters in desert ecosystems. In *Invertebrates as Webmasters in Ecosystems*, eds. D.C. Coleman and P.F. Hendrix, 25–41. CAB International, Wallingford, UK.

Wickings, K., A.S. Grandy, S. Reed, and C. Cleveland. 2010. Management intensity alters decomposition via biological pathways. *Biogeochemistry* DOI 10.1007/s10533-010-9510-x.

chapter 2

Manipulation of beneficial microorganisms in crop rhizospheres

Richard P. Dick
The Ohio State University

Contents

2.1 Introduction

The rhizosphere (including the endophytic habitat) is a unique microbial environment that is a critical interface for the growth and maintenance of plants. It encompasses apoplastic and intracellular root tissue, root surfaces, and the surrounding soil that is influenced by the root. It is significantly different from bulk soil in terms of chemistry and biology. Roots release various organic materials that include sloughed dead cellular material and actively excreted organic compounds. At the same time, they alter the physical environment, which affects gas and water dynamics and ultimately affects microbial communities and processes they mediate (Hinsinger et al., 2009).

The endophytic niche requires special recognition because of its unique habitat and potential to affect plants. Endophytic bacteria can colonize the apoplastic and intracellular locations in roots without causing negative impacts on the host plant (Schulz and Boyle, 2006). Because of their intimate contact or location within roots, they are more protected from adverse changes in the environment than bacteria in the rhizosphere (Beattie, 2006) and would be expected to interact closely with their host and face less competition for nutrients.

The rhizosphere region is a microbially diverse, highly competitive, and complex ecological environment for microorganisms. We now know that they exhibit highly regulated cell-to-cell communication using signaling compounds. The volume of rhizospheric soil is relatively small compared to the bulk soil of the rooting zone of crops but is nonetheless extremely important in affecting plant pathogens, plant growth promotion, and biogeochemical processes that together strongly affect the yield and quality of crops (Gregory, 2006).

There has been long-standing recognition of the importance of root exudates as sources of nutrients and carbon (C) for promoting and sustaining biological organisms. Roots exude ions, free oxygen and water, extracellular enzymes, mucilage, and a wide range of primary and secondary metabolites (Bertin et al., 2003; Uren, 2007), which result in elimination of waste materials of either unknown function or for root lubrication or plant defense (Uren, 2007). These exudates are an array of low molecular weight compounds, such as amino acids, sugars, organic acids, phenolics, and other secondary metabolites or less-diverse high molecular weight exudates of polysaccharides and proteins (Faure et al., 2009). The latter have a much larger proportion of the exudate mass than the low molecular weight compounds. Plants expend a considerable amount of energy and C through exudation, which can include more than 30% of the energy captured by photosynthesis (Morgan and Whipps, 2001).

These exudates primarily affect microbial communities in two ways. First, they provide rich and relatively readily available sources of energy and nutrients. Second, there is growing evidence of a diverse range of chemical signals from plant roots to microorganisms and vice versa that influence community structure and functions. This creates a functionally complex community with a high level of competition for colonization by bacteria and fungi that may be beneficial, neutral, or pathogenic toward plants. Although a general understanding of root exudates and their overall importance relative to plant nutrition, pathogen responses, and beneficial microbial interactions has been established, the role and magnitude of chemical signaling of root to root, root to microorganism/invertebrate and microorganism/invertebrate to root are just beginning to be understood. The mechanisms used by these organisms to select and interpret the signals produced in the rhizosphere are largely unknown (Uren, 2007).

There is a growing body of evidence that the biology of the rhizosphere could be exploited by manipulating root and microbial interactions to improve the productivity and sustainability of agricultural systems. These beneficial rhizospheric organisms have shown the potential to increase nutrient availability to plants, stimulate growth, and protect plants from pests and pathogens. This would be particularly valuable toward the development of sustainable, biologically based agricultural systems because this has potential for reduced or no external inputs—in effect substituting microbial benefits for synthetic inputs.

2.1.1 *Microbial ecology, genomics, and synergisms*

Since 2000, there is increasing evidence for the feasibility of manipulating microorganisms to substitute for pesticides, significantly reducing the need for off-farm fertilizer inputs, and stimulating plant growth. In the following sections, the mechanisms, organisms, and research needs for specific strategies focused on plant protection, disease suppression, plant growth stimulation, and enhanced plant nutrition are discussed. The goal of this section is to provide a brief overview of the areas of basic research in microbial ecology that will be needed to provide the underpinnings for developing biologically based agroecosystems.

The rhizosphere microbial habitat is found within a soil physical structure of aggregates and pore space and is associated with roots (rhizoplane, spaces between cortical cells and nodules) and the intestines of soil animals. On plant or soil surfaces, microorganisms exist in thin water films, and communities may develop biofilms to further survival and withstand stress in soils. Microorganisms occupy a small (1–5%) portion of the soil space and exist in hot spots where environmental conditions allow microbial life due to adequate C and energy sources, nutrients, ionic composition, available water, temperature, pressure, air composition, electromagnetic radiation, pH, and oxidation-reduction potential. Rhizo-organisms live along a root gradient colonizing (1) the soil near roots; (2) the rhizoplane (root surface); (3) root tissue, inhabiting spaces between cortical cells; and (4) intracellular specialized root structures or nodules, where populations less than 50 mm from the root reach 10^9–10^{12} microbial cells g^{-1} soils (Foster et al., 1983; Pinton et al., 2001). Another important characteristic is that almost 80–90% of the microorganisms inhabiting soil are on solid surfaces (Hattori, 1973).

2.1.2 *Gene pools and microbial diversity*

Billions of microorganisms live in a seemingly harsh and competitive soil environment and yet ultimately in harmony for species survival and performance of ecosystem services. Although it seems likely we know of most of the important metabolic pathways, a full understanding of genetic, functional, and spatial diversity of soil microbial communities remains elusive. This is because of the complexity of biogeochemical interactions and communications among and within microbial and plant cells. If this can be understood, the soil and rhizosphere communities could be exploited to develop a truly sustainable agricultural system.

Until 30 years ago, it was estimated that there may be on the order of hundreds of species per gram of soil, which was based on isolation and culturing techniques. Subsequently, landmark reassociation analysis of bacterial community DNA in 1990 and a later study by Torsvik et al. (1990, 1996) estimated there were 4,000–10,000 bacterial species per gram of soil. However, this was superseded by reanalysis of the data using new analytical methods by Gans et al. (2005) and Handelsman and Tiedje (2007), who showed soil could have more than 1 million distinct genomes per gram of soil. It should be pointed out that these estimates are for prokaryotes and do not account for eukaryote species that also must make a significant contribution to soil genomes. It is believed that more than 99% of these are unknown and unculturable (Handelsman and Tiedje, 2007).

2.1.2.1 *Microbial sequencing and metagenomics*

Revolutionary technological developments in high-throughput DNA sequencing have resulted in the publication of many whole-genome sequences. The U.S. Department of Agriculture (USDA) Microbial Genome Sequencing Program, with support from other federal agencies, has sequenced some of the most relevant microorganisms. These sequences are available to the public and can be used by investigators to address issues of scientific and societal importance, including the impact of microorganisms on the productivity and sustainability of agriculture and natural resources.

Extensive sequencing is required to capture the microbial diversity in a complex environment such as soil. Methods used to recover environmental sequences tend to overrepresent the most abundant community members. Newer technologies such as 454 pyrosequencing (also known as massive sequencing) and single-molecule sequencing have the potential to make these projects considerably more cost efficient and productive.

The science of metagenomics involves genome-level characterization of whole microbial communities on a culture-independent and high-throughput scale. While sequencing answers questions about evolutionary relationship (phylogenetics), metagenomics has the potential to determine how communities are functioning in a given environment (Singh et al., 2004).

A promising metagenomics approach is the construction of bacterial artificial chromosome (BAC) clone libraries to define communities based on particular functional genes that are present and potentially active in the community. Screening of BAC libraries for functional genes has identified both novel and previously recognized antibiotics (Rondon et al., 2000), antibiotic resistance genes (Diaz-Torres et al., 2003), enzymes (Henne et al., 2000), and a protein involved in a novel photosynthetic pathway that may play a significant role in the global carbon cycle in the sea (Beja et al., 2000). Current BAC methods rely on the use of an *Escherichia coli* gene expression vector, but advancement of this approach will require the development of additional expression vectors because only a limited number of genes can be expressed in *E. coli.*

Gene arrays using reverse transcribed messenger RNA (mRNA) and designed for a community of interest can assess information about gene diversity and expression. These techniques are still in their infancy, particularly when applied to natural communities. As the database of environmental functional genes grows, habitat-specific microarrays will provide quantitative information about community gene expression. In addition to more database information, enhancement of array technology for greater sensitivity and specificity requires improved methods for extraction of nucleic acids from soil and improvement in data analysis tools.

2.1.2.2 *Mitigating factors for the power of metagenomics*

The high species diversity of soils and rhizospheres seemingly has led to functional redundancy, which would suggest functional characteristics of component species are at least as important or perhaps more important than species diversity per se for maintaining essential processes (Andrén and Balandreau, 1999; Bardgett and Shine, 1999). Undoubtedly, some minimum but likely large number of species living in stable communities is essential for mediating biogeochemical processes to enable ecosystem function in changing environments, the so-called insurance hypothesis (Loreau et al., 2001).

Unlike higher life-forms for which species can be clearly defined, the avalanche of genomic data indicates that defining species diversity or even the concept of species for microorganisms is becoming increasingly difficult. In part this is due to the growing recognition of *horizontal gene flow* (HGF) (nongenealogical transfer of genetic material from one organism to another) among bacteria or between viruses and bacteria. HGF has implications for defining how many species there actually are in soils and suggests that, rather than discrete genomes, there are genetic continuums (Golenfeld and Woese, 2007). In this scenario, microorganisms absorb or abandon genes as needed to adapt to environmental conditions. Furthermore, this would suggest that microorganisms are largely cooperative in terms of survival and function.

Complicating our understanding of species diversity, the role of gene transfer in the environment is a largely unknown entity. It is important to develop experimental tools that can precisely determine the mechanisms and location of gene transfer in soils and rhizospheres to exploit microbial-plant interactions to fully utilize the power of genomics. In addition, it will be important not only to identify putative beneficial microorganisms but also to determine the underlying mode of action at the physiological and genetic levels. This will require improvements in the ability to culture microorganisms from soil in addition to the novel culture-independent methods described in this chapter.

Theories on the role of species diversity are largely based on the ecology of aboveground terrestrial organisms with little understanding of the significance for soil and rhizospheres (Wardle and Giller, 1996; Ohtonen et al., 1997; Griffiths et al., 2000). This is further complicated for soil microbial ecology by the aforementioned limitations of how to define species and implications of gene flow for identifying species. Thus, the links between biodiversity and soil functioning are poorly understood, yet fundamental to exploit microorganisms and their interactions with plants to improve the productivity of crops. Consequently, as research proceeds, there will be a need for in-depth studies of functional genes and gene flow, as well as determining an appropriate definition of microbial species (or discarding the importance of species diversity per se in favor of other, as-yet-unknown means of describing microbial community structure).

2.1.3 Signaling

Besides species diversity and gene flow, another important and poorly understood phenomenon is cell-cell communication by chemical signaling among microorganisms and between microorganisms and plants. There are likely a large number of chemicals that allow microorganisms to coordinate gene expression on a population-wide scale. One such system is quorum sensing (QS), which involves production, secretion, and subsequent detection of small hormone-like signaling molecules known as autoinducers, first discovered over 25 years ago in marine bacteria (Nealson and Hastings, 1979). Signal compounds are produced by bacteria, which on reaching a critical level, induce gene expression encoding for a variety of phenotypical and physiological responses, such as bioluminescence, pathogenicity and pigment induction, cell conjugation, growth regulation, nodulation (e.g., rhizobial-legume symbiosis), biofilm formation, and cell motility (Lithgow et al., 2000; Fray, 2002). The most widely studied QS compounds are N-acylhomoserine lactones (AHLs), which vary in length, oxidation state, and degree of saturation of their acyl side chains to provide a degree of species specificity (Badri et al., 2009). A range of gram-negative bacterial species employs AHLs, whereas gram-positive bacteria typically use peptides as the QS molecule. Signaling can also occur by volatile compounds.

There has been great progress in understanding plant QS compounds that target bacterial receptors for favorable biocontrol responses. In particular, AHLs produced by rhizobacteria have been shown to promote biocontrol agents or interfere with pathogens of biocontrol or affect other rhizosphere processes (Lugtenberg and Leveau, 2007). There is evidence that plants can produce and secrete substances that mimic AHL activity and could therefore influence the density-dependent behavior of biocontrol and other beneficial rhizobacteria (Teplitski et al., 2000). This offers an opportunity through plant breeding or transgenic gene insertion to develop crops with roots that can produce QS AHL-mimicking compounds that stimulate or optimize the application of biocontrol strains (Chin-A-Woeng et al., 2001; Castang et al., 2004) by inducing regulated genes via specific AHL receptors on biocontrol bacteria. This has great potential for engineered crops to deal with both pathogens and symbionts.

To utilize beneficial microorganisms fully, a fundamental understanding of signaling mechanisms and the array of signaling compounds is required. The impacts of signaling are multifaceted, with microorganisms communicating with each other to carry out certain functions and two-way signaling between plants and microbes. One of the better-understood plant-microbe communications is between rhizobia and legumes at the beginning of the symbiotic N_2 fixation process and other stages of symbiosis and involves the exchange of flavonoids and chitin-based compounds.

Another example is fungal elicitors detected by plants such as chitin fragments that cause the plant to have defense reactions to fungal pathogens and to produce phytoalexins (Ebel and Mithöfer, 1998). Most often, these phytoalexins are flavonoids and isoflavonoids, such as genistein.

Protein elicitins are unique to the genus *Phytophthora* and are encoded by a large multigene family divided into at least eight different classes based on their sequence homology and protein motif diversity (Jiang, Tyler, and Govers, 2006; Jiang, Tyler, Whisson, et al., 2006). Class I elicitins became of interest when it was shown that these proteins induced a plant hypersensitive response (HR) when infiltrated on tobacco leaves (Ponchet et al., 1999). Consequently, further research was done to establish whether elicitin genes are avirulence genes that determine the host range (Kamoun, 2001; Ponchet et al., 1999). As a result, studies of elicitins have largely focused on elicitor activity on plants (e.g., Baillieul et al., 2003), characterizing protein structure (e.g., *Pantoea* and *Zoogloea*) (Lascombe et al., 2004), and discovering new elicitins secreted by *Phytophthora* spp. (e.g., Churngchow and Rattarasarn, 2000).

Yousef et al. (2009) showed that sterols trigger the expression of elicitin genes in the plant pathogen *Phytophthora sojae*, and that elicitin secretion provides a protein sterol carrier in the soil/rhizosphere environment. The latter is important as it shows how this eukaryote organism can acquire essential sterols from a plant that it does not produce itself.

Plant growth-promoting (PGP) microorganisms will produce compounds and cause systemic acquired resistance (SAR), which causes a range of defense mechanisms (Metraux et al., 2001). For instance, SAR involves signal transduction, generation of phytoalexins, oxidative stress protection, and lignification (Reymond and Farmer, 1998).

At the same time, there is evidence that plants send signals to microorganisms, which in turn result in stimulation of microorganisms that can have synergistic and positive effects on plants. However, the relative importance of plant signaling versus the increase in nutrients and energy sources provided by plant roots has on promoting beneficial microorganisms is not known. Given that some rhizosphere bacteria are reported to stimulate plant growth through production of plant hormones, it may be that ethylene, salicylic acid, and jasmonic acid are active here as well.

In-depth research is needed to identify the range of signaling compounds produced by rhizosphere microorganisms and by plant roots and the associated response mechanisms. For example, riboflavin and lumichrome, both known to stimulate plant root growth, have been shown to activate the LasR QS receptor, thus acting as QS agonists (J. Robinson, personal communication, 2011). This finding raises the possibility that vitamins and their derivatives might play important roles as QS mimic compounds as well as precursors to essential cofactors. Additional research is needed to exploit the use of activator inducers on beneficial microbial activator production that could further increase plant growth. An improved understanding of this aspect of plant-microbe relationships will allow for manipulation of these compounds to enhance plant growth, increasing crop yield and allowing for efficient and less environmentally damaging production of plant biomass for food and feed production. In summary, microbial and plant signaling is not well understood, and in general, this type of effect remains to be investigated and is strongly needed to fully exploit the potential of beneficial microorganism for crop production. If this breakthrough could be made, it offers the possibility of "programming" plant roots to release compounds to activate beneficial microbial responses as needed at critical growth stages or to withstand pathogenic and environmental stress.

2.1.4 Spatial distributions of microorganisms and biofilms

The communication networking of microorganisms is closely related to spatial distributions and interactions of microbial communities. In soils, there is recognition of the potential role of hyphal organisms such as fungi and actinomycetes versus unicellular organisms (mostly bacteria) relative to exploitation-specific niches. For example, fungi dominate on the surface of larger aggregates (>250 μ) of soils compared to microaggregates and pores where bacteria dominate (Gupta and Germida, 1988). At the same time, cells of singular or multiple species can adhere to each other to form biofilms, which are defined as interphase boundaries (solid/gas, solid/liquid, and liquid/gas).

A biofilm consists of an assemblage of heterogeneous microcolonies (e.g., protozoans, fungi, nematodes, and prokaryotes) that are encapsulated in extracellular polymeric substances (EPSs). Microorganisms embedded in biofilms often have altered phenotypic expression in comparison to the same cells in a planktonic state (Watnick and Kolter, 2000). Each component of biofilms has a unique function: The liquid interface is the source of nutrients and energy; the EPSs provide physical structure, protection from environmental extremes, spatial distribution of organisms, and a medium for diffusion of nutrients and signaling chemicals. The organisms can work in consortia to perform biogeochemical processes or cooperatively to aid individual microorganisms. Interestingly, there is now speculation that biofilms have complex cell-cell signaling and physical structures such as water channels to disperse nutrients, metabolites, and waste products (Sauer et al., 2007) and "nanowires" that transfer energy (Schaudinn et al., 2007). Early biofilm research focused on aquatic systems because they were easily observed. Recently, preliminary research suggests biofilms do exist in soils, but this remains largely unexplored. Biofilms on rhizoplanes have been established and related to rhizodeposits of roots that result in discontinuous nonuniform colonies (Rovira and Campbell, 1974; Hansel et al., 2001).

Clearly, recent advances in molecular biology and microscopy have provided information on the importance of biofilms on roots in performing certain tasks (e.g., plant protection, transforming nutrients to plant-available forms) and hint that biofilms do exist in bulk soil. However, the complexity and intricacies of organismal relationships and biogeochemical processes are largely unknown in these biofilm communities. Indeed, scientists are debating the existence of biofilms in bulk soil and how to define biofilms in terms of cell number or species complexity.

2.2 Emerging components of optimized and sustainable rhizospheres

2.2.1 Phytostimulators

A diverse array of bacteria and some fungi has been identified that can produce PGP substances and increase crop yields (Chen et al., 1994; Amara and Dahdoh, 1997; Biswas et al., 2000a, 2000b; Hilali et al., 2001; Asghar et al., 2002; Khalid et al., 2004; Larkin, 2008). Phytohormones that are known to be produced by PGP microorganisms are auxin, cytokinins, indole-3-acetic acid (IAA), and gibberellin, with auxin of primary interest (Garcia de Salamone et al., 2001; Steenhoudt and Vanderleyden, 2000; Cleland, 1990; Hagen, 1990).

Another beneficial mechanism is the optimization of growth conditions by *Pseudomonas* spp. by removing ethylene (naturally produced by plants during metabolism; inhibits root growth) from the root zone. They do this by producing the enzyme 1-aminocyclopropane-1-carboxylate (ACC)-deaminase, which hydrolyzes the ethylene precursor ACC (Penrose and Glick, 2001).

Some examples of organisms inoculated on crop roots that increase yields or growth include *Azospirillum* on a variety of crops; *Rhizobium leguminosarium* on rice and wheat (but interestingly only in the presence of N fertilizer) (Biswas et al., 2000a, 2000b); *Pseudomonas* on potatoes (Kłoepper et al., 1980); *Bacillus* spp., producing various gibberellins, on alder (Gutierrez Munero et al., 2001); *Pseudomonas fluorescens* strain on radish; and *Bacillus licheniformis* on *Pinus pinea L.* seedlings (Probanza et al., 2002). However, regionality and local environments need to be investigated to determine the universality for using phytohormone-producing microorganisms. This was shown on a strain of *B. japonicum* that grew best under cooler soil conditions, with increased soybean protein (Zhang et al., 2002). Earlier reviews have reported yield increases of inoculated over uninoculated controls of 5–30% in about 70% of inoculation trials reported, mostly with *Azospirillum* (Okon and Labandera-Gonzalez, 1994) and 12–30% on maize at various sites in the United States (Riggs et al., 2001).

Why rhizobacteria have evolved to produce phytohormones is not well understood. It may be that this stimulates greater production of root exudates to increase root development, which provides enhanced nutritional benefits and habitable space for microbial colonization, giving those microorganisms with phytostimulating properties an evolutionary advantage. The possible role of rhizosphere microorganisms that produce plant growth stimulation presents an exciting potential to increase crop yields naturally (Broughton et al., 2003). However, the mechanisms underlying this are poorly understood. These organisms operate within complex communities and often have multiple effects on plants. For example, *Azospirillum* produces phytohormones but also can fix N_2. Commercially available biological inoculation technologies are emerging in Western countries, but most of these products are not based on solid science. Larkin (2008) showed that several commercial biologicals did increase yields but only under certain crop rotations. The research required to fully develop phytostimulating microbial systems will require work at all levels, from ecology to proteomics and metabolomics.

2.2.2 *Disease-suppressive soils and plant-protecting microorganisms*

Since the mid-1990s, there have been surprising and exciting discoveries for natural methods to suppress or eliminate pathogens or protect plants. Intensive studies of disease-suppressive soils have led to the development of new methods of analysis (Gross et al., 2007; Borneman et al., 2007; Bolwerk et al., 2005; Benitez et al., 2007) and new insights into the nature of soilborne disease suppression (Weller et al., 2002; Hoitink and Boehm, 1999). Such advances indicate that active management of soil microbial communities can be an effective approach to develop natural suppression of soilborne diseases and improve crop productivity (Mazzola, 2004; Raaijmakers et al., 2009). Generally, there are two approaches to actively manage crop-associated microbial communities.

The first approach is to develop disease-suppressive soils through manipulation of carbon inputs. This involves adjusting the types and timing of organic inputs, such as cover crops (Widmer et al., 2002), animal manures (Darby et al., 2006), composts (Abbasi et al., 2002; Darby et al., 2006), organic amendments (Rotenberg et al., 2005; Stone et al.,

2003), compost teas (Larkin, 2008), and crop sequencing (Larkin, 2008). Such approaches have been shown to provide site-specific reductions in disease and pest incidence. The advantage of this approach is that it relies on locally available resources to maximize soil health in a sustainable way. The disadvantage is that outcomes tend to be more variable depending on soil type and climate, so knowledge of soils and available inputs is essential. Therefore, for the most part recommendations must be developed and adapted on a regional basis.

The second approach involves inoculation of disease-suppressive microorganisms. All soils harbor detectable populations of disease-suppressive organisms; however, the diversity, relative abundance, and activities of these organisms can vary substantially from site to site. Historically, researchers recovered microbes and then screened them for disease-suppressing activity. However, molecular tools now allow us to identify microbes that are suppressive in situ and recover them in a directed fashion (Borneman et al., 2007). Such an approach has already proven useful at collecting indigenous fungi that acted as effective inoculants to suppress soilborne diseases caused by nematodes (Olatinwo et al., 2006). In addition, analyses of the genetics and genomics of disease-suppressive microbes (Paulsen et al., 2005; Bangera and Thomashow, 1999; Koumoutsi et al., 2004) have led to new methods to isolate disease-suppressive microbes in a directed fashion from any location (McSpadden Gardener et al., 2001; Joshi et al., 2006). This approach also has led to the development of effective and low-cost inoculants (McSpadden Gardener et al., 2005).

Endophytic fungi or bacteria that colonize the host can stimulate systemic resistance to pathogens by producing compounds or causing changes in plant morphology or physiology that can control or prevent plant diseases (Faeth, 2002; Harman et al., 2004; Bailey et al., 2006; Melnick et al., 2008; Hanada et al., 2010). Inoculation with disease-suppressing microorganisms has been extended to harvestable products. This was shown by Shi et al. (2011) with endophytic *Pseudomonas putida* MGY2, which reduced anthracnose (*Colletotrichum gloeosporioides*) infection in harvested papaya fruit. Fruit treated with MGY2 showed a significantly lower disease index, disease incidence, and lesion diameter than the control. Potential control mechanisms were related to enhanced activity of phenylalanine ammonia-lyase, catalase (CAT), and peroxidase and increased the phenolic content. These results suggested that papaya fruit is capable of responding to the endophyte *P. putida.*

In Taita District, Kenya, integrated soil management interventions (cow manure, inorganic fertilizers, and *Trichoderma* inoculant) controlled the *Fusarium* root rot of maize and bean cropping systems (Okoth and Siameto, 2011). The use of mavuno fertilizer (containing 11 nutrients in balanced proportion) recorded the least root infection, followed by triple superphosphate plus calcium ammonium nitrate, mavuno plus *Trichoderma,* and cow manure plus *Trichoderma,* implying that low soil fertility could be the reason for high infection of *Fusarium* in soil (Okoth and Siameto, 2011).

Considerable progress has been made in identifying specific organisms and managing organic inputs to suppress diseases or protect plants from infection in developed countries, but there has been limited progress in developing countries. Methods that transform resident microbial communities in a manner that induces natural soil disease suppression have potential as components of environmentally sustainable systems for management of soilborne plant pathogens to reduce the need for pesticides. A high priority should be devoted to manipulating organic amendments from on-farm sources to naturally suppress diseases and protect plants, particularly for developing countries where farmers operate under infrastructure and economic constraints.

2.2.3 Biological nutrient enhancement and nitrogen fixation

The two of most important plant nutrients are nitrogen (N) and phosphorus (P), for which microbial-based technologies could be developed to greatly reduce the need for off-farm purchased fertilizers. Universally, N is typically the most limiting crop nutrient. Biological N fixation, particularly the development of productive associations of N-fixing bacteria or transgenic N-fixing nonlegume crops, has great potential to transform the productivity and quality of crops produced by sub-Saharan Africa (SSA) and South Asia (SA) farmers. Biofertilization accounts for approximately 65% of the N supply of crops worldwide, primarily via *Rhizobiaceae*-legume symbiosis (Lugtenberg et al., 2002).

Rice, wheat, and maize are the three major staple food crops for the world population. These crops can require as much as 100–200 kg fertilizer-N ha^{-1} to produce optimal yields under ideal moisture and temperature conditions. This is an expensive input, particularly for farmers from developing countries. Thus, development of biological N-fixing systems that they could use directly on the farm would transform their productivity and profitability. Background levels of N fixation by free-living or endophytic bacteria are likely quite low (~5–10 kg N ha^{-1} y^{-1}), which are inadequate for optimum crop growth for most crops. However, it should be pointed out that the standard fertilization method of soil amendments is inefficient, with only 40–60% of the N applied taken up plants because it is lost through leaching or immobilized by soil microorganisms. Here, particularly endophytic N-fixing organisms have potential to overcome this low efficiency because the N they produce presumably would be taken up directly by plants and not be susceptible to losses, such as leaching or immobilization of N, that happen when N fertilizers are applied to soils. This lowers the amount that microbial N fixers would have to provide by about 50% of what is needed when amending soils with N fertilizers.

It should also be pointed out that large areas of the developing world are semiarid, where crops like millet and sorghum are grown under much lower yield potential and thus have much lower N requirements than crops in higher-rainfall regimes. Thus, in this environment optimizing N-fixing organisms would require much lower N requirements, which potentially makes biological N fixation more feasible.

The other important nutrient is phosphorus, particularly in the tropics as many soils in the region have inherent low P levels and high P adsorption capacity. This is exacerbated by the lack of P replacement of P removed by crop harvesting and soil erosion. Unlike N deficiency, which potentially can be corrected by biologically fixing atmospheric N_2, P deficiencies need to be corrected by adding P sources to soils and increasing P uptake efficiency.

In the following sections, specific categories of biological nutrient enhancement opportunities are discussed. Besides biological N_2 fixation and increasing nutrient availability in the rhizosphere, microorganisms can enhance nutrient efficiencies by increasing root surface area, enhancing other beneficial symbioses of the host plant, and having combinations of microbial interactions, which are further discussed in the rhizosphere bioengineering section.

2.2.3.1 Free-living and endophytic nitrogen-fixing bacteria

Free-living diazotrophs require a chemical, nonphotosynthetic energy source, whereas the photosynthetic diazotrophs utilize light energy. Associative diazotrophs inhabit the rhizosphere soil or apoplastic space of roots, can obtain energy materials from the plants, and have the greatest potential to be exploited for fixing N and making it directly available to plants (An et al., 2001). The most notable of these is the legume-*Rhizobia* symbiosis. However, given the long research history of legume N fixation, our emphasis focuses on developing biological N fixation for nonlegume crops because, if successful, this could significantly transform agriculture by enabling on-farm generation of N food crops that are not legumes (e.g., rice, wheat, sorghum, millet).

A wide range of root endophytic bacteria that can fix N_2 have been identified, including *Arhizobium*, several *Burkholderia* spp., *Gluconacetobacter*, *Herbaspirillum*, *Azoarcus*, *Acetobacter*, *Arthrobacter*, *Azospirillum*, *Azotobacter*, *Bacillus*, *Beijerinckia*, *Bradyrhizobium*, *Derxia*, *Enterobacter*, *Erwinia*, *Herbaspirillum*, *Klebsiella*, *Methylobacterium*, *Mesorhizobium*, *Pseudomonas*, *Ralstonia*, *Rhizobium*, *Sinorhizobium*, *Stenotrophomonas*, *Zoogloea*, and a large number of cyanobacterial genera (Werner, 2007; Bergmann et al., 2007; Gillis et al., 1995; Palus et al., 1996; Chelius and Triplett, 2000; Rosenblueth et al., 2004; Jha and Kumar, 2009; Zhang et al., 2002; Lodewyckx et al., 2002; Barraquio et al., 2000; James et al., 2000).

The physical location of diazotrophs seems to be important in controlling rates of N fixation. The interior of plant tissues may be a more favorable niche for N fixation because of lower partial oxygen pressure (pO_2) and more direct plant uptake of N fixed by bacteria (James and Olivares, 1998).

A number of *Azospirillum* diazotrophic species have been shown to fix N (Lodewyckx et al., 2002). This diazotrophic bacterium has been widely studied and found on the rhizoplane, rhizodermis, apoplast, rhizodermal cells, and root hairs of a number of crops using techniques such green fluorescent protein (GFP) labeling, strain-specific antibodies, and fluorescence in situ hybridization (FISH) (Scholter et al., 1997; Assmus et al., 1995; Rothballer et al., 2003). Yield increases of 5–30% with *Azospirillum* inoculation have been found on wheat, sorghum, and maize (Dobbelaere et al., 2001). However, it should be noted that, typically, endophytic organisms produce phytohormones (e.g., auxins), which are likely important for them to colonize plant surfaces and cells and have several implications. This improves early root and shoot development and increases lateral roots, root hair density, and root branching (Dobbelaere et al., 2001). Thus, inoculation with diazotrophic organisms can also cause phytohormone yield responses. However, ^{15}N dilution studies of rice have shown that up to 30% of the N uptake can be attributed to *Azoarcus* sp. BH72 and *Gluconacetobacter iazotrophicus* PAL3 (Hurek et al., 2002; Hurek and Piche, 2002; Selesi et al., 2005).

In another N dilution study, Mia et al. (2007) showed that adding N fertilizer suppressed N_2 fixation N by *Azospirillum* and *Bacillus* spp. inoculated to banana. This same research group showed that inoculation (*Azospirillum brasilense* strain Sp7 and *Bacillus sphaericus* strain UPMB10) of banana root seedlings optimized root and shoot growth, root proliferation, photosynthetic rate, and yield when N fertilizer was applied at 33% of the recommended N rate (Mia et al., 2010). In addition, besides providing a significant level of atmospheric derived N, inoculated plants had increased calcium (Ca) and magnesium (Mg) uptake.

Mia et al. (2007) showed that addition of inorganic N can affect diazotrophic N fixation rates. They found that at 0 ppm N addition 38% of N uptake of 45-day-old seedlings was derived from N fixation compared to only 5% when 50 ppm N was added to soil. This is notable as it showed that addition of inorganic N suppressed N fixation in diazotrophs, which is well established for the symbiotic N fixation of rhizobium in legumes. Therefore, N fertilizer management must be considered in the development of diazotrophic-based N fixation strategies.

Another consideration for optimizing diazotrophic N fixation is plant genotype. This was shown by Naher et al. (2009) on rice. They found that strains of *Rhizobium* sp. (Sb16) and *Corynebacterium* sp. inoculated to rice seedlings significantly increased biomass of 45-day-old seedlings compared to the control, 35 kg N ha^{-1}, but there was an interaction between microbial strain and rice genotype. The N content was highest for the rice cultivar

Mayang Segumpal when colonized by *Rhizobium* sp. (SB16) (4.47% tissue N), while the MR219 rice cultivar had the highest N content when inoculated with *Corynebacterium* sp. (4.30% tissue N). These results were similar to those of Rosenblueth and Martínez-Romero (2006) and Gyaneshwar et al. (2002), which showed an interaction between rice cultivars and microbial species relative to N fixation. Naher et al. (2009) showed that both strains were infecting surface and endophytic locations in both rice cultivars and speculated that the interaction of diazotroph by rice cultivar may be due to differences in root exudates and the plant root tissue architecture. These results demonstrated the importance of multidisciplinary research among microbiologists, plant physiologists, and plant breeders to optimize the N fixation by diazotrophs.

The most notable example for practical utilization of biological N_2 fixation is on sugarcane (*Saccharum* spp.), for which 60–70% of the N comes from endophytic bacteria (Boddey et al., 1995, 2003; Saravanan et al., 2008). An important diazotrophic species on sugarcane (and other crops) is *Gluconacetobacter diazotrophicus*, which is a N-fixing, aerobic α-proteobacterium (Alvarez et al., 1995) isolated from sugarcane that colonizes the plant's apoplast and xylem (Fuentes-Ramírez et al., 1999; Dong et al., 1994) in an endophytic manner (Baldani and Baldani, 2005; Reis et al., 2000).

Typically, the largest plant response to diazotrophic bacteria is on degraded soils and in water-stressed regions when low or intermediate levels of fertilizer are applied (Dobbelaere et al., 2001). This is encouraging for farmers in developing countries because these are exactly the kind of soils and conditions they regularly encounter.

To develop this technology, fundamental research is needed as the detailed mechanisms of this N-fixing symbiosis are not well understood. There needs to be a clear separation of the role of phytohormones in both stimulating plant growth and its role on N fixation. Information is needed on the role of physiological plant responses to endophytic diazotrophs and which plant responses are needed to enhance N fixation. For example, roots treated with natural, synthetic, or precursors of auxin will result in the formation of nodular-like tumors (paranodules) that diazotrophs colonize and exhibit some degree of nitrogenase activity. However, it is unclear whether this is advantageous over other symbiotic relationships between diazotrophs and normal roots (Vessey, 2003).

Another question that needs to be answered is the energy, nutritional, or signaling requirements needed to stimulate N-fixing bacteria. It may well be that considerably more C is needed than is currently produced in modern nonlegume crop rhizospheres. For example, legumes can fix agronomically significant levels of N but do so by utilizing over 30% of the photosynthate produced by the host plant. Modern breeding programs that have developed crops under high N levels that emphasize aboveground yield may have selected against rhizospheres that promote N-fixing microorganisms.

2.2.3.2 *Phosphorus-mineralizing microorganisms*

Most soils are deficient in P readily available to plants. In contrast to N, which can be obtained from the atmosphere through biological fixation, P must be supplied to plants by native P or through the application of fertilizer. The majority of applied P is rapidly "fixed" in plant-unavailable soil fractions (Sanyal and De Datta, 1991). The strong chemical reactivity of phosphate (P) results in very low plant recovery of only 10–20% of the P applied to soils (McLaughlin et al., 1988; Holford, 1997), and global annual consumption of P-based fertilizers exceeds 30 million tons (International Fertilizer Industry Association [IFIA] 2001). It is expected the world's known reserves of high-quality rock phosphate will be consumed within the next 80 years (Isherwood, 2000). However, most farmers in developing countries have used very little P fertilizer because of a lack of infrastructure for delivery or low purchasing power. Consequently, soils in these regions are highly depleted in P (Tiessen, 2005), and any solution to increasing yields in SA and SSA must address P nutrition.

Plant rhizospheres support large populations of soil microorganisms, and a number of bacteria (*Pseudomonas* spp. and *Bacillus* spp.) and fungi (most notably mycorrhizae but also *Aspergillus* and *Penicillium*) have been shown to aid P availability to plants. Microorganisms can enhance P availability by (1) solubilization of inorganic P by acidification (e.g., production of organic acids, carbonic acid, or H^+) or chelation; (2) mineralizing organic P by producing extracellular phosphatases or phytases (phytate constitutes ~50% of soil organic P); and (3) increase root growth or root hair development by production of phytohormones (Jakobsen et al., 2005). A considerable number of organisms have been identified to have P enhancement characteristics. However, if the work was done in the lab via screening or even under field conditions, unless there are specific measurements to determine that P deficiency was reduced or eliminated, a growth response may be due to other factors (Whitelaw, 2000). Nonetheless, there are a significant number of studies that have shown improved P nutrition and yields on a wide range of crops by inoculating with P-enhancing microorganisms (see reviews by Whitelaw, 2000; Leggett et al., 2001; Jakobsen et al., 2005).

Besides the expected response on P-deficient soils, studies have shown P improvement with P-enhancing microorganisms in the presence of rock phosphate (low-soluble material that is used to manufacture P fertilizer) (Barea et al., 2002). This is important for SA and SSA farmers as it may be more cost effective (indigenous sources of rock phosphate exist in SA and SSA that can be mined and used directly), and on high P-fixing soils, it can have longer residual effects that last several years on acid soils (see review by Sahrawat et al., 2001).

Symbiotic fungi that have a long research history that is worthy of special consideration are mycorrhizal fungi. For crop species, endomycorrhizae form arbuscular structures in close association with host cells; the plant provides the carbohydrates to the fungus. The fungus in turn develops an extensive hyphal network that can transport P and other nutrients to the plant (Smith and Read, 1997). This factor greatly increases the volume of soil for the plant to explore; also, mycorrhizae can mobilize P by excreting phosphatases and organic acids (Whitelaw, 2000). It should also be noted that coinoculation of plants with mycorrhizal fungi and phosphate-solubilizing or -mineralizing microorganisms (so-called helper microorganisms, many of the same ones mentioned) can further increase the plant P response and therefore the effectiveness of mycorrhizal fungi (see review by Jakobsen et al., 2005).

In conclusion, it is evident that microorganisms are an integral component of the soil P cycle, and they can mobilize soil P and increase yields. However, exploitation of microbial processes to increase P uptake by crops has had limited success (Jakobson et al., 2005). To achieve optimal benefits, the following are needed:

1. Identification and genetic characterization and manipulation of microbial species that significantly enhance P plant nutrition by either inoculation or general stimulation in the rhizosphere; in particular, there is a need to identify the P-enhancing genes and their role or interactions with plants in affecting P nutrition;
2. Basic understanding of microbial ecology and signaling between plant roots and P-enhancing microorganisms;
3. Development of effective, low-cost, and simple inoculation technologies that can be utilized by SA and SSA farmers;
4. Determination of microbial ecology of P-enhancing organisms and management strategies that enable use of rock phosphate to eliminate P deficiency of SSA and SA crops;

5. Combination of microbial ecology of P-enhancing microorganisms with plant breeding or insertion of transgenic genes to develop roots with architectures, exudates, and signaling that optimize root and microbial responses to increase P uptake by crops.

2.2.4 *Manipulating microorganisms to promote drought tolerance on plants*

Bacteria have been isolated that can enhance the ability of crops to withstand water stress by increasing seedling root elongation (promotes stand establishment) and various crop physiological responses (reduction in cell elasticity and osmotic potential and rise in apoplastic water fractions) (Creus et al., 1998; Alvarez et al., 1996). Also, autochthonous arbuscular mycorrhizae (AM) can increase plant resistance to water stress (Marulanda et al., 2007; Kohler, 2008). Barea and coworkers (2002) have shown that this could be further improved with coinoculation with bacteria (*Bacillus thuringiensis*) because of the reduced amount of water required to produce shoot biomass (Vivas et al., 2003).

Abiotic stress such as water stress causes oxidative damage at the cellular level. To overcome this, plant cells can produce antioxidant enzymes (CAT, various peroxidases, and superoxide dismutase) that inactivate reactive free radicals (Simova-Stoilova et al., 2008). Interestingly, it was shown by Kohler (2008) on lettuce (*Lactuca sativa* L.) that *Pseudomonas mendocina* could assist this response by boosting plant CAT under severe drought conditions. *Paenibacillus polymyxa* was shown by Timmusk and Wagner (1999) to induce drought tolerance in *Arabidopsis thaliana* by induction of a drought-responsive gene. In wheat (*Triticum aestivum*) under drought stress, Creus et al. (2004) showed that inoculation with *Azospirillum brasilense* improved water status and grain yield. Plants inoculated with bacteria producing exopolysaccharide (EPS) can increase resistance to water stress because EPS forms biofilm on the surface of roots (Bensalim et al., 1998). The inoculated seedlings showed improved soil aggregation and root-adhering soil and higher relative water content in the leaves (Sandhya et al., 2010).

Sorghum inoculated with *Azospirillum* (Sarig et al., 1988; Fallik et al., 1994) resulted in a larger root system that improves water uptake in water-deficient soils. These field experiments showed inoculation increased leaf water potential, lowered canopy temperatures, and increased stomatal conductance and transpiration. In addition, inoculated sorghum increased total water extraction from soil over a control (by about 15%) and obtained water from deeper soil layers as compared with noninoculated controls. Similarly, Hamaoui et al. (2001) showed that inoculation of sorghum with *A. brasilense* significantly reduced the negative effects of saline irrigation water, which has been attributed to the stimulation of root development, delayed leaf senescence, and improved water uptake in saline soils (Sarig et al., 1990).

Sandhya et al. (2010) found five *Pseudomonas* spp. (*P. entomophila* strain BV-P13, *P. stutzeri* strain GRFHAP-P14, *P. putida* strain GAP-P45, *P. syringae* strain GRFHYTP52, and *P. monteilli* strain WAPP53) that improved plant biomass, relative water content, leaf water potential, root-adhering soil/root tissue ratio, aggregate stability, and mean weight diameter and decreased leaf water loss. The inoculated plants showed higher levels of proline, sugars, and free amino acids under drought stress. They also showed that inoculation significantly decreased electrolyte leakage and activities of antioxidant enzymes, ascorbate peroxidase (APX), CAT, and glutathione peroxidase (GPX) under drought stress, indicating that inoculated seedlings felt less stress compared to uninoculated seedlings. The strain GAP-P45 was the best in terms of influencing growth and biochemical and physiological status of the seedlings under drought stress.

Microbial-induced drought resistance is a potentially important technology for developing countries that are semiarid and that have crops that are regularly prone to drought stress. However, research on the interactions of microorganisms with plants and drought resistance is in its infancy.

2.2.5 *Companion woody rhizospheres*

Parkland agroforestry to some degree is practiced by farmers in tropical regions by allowing trees that can improve soil quality to grow in their fields. However, trees growing in farmers' fields may compete with crops, and there has been limited success in adoption of such systems (Buresh and Tian, 1998; Rhoades, 1997). However, Dick and coworkers discovered the importance of another prominent woody component in West Africa—native shrubs—that has largely been unrecognized for its potential to improve agroecosystems. The two species found throughout the Sahel are *Piliostigma reticulatum* (PR) and *Guiera senegalensis* (GS), which regrow after the rainy, summer cropping period and are unfortunately cut and burned just prior to the next cropping (rainy) season. Research findings so far have shown shrubs

- Are the single most important source of C inputs (Lufafa, Diédhiou, et al., 2008; Lufafa, Wright, et al., 2008) in farmers' fields;
- Increase soil quality and nutrient availability and could regenerate landscapes (Dossa, 2007; Dossa, Baham, et al., 2009; Dossa, Khouma, et al., 2009; Dossa et al., 2010);
- Are doing "hydraulic lift" (Kizito et al., 2012) (passive movement of water through roots from moist subsoil to dry surface layers [Caldwell et al., 1998]);
- Increase groundwater recharge during high-rainfall periods (Kizito et al., 2007);
- Maintain greater microbial diversity in the rhizosphere year around (presumably due to hydraulic lift) than soil outside the influence of the rhizosphere (Diedhiou et al., 2009);
- Are noncompetitive with crops for water (Kizito et al., 2007);
- Can be managed nonthermally, with unburned residue having positive effects on soils;
- Increased yields of peanut and millet by more than 50% over plots without shrubs (3 years of field testing) (Dossa, 2007).

Developing woody species companion plants has great potential throughout semiarid developing countries to restore degraded landscapes, buffer desert encroachment, and increase crop productivity.

2.3 *Biotechnology within a cultural and agroecosystems context*

It will be important as research proceeds that it is put within a regional and farm-level/cropping systems context. To emphasize this, one example is provided here with regard to manipulation of biocontrol microorganisms.

Larkin (2008) compared various biostimulants and known microbial control agents on potato diseases. Although all of them had measurable effects on soil microbial properties in the field, in no case did any of these reduce disease in continuous potato plots: Rather effective controls only occurred in certain crop rotations. These results indicated that certain rotations were better able to support the added beneficial organisms and amendments to enable more effective biological disease control. Establishment and persistence of microbial inoculants, soil amendments, or engineered rhizospheres likely depend on many edaphic and climatic factors. This emphasizes the need for early testing of promising biological technologies at the farm or cropping systems level.

Besides biological interactions and adapting to local conditions, technology research for developing countries must take into account regional cultural and socioeconomic considerations. Since the 1960s, the developing world has become littered with agricultural technologies that may have been agronomically superior under controlled research conditions but were unacceptable for subsistence farmers because of local culture, lack of technical support, or scales and markets of rural economies.

To attain practical outcomes for microbial technology, it seems that the emphasis should be placed on those approaches that depend on manipulating native populations. For example, there is a body of research for the feasibility of using organic amendments to create conditions and microbial responses to suppress soilborne diseases naturally. Similarly, this appropriate technology consideration makes breeding or insertion of transgenic genes attractive. Such an approach is desirable because attributes like optimized roots for efficient water and nutrient uptake or stimulation of beneficial microbial gene expression would only require the introduction of new cultivars into existing systems—with minimal technical support and reduction of the need for external inputs.

2.4 Conclusions

The rhizosphere (including the endophytic habitat) is a unique microbial environment that has an impact on plant growth positively and negatively. Roots alter the physical environment of soils, release sloughed dead cellular material, and actively excrete organic compounds that promote and sustain biological organisms.

It is now recognized, through the advent of nucleic acid analyses, that soils have a high degree of diversity, with estimates of 1 million distinct bacterial genomes per gram of soil. However, more than 99% of these are uncultured. Nonetheless, there have been important discoveries of microorganisms that can deliver services to plants by stimulating plant growth and protecting plants from pests, diseases, and environmental stress. Metagenomics can be used to construct BAC clone libraries to define these communities based on particular functional genes that are present and potentially active in the community. This can be used to screen BAC libraries for novel and plant-beneficial functional genes. A diverse array of bacteria and some fungi has been identified that can improve crop productivity. One way is that microorganisms can produce PGP substances (e.g., plant hormones auxin, cytokinins, IAA, and gibberellin) that directly increase crop yields. In the recent past, there have been surprising and exciting discoveries for microbial suppression or attack of pathogens or plant protection.

There are two approaches to actively manage crop-associated microbial communities for beneficial services: (1) adjusting the types and timing of organic inputs, such as cover crops to stimulate favorable native populations; and (2) inoculation with known microbial species. In addition, endophytic fungi or bacteria colonizing the host can stimulate morphologically or physiologically systemic resistance in plants to pathogens that can control or prevent plant diseases. Considerable progress has been made in identifying specific organisms and managing organic inputs to suppress diseases or protect plants from infection in developed countries, but this is less so for other microbial-stimulated plant services.

Beneficial microorganisms can increase nutrient uptake and efficiency for crops. The most remarkable process among plant nutrients is biological N fixation because there is an inexhaustible supply from the atmosphere. Outside N fixation by the symbiotic relationship of *Rhizobium* with leguminous plants, for nonsymbiotic microorganisms the most notable example is on sugarcane, for which 40–70% of the required N can be supplied by diazotrophs. However, for other nonlegume crops the rates of N fixation are much lower, typically 5–10% at best. The major challenge for free-living N fixers likely is the large amount of energy that is needed to support diazotrophs to produce N at levels that are agronomically significant. Phosphorus availability can be enhanced by certain microorganisms. This is accomplished by (1) solubilization of inorganic P by acidification (e.g., production of organic acids, carbonic acid, or H^+) or chelation; (2) mineralization of organic P by producing extracellular phosphatases or phytases; and (3) increased root growth or root hair development by production of phytohormones to enable greater exploration of soil for P uptake.

Interestingly, there are certain bacteria that can reduce crop water stress. Bacteria have been isolated that can reduce water stress of crops by increasing seedling root elongation (promotes stand establishment) and various crop physiological responses (reduction in cell elasticity and osmotic potential and rise in apoplastic water fractions). Also, autochthonous AM can increase plant resistance to water stress.

Of special note relative to beneficial microorganisms for crops is the symbiotic fungal species mycorrhizae. For crop species, endomycorrhizae form arbuscular structures and extensive hyphal networks that can transport P and other nutrients to the plant to help relieve drought stress in crops.

This information indicates there is great potential for microorganisms to benefit crop yields and quality and to develop truly sustainable systems that eliminate or greatly reduce external, synthetic inputs. However, considerable research is needed to make these practical and agronomically effective. This will require a multidisciplinary approach of plant scientists and soil microbiologist to bioengineer crops and the associated management systems to utilize beneficial microorganisms effectively. This necessitates fundamental microbial ecology research on microbial-plant interactions.

An emerging and potentially critical area of rhizosphere ecology is cell-to-cell communication using signaling compounds. This communication can be among microorganisms and between microorganisms and plants. There are likely a large number of chemicals that allow microorganisms to coordinate gene expression on a population-wide scale in concert with crop growth responses. One such system is QS, which involves production, secretion, and subsequent detection of small hormone-like signaling molecules known as autoinducers. Fundamental research is needed on the types of communications, cataloguing of signaling compounds and their sources, microbial and plant gene expression in response to signals, genetic controls on signaling, and signaling processes that elicit beneficial crop growth and protection responses. Clearly, the convergence of new research avenues of "-omics," bioinformatics, and micro- or nanoscale biochemistry sensing technologies offers great potential to understand these complex cell-to-cell communications. With such a foundation, one can imagine bioengineered crops with genes that can be turned on when they are experiencing a nutrient, environmental, or pathogenic stress; this results in an appropriate microbial response. If such a breakthrough could be made, it offers the possibility to "program" plant roots to release compounds and activate beneficial microbial responses at a critical growth stage.

To attain practical outcomes for microbial technology, it seems that the emphasis should be placed on those approaches that depend on manipulating native populations.

For example, there is a body of research for the feasibility of using organic amendments to create conditions and microbial responses to suppress soilborne diseases naturally. Similarly, breeding or insertion of transgenic genes in crop plants could, as described, provide the means for a crop to signal native populations for the appropriate response. At the same time, there likely may need to be engineered rhizospheres that provide the habit and energy to enable native subpopulations to proliferate and respond at levels that are agronomically significant. For example, there is evidence from pure culture work that free-living or endophytic diazotrophs have high potential to fix N, but under field conditions these organisms are severely limited by C or energy availability. Such an approach would only require the introduction of new cultivars into existing systems—with minimal technical support and reducing the need for external inputs or inoculations.

Biofilms are another important but poorly understood soil/rhizosphere habitat that could be important and related to this discussion. In this case, an assemblage of heterogeneous microcolonies (e.g., protozoans, fungi, nematodes, and prokaryotes) is encapsulated in EPSs. These organisms can work in consortia to perform biogeochemical processes or cooperatively to aid individual microorganisms. There is now speculation that biofilms have complex cell-cell signaling and physical structures such as water channels to disperse nutrients, metabolites, and waste products and "nanowires" that transfer energy. Understanding biofilms and diversity and potential interactions of their microbial communities could provide the basis for designing or stimulating certain biofilm communities that can exist in more stable and interconnected members with diverse services for plants. Although an engineered plant-microbial system offers great potential, there must be collaborative and parallel research on cropping system management within a regional and farm-level/cropping systems context. This includes the appropriate cultural practices associated with the use of specific organisms for a given agricultural service. If bioengineered technologies are developed in isolation of a social, economic, and agroecosystems context, past experience would suggest this greatly diminishes the potential for implementation and adoption of these systems in farmers' fields. Furthermore, the need for these broader considerations is that biologically based systems are much more sensitive to environmental controls than agricultural systems based on chemical and external input.

In summary, there are exciting research approaches that offer the possibility to develop crop rhizospheres with a consortium of microbial species that support, protect, and enhance the yield and quality of crops. To attain this goal, considerable fundamental research is needed on microbial and rhizosphere ecology, cell-to-cell communications, and microbial genomics and proteomics in combination with cropping systems research.

References

Abbasi, P. A., J. Al-Dahmani, F. Sahin, et al. 2002. Effect of compost amendments on disease severity and yield in organic and conventional tomato production systems. *Plant Dis.* 86:156–161.

Alvarez, B., G. Martínez-Drets, and B. Alvarez. 1995. Metabolic characterization of *Acetobacter diazotrophicus*. *Can. J. Microbiol.* 41:918–924.

Alvarez, M. I., R. J. Sueldo, and C. A. Barassi. 1996. Effect of *Azospirillum* on coleoptile growth in wheat seedlings under water stress. *Cereal Res. Commun.* 24:101–107.

Amara, M. A. T., and M. S. A. Dahdoh. 1997. Effect of inoculation with plant growth promoting rhizobacteria (PGPR) on yield and uptake of nutrients by wheat grown on sandy soil. *Egyptian J. Soil Sci.* 37:467–484.

An, Q. L., X. J. Yang, Y. M. Dong, et al. 2001. Using confocal laser scanning microscope to visualize the infection of rice roots by GFP-labelled *Klebsiella oxytoca* SA2, an endophytic diazotroph. *Acta Bot. Sin.* 43:558–564.

Andrén, O., and J. Balandreau. 1999. Biodiversity and soil functioning—from black box to can of worms? *Appl. Soil Ecol.*13:105–108.

Asghar, H. H., Z. A. Zahir, M. Arshad, and A. Khaliq. 2002. Relationship between in vitro production of auxins by rhizobacteria and their growth-promoting activities in *Brassica juncea* L. *Biol. Fertil. Soils* 35:231–237.

Assmus, B., P. Hutzler, G. Kirchhof, et al. 1995. In situ localization of *Azospirillum brasilense* in the rhizosphere of wheat with fluorescently labeled, ribosomal-RNA targeted oligonucleotide probes and scanning confocal laser microscopy. *Appl. Environ. Microbiol.* 69:6208–6215.

Badri, D. V., T. L. Weir, D. van der Lelie, and J. M. Vivanco. 2009. Rhizosphere chemical dialogues: plant-microbe interactions. *Curr. Opin. Biotechnol.* 20:642–650.

Bailey, B., H. Bae, M. Strem, et al. 2006. Fungal and plant gene expression during the colonization of cacao seedlings by endophytic isolates of four *Trichoderma* species. *Planta* 224:1449–1464.

Baillieul, F., P. de Ruffray, and S. Kauffmann. 2003. Molecular cloning and biological activity of alpha-, beta-, and gamma-megaspermin, three elicitins secreted by *Phytophthora megasperma* H_2O. *Plant Physiol.* 131:155–166.

Baldani, J. I., and V. L. D. Baldani. 2005. History on the biological nitrogen fixation research in graminaceous plants: special emphasis on the Brazilian experience. *An. Acad. Bras. Ciênc.* 77:549–579.

Bangera, M. G., and L. S. Thomashow. 1999. Identification and characterization of a gene cluster for synthesis of the polyketide antibiotic 2,4-diacetylphloroglucinol from *Pseudomonas fluorescens* Q2–87. *J. Bacteriol.* 181:3155–3163.

Bardgett, R. D., and A. Shine. 1999. Linkages between plant litter diversity, soil microbial biomass and ecosystem function in temperate grasslands. *Soil Biol. Biochem.* 31:317–321.

Barea, J. M., M. Toro, M. O. Orozco, et al. 2002. The application of isotopic (^{32}P and ^{15}N) dilution techniques to evaluate the interactive effect of phosphate-solubilizing rhizobacteria, mycorrhizal fungi and *Rhizobium* to improve the agronomic efficiency of rock phosphate for legume crops. *Nutr. Cycl. Agroecosyst.* 63:35–42.

Barraquio, W. L., E. M. Segubre, M. S. Gonzalez, et al. 2000. Diazotrophic enterobacteria: what is their role in rhizosphere of rice? In *The Quest for Nitrogen Fixation in Rice*, eds. J. K. Ladha and P. M. Reddy, 93–118. International Rice Research Institute, Los Banos, Philippines.

Beattie, G. A. 2006. Plant-associated bacteria: survey, molecular phylogeny, genomics and recent advances. In *Plant-Associated Bacteria*, ed. S. S. Gnanamanickam, 1–56. Springer, Dordrecht, the Netherlands.

Beja, O., L. Aravind, E. V. Koonin, et al. 2000. Bacterial rhodopsin: evidence for a new type of phototrophy in the sea. *Science* 289:1902–1906.

Benitez, M.-S., F. B. Tustas, D. Rotenberg, et al. 2007. Multiple statistical approaches of community fingerprint data reveal bacterial populations associated with general disease suppression arising from the application of different organic field management strategies. *Soil Biol. Biochem.* 39:2289–2301.

Bensalim, S., J. Nowak, and S. K. Asiedu. 1998. A plant growth promoting rhizobacterium and temperature effects on performance of 18 clones of potato. *Am. J. Potato Res.* 75:145–152.

Bergmann, B., A. N. Rai, and U. Rasmussen. 2007. Cyanobacterial associations. In *Associative and Endophytic Nitrogen-Fixing Bacteria and Cyanobacterial Associations*, eds. C. Elmerich and W. E. Newton, 257–301. Springer, Dordrecht, the Netherlands.

Bertin, C., X. Yang, and L. A. Weston. 2003. The role of root exudates and allelochemicals in the rhizosphere. *Plant Soil* 256:67–83.

Biswas, J. C., J. K. Ladha, and F. B. Dazzo. 2000a. Rhizobia inoculation improves nutrient uptake and growth of lowland rice. *Soil Sci. Soc. Am. J.* 64:1644–1650.

Biswas, J. C., J. K. Ladha, F. B. Dazzo, et al. 2000b. Rhizobial inoculation influences seedling vigor and yield of rice. *Agron. J.* 92:880–886.

Boddey, R. M., O. C. Oliveira, S. Urquiaga, et al. 1995. Biological nitrogen fixation associated with sugarcane and rice: contributions and prospects for improvement. *Plant Soil* 174:195–209.

Boddey, R. M., S. Urquiaga, B. J. R. Alves, and V. Reis. 2003. Endophytic nitrogen fixation in sugarcane: present knowledge and future applications. *Plant Soil* 252:139–149.

Bolwerk, A., A. L. Lagopodi, B. J. J. Lugtenberg, and G. V. Bloemberg. 2005. Visualization of interactions between a pathogenic and a beneficial *Fusarium* strain during biocontrol of tomato foot and root rot. *Mol. Plant-Microbe Interact.* 18:710–721.

Borneman, J., J. O. Becker, E. Bent, et al. 2007. Identifying microorganisms involved in specific in situ functions: experimental design considerations for rRNA gene-based population studies and sequence-selective PCR assays. In *Manual of Environmental Microbiology*, 3rd ed., eds. C. J. Hurst and R. L. Crawford, 748–757. ASM Press, Washington, DC.

Broughton, W. J., F. Zhang, X. Perret, and C. Staehelin. 2003. Signals exchanged between legumes and *Rhizobium*: agricultural uses and perspectives. *Plant Soil* 252:129–137.

Buresh, R. J., and G. Tian. 1998. *Soil Improvement by Trees in Sub-Saharan Africa*. Kluwer Academic, Dordrecht, the Netherlands.

Caldwell, R. R., R. Dave, and P. J. Steinhardt. 1998. Dwarf elliptical galaxies in the M81 Group—the structure and stellar populations of BK5N and F8D1. *Phys. Rev. Lett.* 80:1582–1585.

Castang, S., B. Chantegrel, C. Deshayes, et al. 2004. N-Sulfonyl homoserine lactones as antagonists of bacterial quorum sensing. *Bioorg. Med. Chem. Lett.* 14:5145–5149.

Chelius, M. K., and E. W. Triplett. 2000. Immunolocalization of dinitrogenase reductase produced by *Klebsiella pneumoniae* in association with *Zea mays* L. *Appl. Environ. Microbiol.* 66:783–787.

Chen, Y., R. Mei, S. Lu, et al. 1994. The use of yield increasing bacteria as plant growth promoting rhizobacteria in Chinese agriculture. In *Management of Soil Borne Diseases*, eds. V. K. Gupta and R. Utkhede, 1–13. Narosa, New Delhi, India.

Chin-A-Woeng, T. F., D. van den Broek, G. de Voer, et al. 2001. Phenazine-1-carboxamide production in the biocontrol strain *Pseudomonas chlororaphis* PCL1391 is regulated by multiple factors secreted into the growth medium. *Mol. Plant-Microbe Interact.* 14:969–979.

Churngchow, N., and M. Rattarasarn. 2000. The elicitin secreted by *Phytophthora palmivora* a rubber tree pathogen. *Phytochemistry* 54:33–38.

Cleland, R. E. 1990. Auxin and cell elongation. In *Plant Hormones and Their Role in Plant Growth and Development*, ed. P. J. Davis, 132–148. Kluwer, Dordrecht, the Netherlands.

Creus, C. M., R. J. Sueldo, and C. A. Barassi. 1998. Water relations in *Azospirillum*-inoculated wheat seedlings under osmotic stress. *Can. J. Bot.* 76:238–244.

Creus, C. M., R. J. Sueldo, and C. A. Barassi. 2004. Water relations and yield in *Azospirillum*-inoculated wheat exposed to drought in the field. *Can. J. Bot.* 82:273–281.

Darby, H. M., A. G. Stone, and R. P. Dick. 2006. Compost and manure mediated impacts on soil borne pathogens and soil quality. *Soil Sci. Soc. Am. J.* 70:347–358.

Diaz-Torres, M. L., R. McNab, D. A. Spratt, et al. 2003. Novel tetracycline resistance determinant from the oral metagenome. *Antimicrob. Agents Chemother.* 47:1430–1432.

Diedhiou, S., A. N. Badiane, I. Diedhiou, et al. 2009. Succession of soil microbial communities during decomposition of native shrub litter of semi-arid Senegal. *Pedobiologia* 52:273–286.

Dobbelaere, S., A. Croonenborghs, A. Thys, et al. 2001. Responses of agronomically important crops to inoculation with *Azospirillum*. *Aust. J. Plant Physiol.* 28:871–879.

Dong, Z., M. J. Canny, M. E. McCully, et al. 1994. A nitrogen-fixing endophyte of sugarcane stems. A new role for the apoplast. *Plant Physiol.* 105:1139–1147.

Dossa, E. 2007. The biogeochemistry of nitrogen and phosphorus cycling in native shrub ecosystems in Senegal. PhD dissertation, Oregon State University, Corvallis.

Dossa, E. L., J. Baham, M. Khouma, et al. 2009. Phosphorus sorption and desorption in semiarid soils of Senegal amended with native shrub residues. *Soil Sci.* 173:669–682.

Dossa, E. L., S. Diedhiou, J. E. Compton, K. B. Assigbetse, and R. P. Dick. 2010. Spatial patterns of P fractions and chemical properties in soils of two native shrub communities in Senegal. *Plant Soil* 327:185–198.

Dossa, E. L., M. Khouma, I. Diedhiou, et al. 2009. Carbon, nitrogen and phosphorus mineralization potential of semiarid Sahelian soils amended with native shrub residues. *Geoderma* 148:251–260.

Ebel, J., and A. Mithöfer. 1998. Early events in the elicitation of plant defense. *Planta* 206:335–348.

Faeth, S. H. 2002. Are endophytic fungi defensive plant mutualists? *Oikos* 98:25–36.

Fallik, E., S. Sarig, and Y. Okon. 1994. Morphology and physiology of plant roots associated with *Azospirillum*. In *Azospirillum/Plant Associations*, ed. Y. Okon, 77–86. CRC Press, Boca Raton, FL.

Faure, D., D. Vereecke, and J. H. J. Leveau. 2009. Molecular communication in the rhizosphere. *Plant Soil* 321:279–303.
Foster, R. C., A. D. Rovira, and T. W. Cock. 1983. *Ultrastructure of the Root–Soil Interface*. American Phytopathological Society, St. Paul, MN.
Fray, R. G. 2002. Altering plant-microbe interaction through artificially manipulating bacterial quorum sensing. *Ann. Bot.* 89:245–253.
Fuentes-Ramírez, L. E., J. Caballero-Mellado, J. Sepúlveda, and E. Martínez-Romero. 1999. Colonization of sugarcane by *Acetobacter diazotrophicus* is inhibited by high N-fertilization. *FEMS Microbiol. Ecol.* 29:117–128.
Gans, J., M. Wolinsky, and J. Dunbar. 2005. Computational improvements reveal great bacterial diversity and high metal toxicity in soil. *Science* 309:1387–1390.
Garcia de Salamone, I. E., R. K. Hynes, and L. M. Nelson. 2001. Cytokinin production by plant growth promoting rhizobacteria and selected-mutants. *Can. J. Microbiol.* 47:404–411.
Gillis, M., V. T. Van, R. Bardin, et al. 1995. Polyphasic taxonomy in the genus *Burkholderia* leading to an emended description of the genus and proposition of *Burkholderia vietnamiensis* sp. nov. for N_2-fixfing isolates from rice in Vietnam. *Int. J. Syst. Bacteriol.* 45:274–289.
Golenfeld, N., and C. Woese. 2007. Biology's next revolution: the emerging picture of microbes as gene-swapping collectives demands a revision of such concepts as organism, species and evolution itself. *Nature* 445:369.
Gregory, P. J. 2006. Roots, rhizosphere and soil: a route to a better understanding of soil science. *Eur. J. Soil. Sci.* 57:2–12.
Griffiths, R. I., A. S. Whiteley, A. G. O'Donnell, and M. J. Bailey. 2000. Rapid method for coextraction of DNA and RNA from natural environments for analysis of ribosomal DNA- and rRNA-based microbial community composition. *Appl. Environ. Microbiol.* 66:5488–5491.
Gross, H., V. O. Stockwell, M. D. Henkels, et al. 2007. The genomisotopic approach: a systematic method to isolate products of orphan biosynthetic gene clusters. *Chem. Biol.* 14:53–63.
Gupta, V. V. S. R., and J. J. Germida. 1988. Distribution of microbial biomass and its activity in different soil aggregate size classes as affected by cultivation. *Soil Biol. Biochem.* 20:777–786.
Gutierrez Munero, F. J., B. Ramos-Solano, A. Probanza, J. Mehouachi, F. R. Tadeo, and M. Talon. 2001. The plant-growth-promoting rhizobacteria *Bacillus pumilus* and *Bacillus licheniformis* produce high amounts of physiologically active gibberellins. *Physiol. Plant* 111:206–211.
Gyaneshwar, P., E. K. James, P. M. Reddy, and J. K. Ladha. 2002. *Herbaspirillum* colonization increases growth and nitrogen accumulation in aluminum-tolerant rice varieties. *New Phytol.* 154:131–145.
Hagen, G. 1990. The control of gene expression by auxin. In *Plant Hormones and Their Role in Plant Growth and Development*, ed. P. J. Davies, 149–163. Kluwer Academic, Dordrecht, the Netherlands.
Hamaoui, B., J. M. Abbadi, S. Burdman, A. Rashid, S. Sarig, and Y. Okon. 2001. Effects of inoculation with *Azospirillum brasilense* on chickpeas (*Cicer arietinum*) and faba beans (*Vicia faba*) under different growth conditions. *Agronomie* 21:553–560.
Hanada, R. E., A. William, V. Pomella, et al. 2010. Endophytic fungal diversity in *Theobroma cacao* (cacao) and *T. grandiflorum* (cupuaçu) trees and their potential for growth promotion and biocontrol of black-pod disease. *Fungal Biol.* 114:901–910.
Handelsman, J., and J. Tiedje. 2007. *The New Science of Metagenomics. Committee on Metagenomics: Changes and Functional Applications, National Research Council of the National Academies.* National Academies Press, Washington, DC.
Hansel, C. M., S. Fendorf, S. Sutton, and M. Newville. 2001. Characterization of Fe plaque and associated metals on the roots of mine-waste impacted aquatic plants. *Environ. Sci. Technol.* 35:3863–3868.
Harman, G. E., C. R. Howell, A. Viterbo, I. Chet, and M. Lorito. 2004. *Trichoderma* species—opportunistic, avirulent plant symbionts. *Nature Rev. Microbiol.* 2:43–56.
Hattori, T. 1973. *Microbial Life in Soil*. Dekker, New York.
Henne, A., R. A. Schmitz, M. Bomeke, et al. 2000. Screening of environmental DNA libraries for the presence of genes conferring lipolytic activity on *Escherichia coli*. *Appl. Environ. Microbiol.* 66:3113–3116.

Hilali, A., D. Prevost, W. J. Broughton, and H. Antoun. 2001. Effects of inoculation with strains of *Rhizobium leguminosarum* biovar *trifolii* on the growth of wheat in two different soils from Morocco. *Can. J. Microbiol.* 47:590–593.

Hinsinger, P., A. G. Bengough, D. Vetterlein, and I. M. Young. 2009. Rhizosphere: biophysics, biogeochemistry and ecological relevance. *Plant Soil* 321:117–152.

Hoitink, H. A. J., and M. J. Boehm. 1999. Biocontrol within the context of soil microbial communities: a substrate-dependent phenomenon. *Annu. Rev. Phytopathol.* 37:427–446.

Holford, I. C. R. 1997. Soil phosphorus, its measurements and its uptake by plants. *Aust. J. Soil Res.* 35:227–239.

Hurek, T., L. L. Handley, B. Reinhold-Hreck, and Y. Piche. 2002. *Azoarcus* grass endophytes contribute fixed nitrogen to the plant in an unculturable state. *Mol. Plant-Microbe Interact.* 15:233–242.

Hurek, B., and Y. Piche. 2002 *Azoarcus* grass endophytes contribute fixed nitrogen to the plant in an unculturable state. *Mol. Plant Microbe Interact.* 15:233–242.

International Fertilizer Industry Association (IFIA). 2001. International Fertilizer Industry Association home page. http://www.fertilizer.org.

Isherwood, K. F. 2000. *Mineral Fertilizer Use and the Environment.* International Fertilizer Industry Association/United Nations Environment Programme, Paris.

Jakobsen, I., M. E. Leggett, and A. E. Richardson. 2005. Rhizosphere microorganisms and plant phosphorus uptake. In *Phosphorus: Agriculture and the Environment*, eds. J. T. Sims and A. N. Sharpley, 437–492. Agronomy Monograph 46. ASA, CSA, SSSA, Madison, WI.

James, E. K., P. Gyaneshwar, W. L. Barraquio, et al. 2000. Endophytic diazotrophs associated with rice. In *The Quest for Nitrogen Fixation in Rice,* eds. J. K. Ladha and P. M. Reddy, 119–140. International Rice Research Institute, Los Banos, Philippines.

James, E. K., and F. L. Olivares. 1998. Infection and colonization of sugar cane and other graminaceous plants by endophytic diazotrophs. *Crit. Rev. Plant Sci.* 17:77–119.

Jha, P., and A. Kumar. 2009. Characterization of novel plant growth promoting endophytic bacterium *Achromobacter xylosoxidans* from wheat plant. *Microbial Ecol.* 58:179–188.

Jiang, R. H. Y., B. M. Tyler, and F. Govers. 2006. Comparative analysis of *Phytophthora* genes encoding secreted proteins reveals conserved synteny and lineage-specific gene duplications and deletions. *Mol. Plant Microbe Interact.* 19:1311–1321.

Jiang, R. H. Y., B. M. Tyler, S. C. Whisson, et al. 2006. Ancient origin of elicitin gene clusters in *Phytophthora* genomes. *Mol. Biol. Evol.* 23:338–351.

Joshi, R., and B. McSpadden Gardener. 2006. Identification of genes associated with pathogen inhibition in different strains *B. subtilis. Phytopathology* 96:145–154.

Kamoun, S. 2001. Nonhost resistance to *Phytophthor*a: novel prospects for a classical problem. *Curr. Opin. Plant Biol.* 4:295–300.

Khalid, A., M. Arshad, and Z. A. Zahir. 2004. Screening plant growth-promoting rhizobacteria for improving growth and yield of wheat. *J. Appl. Microbiol.* 96:473–480.

Kizito, M., I. Dragila, R. Brooks, et al. 2012. Hydraulic redistribution by two semi-arid shrubs: implications on agro-ecosystems. *J. Arid Environ.* in press.

Kizito, F., M. Senè, M. I. Dragila, et al. 2007. Soil water balance of annual crop-native shrub systems in Senegal's Peanut Basin. *Agric. Water Manag.* 90:137–148.

Kloepper, J. W., J. Leong, M. Teintze, and M. N. Schroth. 1980. Enhanced plant growth by siderophores produced by plant growth promoting rhizobacteria. *Nature* 286:885–886.

Kohler, J. 2008. Plant-growth-promoting rhizobacteria and arbuscular mycorrhizal fungi modify alleviation biochemical mechanisms in water-stressed plants. *Funct. Plant Biol.* 35:141–151.

Koumoutsi, A., X.-H. Chen, A. Henne, et al. 2004. Structural and functional characterization of gene clusters directing nonribosomal synthesis of bioactive cyclic lipopeptides in *Bacillus amyloliquefaciens* strain FZB42. *J. Bacteriol.* 186:1084–1096.

Larkin, R. P. 2008. Relative effects of biological amendments and crop rotations on soil microbial communities and soilborne diseases of potato. *Soil Biol. Biochem.* 40:1341–1351.

Lascombe, M. B., M. Ponchet, L. Cardin, et al. 2004. Purification, crystallization and preliminary X-ray studies of sylvactin, an elicitin-like protein from *Pythium sylvaticum. Acta Cryst.* D60:362–364.

Leggett, M. E., S. Gleddie, and G. Holloway. 2001. Phosphate-solubilizing microorganisms and their use. In *Plant Nutrient Acquisition: New Perspectives,* eds. N. Ae, J. Arihara, K. Okada, and A. Srinivasan, 299–318. Springer-Verlag, Tokyo.

Lithgow, J. K., A. Wilkinson, A. Hardman, et al. 2000. The regulatory locus cinR1 in *Rhizobium leguminosarum* controls a network of quorum-sensing loci. *Mol. Microbiol.* 37:81–97.

Lodewyckx, C., J. Vangronsveld, F. Porteous, et al. 2002. Endophytic bacteria and their potential applications. *Crit. Rev. Plant Sci.* 21:583–606.

Loreau, M., S. Naeem, P. Inchausti, et al. 2001. Biodiversity and ecosystem functioning: current knowledge and future challenges. *Science* 294:804–808.

Lufafa, A., I. Diédhiou, S. Ndiaye, et al. 2008. Carbon stocks and patterns in native shrub communities of Sénégal's Peanut Basin. *Geoderma* 146:75–82.

Lufafa, A., D. Wright, J. Bolte, et al. 2008. Regional carbon stocks and dynamics in native woody shrub communities of Senegal's Peanut Basin. *Agric. Ecosyst. Environ.* 128:1–11.

Lugtenberg, B. J. J., T. F. C. Chin-A-Woeng, and G. V. Bloemberg. 2002. Microbe-plant interactions: principles and mechanisms. *Anton. van Leeuwen.* 81:373–383.

Lugtenberg, B., and J. Leveau. 2007. Biocontrol of plant pathogens: principles, promises, and pitfalls. In *The Rhizosphere: Biochemistry and Organic Substances at the Soil-Plant Interface*, 2nd ed., eds. R. Pinton, Z. Varanini, and P. Nannipieri, 267–296. CRC Press, New York.

Marulanda, A., R. Porcel, J. M. Barea, and R. Azcón. 2007. Drought tolerance and antioxidant activities in lavender plants colonized by native drought-tolerant or drought-sensitive *Glomus* species. *Microbial Ecol.* 54:543–552.

Mazzola, M. 2004. Assessment and management of soil microbial community structure for disease suppression. *Annu. Rev. Phytopathol.* 42:35–59.

McLaughlin, M. J., A. M. Alston, and J. K. Martin. 1988. Phosphorus cycling in wheat-pasture rotations. III. Organic phosphorus turnover and phosphorus cycling. *Aust. J. Soil Res.* 26:343–353.

McSpadden Gardener, B. B., L. Gutierrez, R. Joshi, et al. 2005. Distribution and biocontrol potential of *phlD+* pseudomonads in corn and soybean fields. *Phytopathology* 95:715–724.

McSpadden Gardener, B. B, D. Mavrodi, L. Thomashow, and D. Weller. 2001. A polymerase chain reaction-based assay characterizing rhizosphere populations of 2,4-DAPG-producing-bacteria. *Phytopathology* 91:44–54.

Melnick, R. L., N. K. Zidack, B. A. Bailey, S. N. Maximova, M. Guiltinan, and P. A. Backman. 2008. Bacterial endophytes: *Bacillus* spp. from annual crops as potential biological control agents of black pod rot of cacao. *Biol. Control* 46:46–56.

Metraux, J.-P., M. R. McAinsh, and K. Palme. 2001. Plant biology. *Curr. Opin. Plant Biol.* 4:169–176.

Mia, M. A. B., Z. H. Shamsuddin, and M. Mahmood. 2010. Use of plant growth promoting bacteria in banana: a new insight for sustainable banana production. *Int. J. Agric. Biol.* 12:459–467.

Mia, M. A. B., Z. H. Shamsuddin, W. Zakaria, and M. Marziah. 2007. Associative nitrogen fixation by *Azospirillum* and *Bacillus* spp. in bananas. *Infomusa* 16:11–15.

Morgan, J. A., and J. M. Whipps. 2001. Methodological approaches to the study of rhizosphere carbon flow and microbial population dynamics. In *The Rhizosphere-Biochemistry and Organic Substrates at the Soil-Plant-Interface,* eds. R. Pinton, Z. Varanini, and P. Nannipieri, 373–409. Dekker, New York.

Naher, U. A., R. Othman, Z. H. J. Shamsuddin, et al. 2009. Growth enhancement and root colonization of rice seedlings by *Rhizobium* and *Corynebacterium* spp. *Int. J. Agric. Biol.* 11:586–590.

Nealson, K. H., and J. W. Hastings. 1979. Bacterial bioluminescence: its control and ecological significance. *Microbiol. Rev.* 43:496–518.

Ohtonen, R., S. Aikio, and H. Väre. 1997. Ecological theories in soil biology. *Soil Biol. Biochem.* 29:1613–1619.

Okon, Y., and C. A. Labandera-Gonzalez. 1994. Agronomic applications of *Azospirillum*: an evaluation of 20 years world-wide field inoculation. *Soil Biol. Biochem.* 26:1591–1601.

Okoth, S. A., and E. Siameto. 2011. Evaluation of selected soil fertility management interventions for suppression of *Fusarium* spp. in a maize and beans intercrop. *Tropic. Subtropic. Agroecosyst.* 13:73–80.

Olatinwo, R., B. Yin, J. O. Becker, and J. Borneman. 2006. Suppression of the plant-parasitic nematode *Heterodera schachtii* by the fungus *Dactylella oviparasitica*. *Phytopathology* 96:111–114.

Palus, J. A., J. Borneman, P. W. Ludden, and E. W. Triplett. 1996. Isolation and characterization of endophytic diazotrophs from *Zea mays* L., and *Zea luxurians* Iltis and Doebley. *Plant Soil* 186:135–142.

Paulsen, I. T., C. M. Press, J. Ravel, et al. 2005. Complete genome sequence of the plant commensal *Pseudomonas fluorescens* Pf-5. *Nat. Biotechnol.* 23:873–878.

Penrose, D. M., and B. R. Glick. 2001. Levels of 1-aminocyclopropane-1-carboxylic acid (ACC) in exudates and extracts of canola seeds treated with plant growth-promoting bacteria. *Can. J. Microbiol.* 47:368–372.

Pinton, R., Z. Varanini, and P. Nannipieri. 2001. The rhizosphere as a site of biochemical interactions among soil components, plants, and microorganisms. In *The Rhizosphere,* eds. R. Pinton, Z. Varanini, and P. Nannipieri, 1–18. Dekker, New York.

Ponchet, M., F. Panabiéres, M. I. Milat, et al. 1999. Are elicitins cryptograms in plant-oomycete communications? *Cell. Mol. Life Sci.* 56:1020–1047.

Probanza, A., J. A. Lucas Garcia, M. Ruiz Palomino, et al. 2002. *Pinus pinea L.* seedling growth and bacterial rhizosphere structure after inoculation with PGPR *Bacillus* (*B. licheniformis* CET 5106 and *B. pumilus* CECT 5105). *Appl. Soil Ecol.* 20:75–84.

Raaijmakers, J. M., T. C. Paulitz, C. Steinberg, et al. 2009. The rhizosphere: a playground and battlefield for soilborne pathogens and beneficial microorganisms. *Plant Soil* 321:341.

Reis, F. B., Jr., V. M. Reis, S. Urquiaga, and J. Döbereiner. 2000. Influence of nitrogen fertilization on the population of diazotrophic *Herbaspirillum* spp. and *Gluconacetobacter diazotrophicus* in sugar cane (*Saccharum* spp.). *Plant Soil* 219:153–159.

Reymond, P., and E. E. Farmer. 1998. Jasmonate and salicylate as global signals for defense gene expression. *Curr. Opin. Plant Biol.* 5:404–411.

Rhoades, R. E. 1997. *The Incredible Potato.* National Geographic Society, Washington, DC.

Riggs, P. J., M. K. Chelius, A. L. Iniguez, et al. 2001. Enhanced maize productivity by inoculation with diazotrophic bacteria. *Aust. J. Plant Physiol.* 28:829–836.

Rondon, M. R., P. R. August, A. D. Bettermann, et al. 2000. Cloning the soil metagenome: a strategy for accessing the genetic and functional diversity of uncultured microorganisms. *Appl. Environ. Microbiol.* 66:2541–2547.

Rosenblueth, M., L. Martinez, J. Silva, and E. Martinez-Romero. 2004. *Klebsiella variicola,* a novel species with clinical and plant-associated isolates. *Syst. Appl. Microbiol.* 27:27–35.

Rosenblueth, M., and E. Martínez-Romero. 2006. Bacterial endophytes and their interactions with hosts. *Mol. Plant Microbe. Interact.* 19:827–837.

Rotenberg, D., L. Cooperband, and A. Stone. 2005. Dynamic relationships between soil properties and foliar disease as affected by annual additions of organic amendment to a sandy-soil vegetable production system. *Soil Biol. Biochem.* 37:1343–1357.

Rothballer, M., M. Schmid, and A. Hartmann. 2003. In situ localization and PGPR-effect of *Azospirillum brasilense* strains colonizing roots of different wheat varieties. *Symbiosis* 34:261–279.

Rovira, A. D., and R. Campbell. 1974. Scanning electron microscopy of micro-organisms on the roots of wheat. *Microbial Ecol.* 1:15–23.

Sahrawat, K. L., M. K. Abekoe, and S. Diatta. 2001. Application of inorganic phosphorus fertilizer. In *Sustaining Soil Fertility in West Africa,* ed. G. Tian, 225–246. Soil Science Society of America, Madison, WI.

Sandhya, V., Sk. Z. Ali, M. Grover, et al. 2010. Effect of plant growth promoting *Pseudomonas* spp. on compatible solutes, antioxidant status and plant growth of maize under drought stress. *Plant Growth Regul.* 62:21–30.

Sanyal, S. K., and S. K. De Datta. 1991. Chemistry of phosphorus transformations in soil. *Adv. Soil Sci.* 16:1–120.

Saravanan, V. S., M. Madhaiyan, J. Osborne, M. Thangaraju, and T. M. Sa. 2008. Ecological occurrence of *Gluconacetobacter diazotrophicus* and nitrogen-fixing *Acetobacteraceae* members: their possible role in plant growth promotion. *Microbial Ecol.* 55:130–140.

Sarig, S., A. Blum, and Y. Okon. 1988. Improvement of the water status and yield of field-grown grain sorghum (*Sorghum bicolor*) by inoculation with *Azospirillum brasilense*. *J. Agric. Sci.* 110:271–277.

Sarig, S., Y. Okon, and A. Blum. 1990. Promotion of leaf area development and yield in *Sorghum bicolor* inoculated with *Azospirillum brasilense*. *Symbiosis* 9:235–245.

Sauer, K., A. H. Rickard, and D. G. Davies. 2007. Biofilms and biocomplexity. *Microbe* 2:347–353.
Schaudinn, C., P. Stoodley, A. Kainović, et al. 2007. Bacterial biofilms, other structures seen as mainstream concepts. *Microbe* 2:231–237.
Scholter, M., W. Wiehe, B. Assmus, et al. 1997. Root colonization of different plants by plant-growth-promoting *Rhizobium leguminosarum bv. Trifolii* R39 studied with monospecific polyclonal antisera. *Appl. Environ. Microbiol.* 63:2038–2046.
Schulz B., and C. Boyle. 2006. What are endophytes? In *Microbial Root Endophytes*, eds. B. J. E. Schulz, C. J. C. Boyle, and T. N. Sieber, 1–13. Springer-Verlag, Berlin.
Selesi, D., M. Schmid, and A. Hartmann. 2005. Diversity of "green-like" and "red-like" ribulose 1,5 bisphosphate carboxylase/oxygenase large subunit genes (*cbbL*) in differently managed agricultural soils. *Appl. Environ. Microbiol.* 71:175–184.
Shi, J., A. Liu, X. Li, S. Feng, and W. Chen. 2011. Inhibitory mechanisms induced by the endophytic bacterium MGY2 in controlling anthracnose of papaya. *Biol. Control* 56:2–8.
Simova-Stoilova, L., K. Demirevska, T. Petrova, N. Tsenov, and U. Feller. 2008. Antioxidative protection in wheat varieties under severe recoverable drought at seedling stage. *Plant Soil Environ.* 54:529–536.
Singh, B. K., P. Millard, A. S. Whiteley, and J. C. Murrell. 2004. Unraveling rhizosphere-microbial interactions: opportunities and limitations. *Trends Microbiol.* 12:386–393.
Smith, S. E., and D. J. Read. 1997. *Mycorrhizal Symbiosis.* Academic Press, San Diego, CA.
Steenhoudt, O., and J. Vanderleyden. 2000. *Azospirillum*, a free-living nitrogen-fixing bacterium closely associated with grasses: genetic, biochemical and ecological aspects. *FEMS Microbiol. Rev.* 24:487–506.
Stone, A. G., G. E. Vallad, L. R. Cooperband, et al. 2003. Impact of annual organic amendment on disease incidence in a three year vegetable rotation. *Plant Dis.* 87:1037–1042.
Teplitski, M., J. B. Robinson, and W. D. Bauer. 2000. Plants secrete substances that mimic bacterial *N*-acyl homoserine lactone signal activities and affect population density-dependent behaviors in associated bacteria. *Mol. Plant Microbe Interact.* 13:637–648.
Tiessen, H. 2005. Phosphorus dynamics in tropical soils. In *Phosphorus: Agriculture and the Environment*, eds. J. T. Sims and A. N. Sharpley, 253–262. Agronomy Monograph 46. ASA, CSA, SSSA, Madison, WI.
Timmusk, S., and E. G. H. Wagner. 1999. The plant-growth-promoting rhizobacterium *Paenibacillus polymyxa* induces changes in *Arabidopsis thaliana* gene expression: a possible connection between biotic and abiotic stress responses. *Mol. Plant Microbe Interact.* 12: 951–959.
Torsvik, V., J. Goksøyr, and F. L. Daae. 1990. High diversity in DNA of soil bacteria. *Appl. Environ. Microbiol.* 56:782.
Torsvik, V., R. Sorheim, and J. Goksoyr. 1996. Total bacterial diversity in soil and sediment communities—a review. *J. Indust. Microbiol.* 17:170–178.
Uren, N. C. 2007. Types, amounts, and possible functions of compounds released into the rhizosphere by soil-grown plants. In *The Rhizosphere: Biochemistry, and Organic Substances at the Soil-Plant Interface*, eds. R. Pinton, Z. Varanini, and P. Nannipieri, 1–21. CRC Press/Taylor & Francis Group, Boca Raton, FL.
Vessey, J. K. 2003. Plant growth promoting rhizobacteria as biofertilizers. *Plant Soil* 255:571–586.
Vivas, A., I. S. Voro, B. Biro, E. Camposa, J. M. Bareaa, and R. Azco. 2003. Symbiotic efficiency of autochthonous arbuscular mycorrhizal fungus (*G. mosseae*) and *Brevibacillus* sp. isolated from cadmium polluted soil under increasing cadmium levels. *Environ. Pollut.* 126:179–189.
Wardle, D. A., and K. E. Giller. 1996. The quest for a contemporary ecological dimension to soil biology. *Soil Biol. Biochem.* 28:1549–1554.
Watnick, P., and R. Kolter, 2000. Biofilm, city of microbes. *J. Bacteriol.* 182:2675–2679.
Weller, D. M., J. Raaijmakers, B. B. McSpadden Gardener, and L. S. Thomashow. 2002. Microbial populations responsible for specific soil suppressiveness. *Annu. Rev. Phytopathol.* 40:309–348.
Werner, D. 2007. Molecular biology and ecology of the rhizobia-legume symbiosis. In *The Rhizosphere*, 2nd ed., eds. R. Pinton, Z. Varanini, and P. Nannipieri, 237–266. CRC Press, Boca Raton, FL.
Whitelaw, M. A. 2000. Growth promotion of plants inoculated with phosphate-solubilizing fungi. *Adv. Agron.* 69:99–151.
Widmer, T. L., N. A. Mitkowski, and G. S. Abawi. 2002. Soil organic matter and management of plant-parasitic nematodes. *J. Nematol.* 34:289–295.

Yousef, L. F., A. F. Yousef, J. S. Mymryk, W. A. Dick, and R. P. Dick. 2009. Stigmasterol and cholesterol regulate the expression of elicitin genes in *Phytophthora sojae*. *J. Chem. Ecol.* 35:824–832.

Zhang, H., B. Prithiviraj, A. Souleimanov, et al. 2002. The effect of temperature and genistein concentration on lipo-chitooligosaccharide (LCO) production by wild-type and mutant strains of *Bradyrhizobium japonicum*. *Soil Biol. Biochem.* 34:1175–1180.

chapter 3

The influence of heterogeneity on soil microbial processes in agroecosystems

Theory, evidence, and opportunities

Terry D. Loecke
University of Nebraska–Lincoln

Contents

3.1 Introduction

Soil heterogeneity is the norm. The complex physical structure of soil is a product of and causal agent of interacting biological diversity and chemical gradients at widely ranging spatial (nanometers to kilometers) and temporal (seconds to centuries) scales. Heterogeneity creates the conditions to select for the myriad biological species present in soil (Brussaard et al., 1997). Chemical gradients (e.g., O_2, CO_2, NO_3^-, metals, and H^+) persist in agricultural soils despite frequent mixing through cultivation (e.g., Robertson et al., 1993). Understanding heterogeneity of soil resources is clearly an important consideration from an agronomic perspective, but the role of heterogeneity as a driver of soil microbial processes and biogeochemical transformations remains unclear.

This chapter describes the theoretical underpinnings of how heterogeneity influences soil microbial processes and agroecosystem functioning. Throughout this analysis, I highlight continuing challenges and opportunities for advancing our understanding of soil microbial processes in light of heterogeneous conditions. I do this through examples of how heterogeneity allows or promotes soil microbial processes to proceed that do not occur at average resource concentrations or constant environmental conditions.

The study of soil heterogeneity cuts across many subdisciplines, the synthesis of which could fill entire volumes. To narrow the scope of this chapter, the examples presented are limited to soil microbial processes affecting carbon (C) and nitrogen (N) cycling in temperate agroecosystems. The goal is not to provide an exhaustive review of the forms and drivers of soil heterogeneity but to couple the theoretical underpinnings to process rates using biogeochemically relevant examples. An emphasis is placed on processes associated with the presence of newly incorporated organic matter (e.g., decomposing plant litter, crop residue, and livestock feces and urine). Furthermore, I do not attempt to review the many techniques for characterizing heterogeneous systems (e.g., geospatial statistics) as there are several excellent summaries already available (e.g., Robertson, 1987; Goovaerts, 1998; Ettema and Wardle, 2002; Webster and Oliver, 2007).

3.2 Theoretical underpinnings

A simple mathematical property of nonlinear functions known as Jensen's inequality provides the theoretical underpinnings for how heterogeneity affects soil microbial process rates and biological systems in general (Jensen, 1906; Ruel and Ayres, 1999). This function was defined over a century ago by Danish mathematician Johan Jensen. Jensen's inequality describes the effect of averaging independent variables (i.e., predictor variables) prior to applying nonlinear functions to explain patterns and contrasts that to applying the function to each individual datum and then summarizing the resulting functions. Jensen's inequality is important to soil process rates for two reasons: (1) Most biochemical processes are nonlinear functions of fundamental abiotic controls on soil biology, such as temperature, moisture availability, substrate abundance, and pH (e.g., Webster et al., 2009). (2) When spatial or temporal heterogeneity of these controls is measureable (the norm), then aggregating across spatial or temporal scales can introduce bias in process rate estimations (Robertson, 1987; Ruel and Ayres, 1999). Because this inequality is inconsistently applied when interpolating process rates from point measurements to greater scales in time or space, these biases can systematically or randomly alter ecosystem-level rate estimation.

As a mathematical expression, Jensen's inequality states that for any nonlinear function, $f(x)$ with variation around the mean predictor variable $\overline{x}$ that

$$f(\overline{x}) \neq \overline{f(x)} \tag{3.1}$$

This gives rise to the memorable phasing of the inequality that the function of the mean ($f(\overline{x})$) does not equal the mean of the functions ($\overline{f(x)}$). To illustrate Jensen's inequality, I use the nonlinear influence of soil temperature on relative metabolic activity of soil organisms (Figure 3.1). In particular, I apply the commonly used van't Hoff function (Webster et al., 2009):

$$\text{Relative Activity} = Q_{10} * e_{T/10} \tag{3.2}$$

where Q_{10} is the fitted constant by which respiration increases when the temperature rises by 10°C, and T is the increase in temperature above 0°C. A Q_{10} of 2.0 is illustrated in Figure 3.1. In this simple example, the heterogeneous system is represented by a soil system with two discrete temperatures (dashed lines at 20°C and 30°C). The activity levels at the two temperatures are then estimated at 15% and 40% for the 20°C and 30°C systems, respectively. The mean activity level of these two systems (i.e., the heterogeneous system) is about 27.5% (represented by the short horizontal dashed line). In contrast, when the function is applied to the mean temperature of these two components (25°C), the estimated activity level is only 24.4%. The approximately 13% difference between the function of the mean and the mean of the functions constitutes an underestimation bias.

My own first encounter with Jensen's inequality was not in a statistics course or reading of it in the primary literature, but an accidental one that is likely common to many scientists.

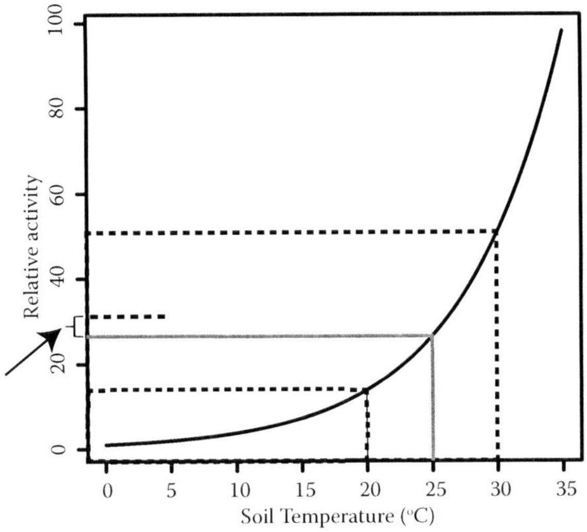

Figure 3.1 A graphical representation of Jensen's inequality using the example of the nonlinear relationship between soil temperature (°C) and relative soil microbial activity (arbitrary units). Here, the simplest case of a heterogeneous system is represented by two states (conditions), 20°C and 30°C (the dashed black lines). Applying the nonlinear function to each state and then finding the average relative activity level (27.5, indicated by the short dashed black horizontal line) yields a approximately 13% higher activity level than when the function is evaluated at the mean soil temperature (25°C, indicated by the solid gray line). The difference between the activity level estimates for the heterogeneous and uniform soil temperatures (~13%, indicated by the bracket and arrow) is referred to as the aggregation bias.

I was applying log transformations to data to meet the homogeneity of variance assumption of an analysis of variance (ANOVA) test and then transforming the means back into a linear scale for visualization purposes. However, I realized that the back-transformed means were not equal to the original means. Not understanding this difference, I sought guidance and was eventually pointed to Jensen's inequality to learn why the transformation process had biased my results. For lack of a more precise term, I refer to the estimation bias due to Jensen's inequality (either an over- or an underestimate) as an aggregation bias. It is important to note that the term *aggregation bias* is also used in a similar but more specific manner in statistics when discussing the level of inference appropriate for a given data set (Freedman et al., 1998).

Aggregation bias observed when studying nonlinear functions is influenced by several components of a functional relationship between environmental variables (or resource abundance) and process rates, including (1) the type of function used to describe the relationship, (2) the sign and magnitude of the function constants, (3) the variance around the mean predictor level (e.g., diel oscillations in soil temperature), and (4) the range in which the function is evaluated. For simple monotonic nonlinear functions (e.g., quadratic, single-parameter power, exponential, logarithm), the direction of the aggregation bias is systematic (i.e., predictable). For example, the exponential relationship between soil temperature and soil metabolic (or enzymatic) activity is an accelerating monotonic function for which the aggregation bias is always negative (i.e., the mean soil temperature will always underpredict the soil microbial activity relative to the unaggregated estimate). For monotonic functions, I recall the direction of the aggregation bias because it is opposite of the sign of the second derivative (or a close approximation thereof). For decelerating monotonic functions (e.g., Michaelis-Menten kinetics), the second derivative is negative, and the aggregation bias is positive (i.e., aggregation causes an overestimation of activity).

The magnitude of the aggregation bias for monotonic response curves is a function of the second derivative (i.e., the greater the acceleration or deceleration, the greater the absolute aggregation bias) and the variance around the mean predictor variable. For example, if the second derivative is a constant function of the predictor variable, then the magnitude of the aggregation bias is a predictable function of the variance around the mean predictor variable. This has implications for many processes in soil systems.

Diel soil temperature oscillations provide a good heuristic for demonstrating the effect of variance on the magnitude of an aggregation bias (Figure 3.2). In general, daily mean temperature is nearly constant with depth, whereas the diel variance in soil temperature varies with depth (Parton and Logan, 1981). In this example, the daily mean soil temperature is constant (25°C) throughout the soil prolife to a depth of 1 m; however, the diel amplitude of the soil temperature produces standard deviations of 3.6, 1.8, 0.9, and 0.1 at 2, 10, 20, and 100 cm of soil depth, respectively. Using a numerical simulation of the influence of variance in temperature on aggregation bias for metabolic rates (using Q_{10}s of 1.5, 2.0, and 3.0), Ruel and Ayres (1999) found aggregation bias to increase monotonically with standard deviation (Figure 3.2b). Comparing the results of Ruel and Ayres's simulation to the example soil profile (Figure 3.2b) at a Q_{10} of 2.0, the aggregation bias is approximately 3%, 1%, 0.4%, and 0.04% at the four depths, respectively. If soil organismal activity is more sensitive to temperature, for example, the Q_{10} is raised to 3.0, the aggregation bias increases to approximately 7%, 2%, 0.7%, and less than 0.1% at 2, 10, 20, and 100 cm of soil depth, respectively. The reported range of Q_{10}s is from less than 1.5 to more than 3.5 (Wang et al., 2010); thus, the aggregation bias can exceed more than 10% for soil microbial communities that are sensitive to temperature fluctuations and vary in temperature.

Use of the Q_{10} function is a common method to scale point measurements of soil processes to longer time scales (e.g., mean annual temperature) (Wang et al., 2010) or larger spatial scales (Raich and Schlesinger, 1992). There has been criticism of this approach for other mechanistic reasons (Davidson et al., 2006); however, the systematic aggregation bias that is introduced into Q_{10} interpolations (e.g., to obtain annual process rate estimates from mean annual temperatures) appears inconsistently acknowledged in soil and agricultural sciences. In my experience, modeling groups generally account for systematic aggregation bias when working with raw data, but scientists who are more field oriented often ignore this bias when interpolating their results.

The systematic aggregation bias that occurs when estimating soil process rates can be minimized through a variety of methods (Ruel and Ayres, 1999; Inouye, 2005; Webster and Oliver, 2007). For example, aggregation bias can be minimized if first the Q_{10} is derived from a fairly continuous measurement of soil processes and soil temperature (i.e., at least hourly, such that the entire diel temperature range is included in the calculation). Next, the process rate can be interpolated without further correction to obtain minimally biased mean annual process rates using this Q_{10} and the mean annual temperature. In contrast, biased Q_{10} functions are often derived from a subset of the diel range, such as weekly measures of soil surface CO_2 flux measurements taken during "business hours." In this case, a correction factor may be developed from the relationship between diel soil temperature variance and the biased Q_{10} factor to minimize bias in the annual process rate estimate.

A synthetic review of diel soil temperature variation with depth across a wide range of agroecosystems does not appear to be easily accessible in the literature, so the overall relevance of diel soil temperature variance on soil metabolic rates is unclear. In general, soil surface albedo, soil moisture content, soil texture, incoming solar radiation, and surface air

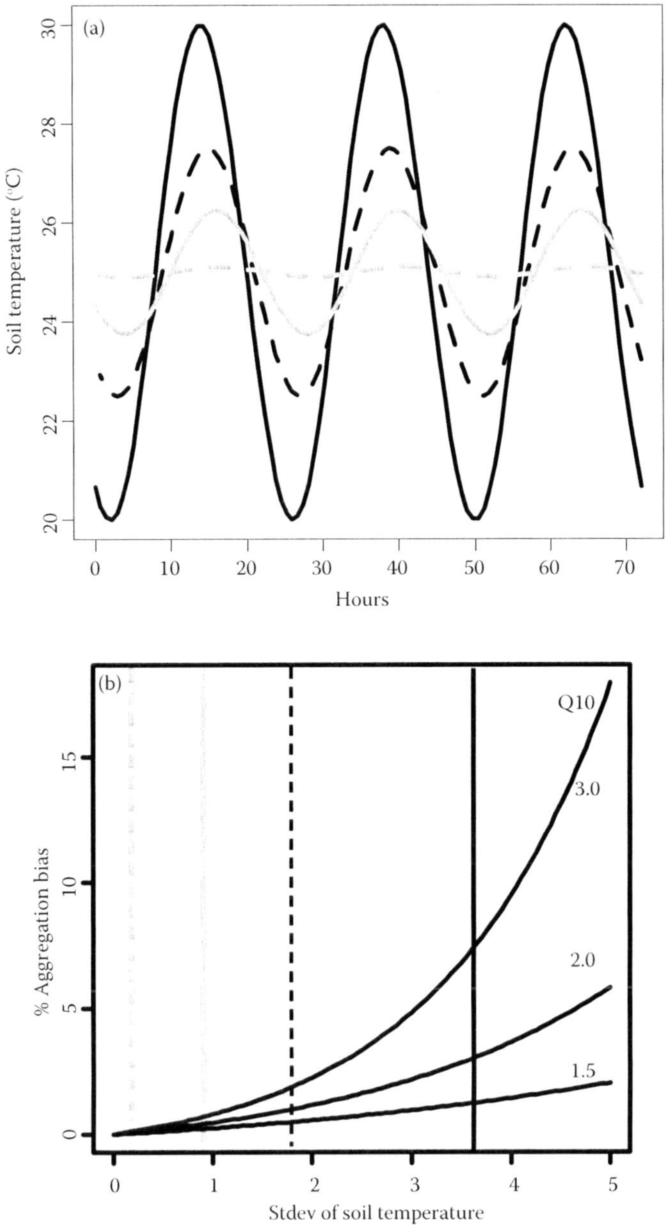

Figure 3.2 An illustration of how variance in biogeochemical controls (e.g., soil temperature) influences aggregation bias in the case of a simple monotonic nonlinear function. (a) Diel soil temperature oscillations over 3 days at four depths; 2 cm (solid black line), 10 cm (dashed black line), 20 cm (solid gray line), and 100 cm (dashed gray line). In this example, the amplitude of the diel oscillation decreased with soil depth, and the daily mean soil temperature was constant with depth. (b) A numerical simulation of the influence of variance on aggregation bias with three levels of Q_{10}: 1.5, 2.0, and 3.0 (as redrawn from Ruel, J. J., and M. P. Ayres. 1999. Jensen's inequality predicts effects of environmental variation. *Trends Ecol. Evol.* 14:361–366). Here, the vertical lines correspond to those from Figure 3.2a and represent the standard deviation of each soil temperature oscillation with soil depth.

temperature are the primary controls on diel soil temperature variation (Hillel, 1998). This indicates that aggregation bias should decrease as soil albedo and moisture content rise. In contrast, as agronomic management practices are implemented to promote soil drying (e.g., to increase seedbed temperature ahead of row crop planting in humid regions), the diel soil temperature variance becomes increasingly important.

Several biogeochemically important soil process-environmental driver functions are more complex than monotonic functions and require individual consideration for inclusion in process-based modeling. For instance, the relationship between soil moisture and soil microbial activity is controlled by two separate factors. At low soil moisture content (approximately less than field capacity), the diffusion of soluble substrates such as simple organic compounds and enzymes limits microbial activity, whereas at soil moisture contents greater than field capacity microbial activity tends to be limited by the diffusion of soil O_2. These two opposing controls on microbial activity result in a discontinuous microbial response function to soil moisture content (Linn and Doran, 1984) with an optimum near field capacity (~60% water-filled pore space) and minimal activity at the extreme soil moisture contents (Figure 3.3a).

For complex functional responses, the question then becomes, how does aggregating soil moisture content across space or time influence estimation bias (i.e., aggregation bias)? To explore this question, I use soil moisture and ammonium (NH_4^+) concentration data and a simple model to describe soil nitrification rates (Parton et al., 1996):

$$\text{Nitrification Rate} = [\text{NH4+}] * \text{WFPS}_f * \text{Temperature}_f * \text{pH}_f \tag{3.3}$$

where $[NH_4^+]$ is the micrograms of 1*N* KCl extractable NH_4^+ as nitrogen per gram of dry soil; $WFPS_f$ is the water-filled pore space activity coefficient (Figure 3.3a); $temperature_f$ is the temperature activity coefficient; and pH_f is an activity coefficient with respect to soil pH. To contrast the influence of a complex functional soil process response with that of a simple monotonic response (e.g., soil temperature), I limit the nitrification model to include soil moisture and NH_4^+ concentration only and assume a constant temperature and pH for simplicity.

For the soil data, I use data collected from the Kellogg Biological Station's Long-Term Ecological Research (KBS-LTER) site located in southwest Michigan in the United States. Specifically, I use data from an experiment comparing four annual row crop management systems that vary in management intensity (e.g., tillage, fertilizer, and pesticide inputs). Soil cores (five 2.5-cm inner diameter [i.d.] × 30 cm deep) have been collected monthly (1989–2001) from each replicate ($r = 6$) of each treatment ($n = 4$). Within-replicate soil samples are composited and subsampled for moisture and inorganic N (NO_2^-, NO_3^-, and NH_4^+).

For modeling simplicity, one might desire to average soil data across replicates to estimate nitrification rates for each treatment on each sampling date (i.e., the aggregated estimate) rather than using the individual replicate data to predict nitrification and then average plot-level estimates to obtain treatment-level nitrification rates (i.e., the unbiased estimate). Doing so creates a treatment-level aggregation bias that is not related to treatment-level soil moisture variance (Figure 3.3b). The aggregation bias is also not simply related to soil moisture content (data not shown). This random aggregation bias precludes any simple correction for variance like that available with simple monotonic functions (e.g., soil temperature vs. microbial activity). The importance of the aggregation bias observed in the nitrification example presented is relative to the desired precision of estimate. The long-term (12 years of data) aggregation bias has a distinct central tendency of no net bias (Figure 3.3b); however, the aggregation bias over shorter periods of time (e.g.,

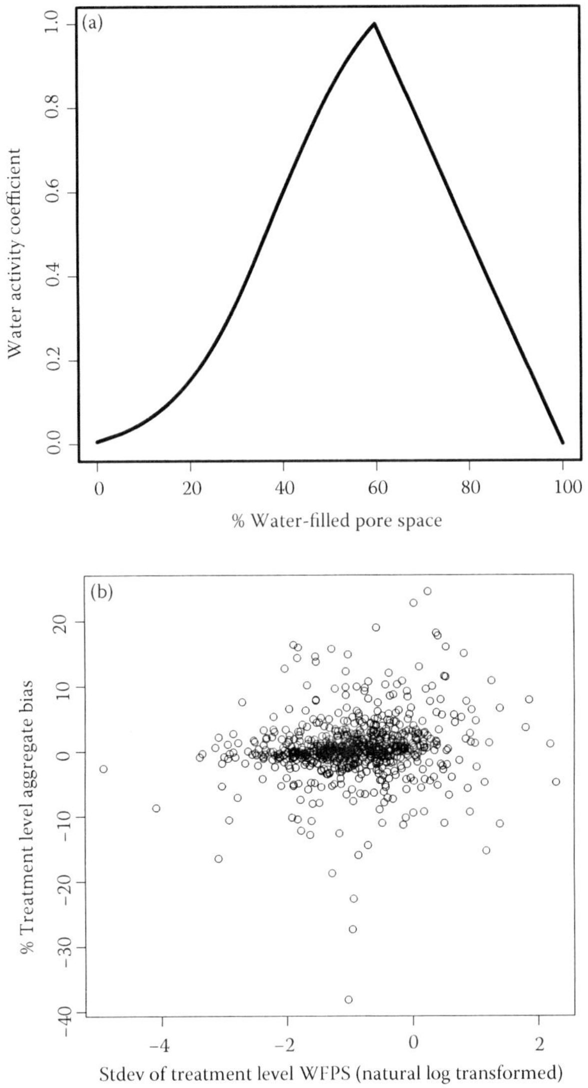

Figure 3.3 An illustration of how variance in biogeochemical controls (e.g., water-filled pore space) influences aggregation bias in the case of a complex nonlinear function. (a) A complex nonlinear function describing how the percentage of soil pore space filled with water (i.e., %WFPS) affects soil microbial activity (Linn and Doran, 1984; Stehfest et al., 2007). (b) The aggregation bias of nitrification rate estimates introduced when aggregating %WFPS data across treatment replicates (for monthly sampling from 1989 to 2001) before applying the %WFPS-microbial activity function (a) versus obtaining the unbiased estimate by deriving the mean nitrification rate for each treatment-sampling date after evaluating each replicate-sampling date separately. (Data for this comparison are from the KBS-LTER.)

within a growth season) may show quite different patterns. For growing seasons with higher-than-average soil moisture contents (i.e., a greater proportion of soil moisture measurements near field capacity), there will be a greater probability of significant aggregation bias because soils will more often be in the nonlinear soil moisture range (i.e., near the optimal soil moisture). In contrast, seasons with drier soils will have less aggregation bias because the soil nitrification rates will more often stay within the linear response range to available moisture.

The prevalence and magnitude of systematic and random (or unpredictable) aggregation bias of microbial processes in agroecosystems due to Jensen's inequality is largely unknown. The opportunities for including correction factors (e.g., variance-weighted interpolation) for systematic bias may help reconcile mass balance and interpolated estimates from point process rate measurements (Webster and Oliver, 2007). When the aggregation bias is unpredictable, however, it is important to minimize aggregation across the predictor variable and perform process-based modeling on the smallest spatial and temporal scales that the data allow to minimize bias (Parkin, 2008).

Jensen's inequality does not appear to be explicitly recognized in the ecological (Ruel and Ayres, 1999), soil, or agricultural sciences. A search of ISI Web of Science for the topic keywords "Jensen's inequality" and "soil" or "ecosystem" or "agriculture" resulted in only 3 articles (Aikio and Ruotsalainen, 2002; Kon, 2004; Benedetti-Cecchi, 2005b) compared to 251 articles with the keyword "Jensen's inequality" alone. The theory is well established, but it is not being communicated to the natural sciences student (Ruel and Ayes, 1999). Ruel and Ayes (1999) reported finding no mention of Jensen's inequality in several biometry textbooks as of 1999. I do not report these findings as proof that Jensen's inequality is not considered in these fields; however, it appears that this principle is used inconsistently in agroecological research.

3.3 *Processes proceed under heterogeneous, not uniform, conditions*

A potentially significant corollary of Jensen's inequality is that soil heterogeneity should allow microbial processes to proceed that would not occur under uniform or constant conditions. Here, I explore the evidence for rare processes occurring in agricultural soils despite attempts at uniformity.

One widely known illustration of soil processes occurring under field conditions at small spatial scales that is unlikely to occur under uniform conditions is Parkin's denitrifying pigweed (*Amaranthus retroflexus* L.) leaf (Parkin, 1987) (Figure 3.4). Within a single soil core (2-cm i.d. × 15 cm in length) taken from an agricultural field in Michigan, Parkin found that about 84% of the denitrification activity (determined by the acetylene block technique) occurring in this core was near a small pigweed leaf that had been incorporated into the soil. This was despite the disproportionately small mass of the leaf relative to the whole soil core (less than 0.01%) and the overall aerobic soil conditions of this field. This work demonstrated the potential for hot spots in the soil to dominate process rates. Furthermore, this observation stimulated the hypothesis that if the components of this leaf (e.g., sugars, proteins, etc.) were evenly dispersed throughout the soil core it would not have provided a sufficiently concentrated resource to deplete soil O_2 concentrations to the point at which denitrification would have been favorable.

Concurrent with Parkin's (1987) work, the role of oxygen diffusion in soils was being investigated. For example, Sexstone et al. (1985) demonstrated that O_2 gradients within individual soil aggregates could range from near-atmospheric concentrations (~20,000 μL L^{-1}) on the surface of the aggregate to less than 200 μL L^{-1} in the interior of a soil aggregate, a distance of less than 2 mm. This novel demonstration helps explain why the products of anaerobic metabolism (e.g., denitrification) are produced in soils that on average were well aerated.

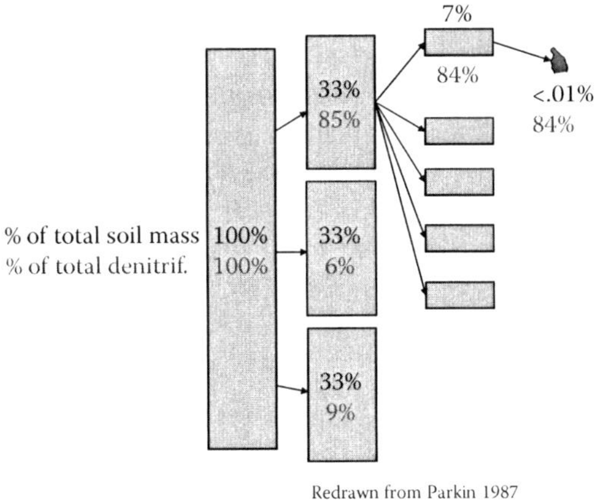

Figure 3.4 A depiction of Parkin's soil core dissection that traced the origin of most of the denitrification activity (84% of the total activity) within a single core (2-cm i.d. × 15 cm in length) back to a single pigweed leaf. The three columns of the soil core represent the sequential process of dissection and acetylene block assessment of denitrification activity in each divided core section. (Redrawn from Parkin, T. B. 1987. Soil microsites as a source of denitrification variability. *Soil Sci. Soc. Am. J.* 51:1194–1199.)

At even smaller scales (e.g., within soil pores, << mm), several efforts to characterize the relationships between the spatial distribution of microbial communities, the processes they perform, and the physical structure of soil habitats have yielded informative insights. For instance, using hypodermic needles as soil corers, Grundmann et al. (2001) dissected a soil aggregate and sampled for the spatial distribution of ammonia- and nitrite-oxidizing bacteria (i.e., nitrifiers). They found the nitrifier communities were spatially clustered, and this spatial pattern was strongly associated with soil micropore size. Through a three-dimensional simulation of microhabitat, Grundmann et al. (2001) were able to conclude that the diffusional properties of micropores likely controlled the spatial distribution of nitrification. This provides evidence of soil spatial structure controlling the spatial distribution of a specific biogeochemical process, nitrification. At a slightly larger scale (2–5 cm), the soil surrounding decomposing plant litter (e.g., *Trifolium repens* L.) can be a hot spot for nitrification within a few weeks of soil incorporation given sufficient soil moisture (Hesselsoe et al., 2001).

Many agronomic management practices intentionally and unintentionally manipulate the spatial distribution of crop and cover crop residue. In temperate humid annual cropping systems (e.g., the U.S. cornbelt), during or prior to planting the soil surface immediately above the seed is intentionally maintained free of crop residue, thereby decreasing surface albedo to promote seedbed warming and drying (e.g., strip tillage and no-till planting) (Licht and Al-Kaisi, 2005). During cool and wet springs, the influence on the seed germination and seedling survival can be dramatic to nonexistent in mild springs (as reviewed in Guerif et al., 2001). Knowledge of how seedbed preparation (apart from total soil surface tillage) alters soil microbial processes appears limited (Guerif et al., 2001; Lee et al., 2009); however, these spatially limited tillage systems (e.g., strip tillage and no till) can substantially increase soil C and N content (e.g., Al-Kaisi et al., 2005), likely a product of altered microbial processes.

Farmers have a range of choices for handling the crop residue that is processed through a modern combination harvester. The residue can be chopped into a range of particle sizes and spread uniformly from where it came or deposited more narrowly directly behind the harvester. The influence of this spatial distribution is often discussed in terms of trade-offs for soil erosion and moisture management. Considerations into how this spatial distribution may affect soil C and N processes is less often mentioned. Few peer-reviewed articles quantitatively described the spatial distribution of crop residue before or after soil tillage (Staricka et al., 1991; Muller et al., 2009), so the influence of crop residue spatial heterogeneity on soil processes remains poorly known (see Grigera et al., 2007).

In humid regions where leaching of field-applied animal manure is of a concern for ground- and surface-water quality, a relatively novel manure applicator (e.g., AerWay) has been designed to apply liquid manure (e.g., swine manure) intentionally in a spatially heterogeneous pattern. This manure applicator is designed to rotate a tool into the surface 8 cm of soil, which creates a cone-shaped void (~10 cm at maximum diameter and 8 cm in depth), and then the resulting depression is filled with a similar volume of liquid manure. The rotating tool is intended to shear and compress soil macropores. This minimizes macropore flow of liquid manure to below the root zone and results in the manure absorbing into the soil directly surrounding the depression. The efficacy of this tool for minimizing manure leaching and promoting crop growth is being studied; however, the spatially heterogeneous distribution of manure that this implement creates raises many questions. For example, is this manure spreader simulating urine and feces patches?

Patches of livestock manure and urine are probably the most well-studied heterogeneous soil resource. At the pasture scale, the distribution of feces and urine patches is a function of topography and animal species, size, behavior, and husbandry (Auerswald et al., 2010). On the individual patch scale, N loss (via NH_4^+ volatilization, denitrification, and nitrate leaching) increases with urine patch size, N concentration, and urine salt concentration (Orwin et al., 2009). Unique soil microbial communities form around newly deposited urine (Orwin et al., 2009) and feces patches (Meyer et al., 2002), similar to that surrounding decomposing plant litter (Hesselsoe et al., 2001). Specifically, the abundance (or activity) of genes involved in nitrification (*AmoA*) and denitrification (*NirS* and *NirK*) increases within days in the soil surrounding feces and urine patches (Orwin et al., 2010) relative to background levels. This indicates that the soil microbial community is poised to respond to resource additions regardless of the spatial distribution.

Naturally occurring heterogeneity of soil properties is often confounded with changes in the mean resource level or environmental control. In that the two factors, heterogeneity and mean levels, often covary in soils, this leads to a difficult interpretation of soil heterogeneity as a property of the system by itself (Benedetti-Cecchi, 2005a). Because of this confounding, several investigators have been motivated to manipulate heterogeneity independently of mean conditions (resources or environmental controls) in an attempt to separate heterogeneity effects from mass effects.

Breland (1994) amended a clay-loam soil with the same mass of white clover (*Trifolium repens*) leaves in a uniform distribution or as a single layer and found the layered clover decomposed and mineralized N more rapidly than the uniform distribution. This layering effect also resulted in greater denitrification (assessed through the acetylene block technique). Breland attributed the process rate differences to a soil-to-litter contact effect. Because the uniformly distributed litter had greater contact with soil particles, there was a greater chance of the litter-soil interaction, leading to the physical protection of the litter C and making it less available for enzymatic degradation. In contrast, uniformly distributed red clover leaves initially decomposed more rapidly than when the litter was aggregated into clumps varying in mass from 0.5 to 4.5 g in a sandy-loam soil (Loecke and Robertson, 2009b).

The contrast between the results of these two studies is likely due to (1) soil mineral-plant litter physical interactions (clay-loam vs. sandy-loam) and (2) microsite O_2 consumption and diffusion rates. Loecke and Robertson (2009b) tested the second part of this hypothesis by incubating uniform and clumped distributions of clover litter under 50% and 80% WFPS to alter the diffusion of soil O_2. Under the more O_2-restrictive environment (80% WFPS), the uniformly distributed litter continued to decompose more rapidly than the clumped distribution, whereas at 50% WFPS decomposition of the clumped litter was only temporarily dampened and exceeded the uniform decomposition rate by the end of the first week of the incubation. Further support for the hypothesis is provided by the sevenfold greater N_2O emissions from the clumped relative to the uniformly distributed litter at 50% WFPS, presumably from denitrification. This largely confirmed Parkin's assertion that if the pigweed leaf had been uniformly distributed in the soil denitrification would have been dramatically lowered.

The *Trifolium* litter used in the two examples (Breland, 1994; Loecke and Robertson, 2009b) is known to decompose rapidly and is relatively rich in N. In contrast, N-poor plant litter (high C:N ratio, *Brassica napus* L.) has been found to decompose more rapidly when aggregated in a sandy-loam soil (Magid et al., 2006). This suggests that heterogeneity affects soil processes independently of mean resource levels; however, the exact processes that are affected will vary with the characteristics of the resource. For example, an N-poor litter may decompose more rapidly within a clump, yet it may be less likely to promote denitrification due to greater microbial assimilatory demand for N in the clump versus uniform distribution. Largely, these ideas are yet to be tested under field conditions.

Given the supposition that heterogeneity is the norm, the utility of including a uniform distribution in comparison to heterogeneous resource distributions is questionable. I am aware of only a single study that has included multiple heterogeneous distributions (Loecke and Robertson, 2009b), so the influence of the spatial distribution of soil resources on biogeochemical processes is still limited. Using a gradient approach to spatial heterogeneity, Loecke and Robertson found large changes in decomposition and N_2O emission between a uniform distribution and 0.5-g clumps of red clover (as mentioned) but few differences between 0.5-, 1.5-, and 4.5-g clumps. At least for red clover litter (either ground or whole leaf), this suggests a threshold between tiny patches (<1 mg) composed of individual ground red clover particles (i.e., the uniform distribution) and the 0.5-g clumps.

3.4 *Opportunities to advance the understanding and management of heterogeneity in agroecosystems*

I see two general areas in which explicit consideration of heterogeneity in agroecosystems can improve management and understanding. First, manipulating the spatial heterogeneity of organic matter (e.g., crop residue or organic amendments) has the potential to improve agronomic efficiency and environmental protection. Second, conceptual and process-based modeling may benefit from considering the role of heterogeneity in nutrient turnover, C sequestration, and processes rate estimation. These two general opportunity areas are synergistic as the former will help test and parameterize the latter.

The interface of hydrologic flow paths from terrestrial and aquatic ecosystems is a well-known area of intense biogeochemical transformations; in particular, this habitat interface (i.e., riparian zone) is a hot spot for denitrification of nitrate-rich agricultural drainage waters (Hedin et al., 1998; McClain et al., 2003; David et al., 2010). The thermodynamic conditions are not always favorable for riparian zone denitrification to deplete NO_3^- before it enters surface waters (Hedin et al., 1998). This has prompted several researchers to investigate methods of promoting zones of denitrification higher in agricultural watersheds (e.g., Moorman et al., 2010; Woli et al., 2010). One method introduces a layer of wood chips (i.e., a source of substrate and habitat) directly surrounding drainage tiles to provide an opportunity for nitrate removal (i.e., denitrification) prior to entering the engineered drainage system (Moorman et al., 2010). In other words, this method spatially manipulates organic amendments into agricultural systems explicitly for environmental protection.

I see several opportunities to spatially manipulate organic amendments and crop residues to promote various soil microbial processes for agronomic and environmental benefits. As highlighted, the spatial distribution of N-rich crop residue can alter the rate of N_2O emissions and N mineralization; however, these few studies were conducted under simplified conditions (e.g., lab incubations and container experiments). More work is warranted on the spatial manipulation of organic amendments and crop residues under field conditions. It is likely that the influence of spatial manipulation on environmental protections (i.e., greenhouse gas emissions) will vary with amendment chemistry, particular size, and edaphic factors. The influence of such manipulations on crop yields and nutrient use efficiencies has received even less attention (Loecke and Robertson, 2009a). See Chapter 6 by Grandy et al. in this volume for a discussion of the interactions between spatial heterogeneity and N use efficiency in agroecosystems.

Conceptual models of plant litter decomposition and soil C sequestration (e.g., Moorhead and Sinsabaugh, 2006; Manzoni and Porporato, 2009) often detail the role of spatial heterogeneity in the form of soil structure or litter particle size and vertical distribution as well as larger landscape heterogeneous factors (e.g., soil texture, topography, and exposure); however, to date the spatial aggregation of plant litter has not explicitly been considered in these models. Empirical work from forested systems suggested litter aggregation significantly affects soil micro- and mesofauna activity (Sulkava and Huhta, 1998). In agricultural systems, field testing of the influence of crop residue spatial manipulations on soil microbial processes and plant-microbe interactions will inform the necessity of including litter aggregation into these conceptual models.

Introducing bias into soil microbial process rate estimates due to Jensen's inequality (i.e., aggregation bias) is unavoidable and unpredictable in some cases and yet systematic and easily correctable in others (e.g., Q_{10} interpolations) (Ruel and Ayres, 1999). Methods for interpolating with known variance are well appreciated in geospatial statistics (Webster and Oliver, 2007); however, too often interpolation of process rates across time does not consider aggregation bias. Using correction factors when interpolating process rates derived from simple monotonic nonlinear functions (e.g., soil temperature effect on metabolic rate) should be a standard practice for scaling point measures to longer temporal scales in agroecological research.

References

Aikio, S., and A. L. Ruotsalainen. 2002. The modelled growth of mycorrhizal and non-mycorrhizal plants under constant versus variable soil nutrient concentration. *Mycorrhiza* 12:257–261.

Al-Kaisi, M. M., X. H. Yin, and M. A. Licht. 2005. Soil carbon and nitrogen changes as affected by tillage system and crop biomass in a corn-soybean rotation. *Appl. Soil Ecol.* 30:174–191.

Auerswald, K., F. Mayer, and H. Schnyder. 2010. Coupling of spatial and temporal pattern of cattle excreta patches on a low intensity pasture. *Nutr. Cycl. Agroecosyst.* 88:275–288.

Benedetti-Cecchi, L. 2005a. The importance of the variance around the mean effect size of ecological processes: reply. *Ecology* 86:265–268.

Benedetti-Cecchi, L. 2005b. Unanticipated impacts of spatial variance of biodiversity on plant productivity. *Ecol. Lett.* 8:791–799.

Breland, T. A. 1994. Enhanced mineralization and denitrification as a result of heterogeneous distribution of clover residues in soil. *Plant Soil* 166:1–12.

Brussaard, L., V. M. Behan-Pelletier, D. E. Bignell, et al. 1997. Biodiversity and ecosystem functioning in soil. *Ambio.* 26:563–570.

David, M. B., L. E. Drinkwater, and G. F. McIsaac. 2010. Sources of nitrate yields in the Mississippi River Basin. *J. Environ. Qual.* 39:1657–1667.

Davidson, E. A., I. A. Janssens, and Y. Luo. 2006. On the variability of respiration in terrestrial ecosystems: moving beyond Q10. *Glob. Change Biol.* 12:154–164.

Ettema, C. H., and D. A. Wardle. 2002. Spatial soil ecology. *Trends Ecol. Evol.* 17:177–183.

Freedman, D. A., S. P. Klein, M. Ostland, M. R. Roberts. 1998. Review of a solution to the ecological inference problem. *J. Am. Stat. Assoc.* 93:1518–1522.

Goovaerts, P. 1998. Geostatistical tools for characterizing the spatial variability of microbiological and physio-chemical soil properties. *Biol. Fertil. Soils* 27:315–334.

Grigera, M. S., R. A. Drijber, and B. J. Wienhold. 2007. Redistribution of crop residues during row cultivation creates a biologically enhanced environment for soil microorganisms. *Soil Tillage Res.* 94:550–554.

Grundmann, G. L., A. Dechesne, F. Bartoli, J. P. Flandrois, J. L. Chasse, and R. Kizungu. 2001. Spatial modeling of nitrifier microhabitats in soil. *Soil Sci. Soc. Am. J.* 65:1709–1716.

Guerif, J., G. Richard, C. Durr, J. M. Machet, S. Recous, and J. Roger-Estrade. 2001. A review of tillage effects on crop residue management, seedbed conditions and seedling establishment. *Soil Tillage Res.* 61:13–32.

Hedin, L. O., J. C. von Fischer, N. E. Ostrom, B. P. Kennedy, M. G. Brown, and G. P. Robertson. 1998. Thermodynamic constraints on nitrogen transformations and other biogeochemical processes at soil-stream interfaces. *Ecology* 79:684–703.

Hesselsoe, M., A. Pedersen, K. Bundgaard, K. K. Brandt, and J. Sorensen. 2001. Development of nitrification hot-spots around degrading red clover (*Trifolium pratense*) leaves in soil. *Biol. Fertil. Soils* 33:238–245.

Hillel, D. 1998. *Environmental Soil Physics.* Academic Press, San Diego, CA.

Inouye, B. D. 2005. The importance of the variance around the mean effect size of ecological processes: comment. *Ecology* 86:262–265.

Jensen, J. L. W. V. 1906. Sur les fonctions convexes et les inegalites entre les valeurs moyennes. *Acta Math.* 30:175–193.

Kon, R. 2004. Permanence of discrete-time Kolmogorov systems for two species and saturated fixed points. *J. Math. Biol.* 48:57–81.

Lee, J., J. W. Hopmans, C. vanKessel, A. P. King, K. J. Evatt, D. Louie, D. E. Rolston, and J. Six. 2009. Tillage and seasonal emissions of CO_2, N_2O, and NO across a seed bed and at the field scale in a Mediterranean climate. *Agric. Ecosyst. Environ.* 129:378–390.

Licht, M. A., and M. M. Al-Kaisi. 2005. Strip-tillage effect on seedbed soil temperature and other soil physical properties. *Soil Tillage Res.* 80:233–249.

Linn, D. M., and J. W. Doran. 1984. Effect of water-filled pore-space on carbon-dioxide and nitrous-oxide production in tilled and nontilled soils. *Soil Sci. Soc. Am. J.* 48:1267–1272.

Loecke, T. D., and G. P. Robertson. 2009a. Soil resource heterogeneity in the form of aggregated litter alters maize productivity. *Plant Soil* 325:231–241.

Loecke, T. D., and G. P. Robertson. 2009b. Soil resource heterogeneity in terms of litter aggregation promotes nitrous oxide production and slows decomposition. *Soil Biol. Biochem.* 41:228–235.

Magid, J., A. De Neergaard, and M. Brandt. 2006. Heterogeneous distribution may substantially decrease initial decomposition, long-term microbial growth and N-immobilization from high C-to-N ratio resources. *European J. Soil Sci.* 57:517–529.

Manzoni, S., and A. Porporato. 2009. Soil carbon and nitrogen mineralization: theory and models across scales. *Soil Biol. Biochem.* 41:1355–1379.

McClain, M. E., E. W. Boyer, C. L. Dent, S. E. Gergel, N. B. Grimm, P. M. Groffman, S. C. Hart, J. W. Harvey, C. A. Johnston, M. Emilio, W. H. McDowell, and G. Pinay. 2003. Biogeochemical hot spots and hot moments at the interface of terrestrial and aquatic ecosystems. *Ecosystems* 6:301–312.

Meyer, R. L., T. Kjaer, and N. P. Revsbech. 2002. Nitrification and denitrification near a soil-manure interface studied with a nitrate-nitrite biosensor. *Soil Sci. Soc. Am. J.* 66:498–506.

Moorhead, D. L., and R. L. Sinsabaugh. 2006. A theoretical model of litter decay and microbial interaction. *Ecol. Monogr.* 76:151–174.

Moorman, T. B., T. B. Parkin, T. C. Kaspar, and D. B. Jaynes. 2010. Denitrification activity, wood loss, and N_2O emissions over 9 years from a wood chip bioreactor. *Ecol. Eng.* 36:1567–1574.

Muller, E., H. Wildhagen, M. Quintern, J. Hess, F. Wichern, and R. G. Joergensen. 2009. Spatial patterns of soil biological and physical properties in a ridge tilled and a ploughed luvisol. *Soil Tillage Res.* 105:88–95.

Orwin, K. H., J. E. Bertram, T. J. Clough, L. M. Condron, R. R. Sherlock, and M. O'Callaghan. 2009. Short-term consequences of spatial heterogeneity in soil nitrogen concentrations caused by urine patches of different sizes. *Appl. Soil Ecol.* 42:271–278.

Orwin, K. H., J. E. Bertram, T. J. Clough, L. M. Condron, R. R. Sherlock, M. O'Callaghan, J. Ray, and D. B. Baird. 2010. Impact of bovine urine deposition on soil microbial activity, biomass, and community structure. *Appl. Soil Ecol.* 44:89–100.

Parkin, T. B. 1987. Soil microsites as a source of denitrification variability. *Soil Sci. Soc. Am. J.* 51:1194–1199.

Parkin, T. B. 2008. Effect of sampling frequency on estimates of cumulative nitrous oxide emissions. *J. Environ. Qual.* 37:1390–1395.

Parton, W. J., and J. A. Logan. 1981. A model for diurnal-variation in soil and air-temperature. *Agric. Meteorol.* 23:205–216.

Parton, W. J., A. R. Mosier, D. S. Ojima, D. W. Valentine, D. S. Schimel, K. Weier, and A. E. Kulmala. 1996. Generalized model for N_2 and N_2O production from nitrification and denitrification. *Global Biogeochem. Cycles* 10:401–412.

Raich, J. W., and W. H. Schlesinger. 1992. The global carbon-dioxide flux in soil respiration and its relationship to vegetation and climate. *Tellus Series B Chem. Phys. Meteorol.* 44:81–99.

Robertson, G. P. 1987. Geostatistics in ecology—interpolating with known variance. *Ecology* 68:744–748.

Robertson, G. P., J. R. Crum, B. G. Ellis. 1993. The spatial variability of soil resources following long-term disturbance. *Oecologia* 96:451–456.

Ruel, J. J., and M. P. Ayres. 1999. Jensen's inequality predicts effects of environmental variation. *Trends Ecol. Evol.* 14:361–366.

Sexstone, A. J., N. P. Revsbech, T. B. Parkin, and J. M. Tiedje. 1985. Direct measurement of oxygen profiles and denitrification rates in soil aggregates. *Soil Sci. Soc. Am. J.* 49:645–651.

Staricka, J. A., R. R. Allmaras, and W. W. Nelson. 1991. Spatial variation of crop residue incorporated by tillage. *Soil Sci. Soc. Am. J.* 55:1668–1674.

Stehfest, E., M. Heistermann, J. A. Priess, D. S. Ojima, and J. Alcamo. 2007. Simulation of global crop production with the ecosystem model DayCent. *Ecol. Model.* 209:203–219.

Sulkava, P., and V. Huhta. 1998. Habitat patchiness affects decomposition and faunal diversity: a microcosm experiment on forest floor. *Oecologia* 116:390–396.

Wang, X. H., S. L. Piao, P. Ciais, I. A. Janssens, M. Reichstein, S. S. Peng, and T. Wang. 2010. Are ecological gradients in seasonal Q(10) of soil respiration explained by climate or by vegetation seasonality? *Soil Biol. Biochem.* 42:1728–1734.

Webster, K. L., I. F. Creed, M. D. Skowronski, and Y. H. Kaheil. 2009. Comparison of the performance of statistical models that predict soil respiration from forests. *Soil Sci. Soc. Am. J.* 73:1157–1167.

Webster, R., and M. A. Oliver. 2007. *Geostatistics for Environmental Scientists.* Wiley, West Sussex, UK.

Woli, K. P., M. B. David, R. A. Cooke, G. F. McIsaac, and C. A. Mitchell. 2010. Nitrogen balance in and export from agricultural fields associated with controlled drainage systems and denitrifying bioreactors. *Ecol. Eng.* 36:1558–1566.

chapter 4

Soil food webs in agricultural ecosystems

John C. Moore
Colorado State University

Peter C. de Ruiter
Wageningen University

Contents

4.1 *Introduction*

Odum (1969) provided a framework for viewing the relationships between plant communities and biogeochemical cycles in his seminal article, "The Strategy of Ecosystem Development." The framework is a compilation of ecosystem thinking of that time that juxtaposes ecosystem properties of early development communities with mature communities, revealing an end game of homeostasis and balance between plants and nutrients (Table 4.1). The transition from early developmental communities to mature communities is characterized structurally and functionally by a transition from *r*-selected to *k*-selected plant species, from food chains that rely primarily on primary production to ones that rely on both primary production and detritus, and from biogeochemical cycles that are open with exogenous inputs and high exports of plant-limiting nutrients to ones that are more closed and dependent on symbiotic interactions that promote internal cycling of nutrients and nutrient exchanges (Wall and Moore, 1999).

Agricultural practices affect many of these key structural and functional attributes of ecosystems in ways that transform mature ecosystems to ones that are in a managed developmental state. The regular additions of inorganic or organic fertilizers and amendments and harvesting of crops have the potential to shift the elemental balance of a system.

Table 4.1 Summary of Ecological Succession from Early Developmental Stages to a Mature Community

Ecosystem attributes	Developmental stages	Mature stages
	Community energetics	
Gross production/community respiration (P/R ratio)	Greater or less than 1	Approaches 1
Gross production/standing crop biomass (P/B ratio)	High	Low
Biomass supported/unit energy flow (B/E ratio)	Low	High
Net community production	High	Low
Food chains	Linear, predominantly grazing	Weblike, predominantly detritus
	Community structure	
Total organic matter	Small	Large
Inorganic nutrients	Extrabiotic	Intrabiotic
Species diversity: richness	Low	High
Species diversity: evenness	Low	High
Biochemical diversity	Low	High
Stratification and spatial diversity (pattern diversity)	Poorly organized	Well organized
	Life history	
Niche specialization	Broad	Narrow
Size of organisms	Small	Large
Life cycles	Short, simple	Long, complex
Nutrient cycling		
Mineral cycles	Open	Closed
Nutrient exchange rate, between organisms and environment	Rapid	Slow
Role of detritus in nutrient regeneration	Unimportant	Important
	Selective pressure	
Growth form	For rapid growth ("*r*-selection")	For feedback control ("*K*-selection")
Production	Quantity	Quality
	Overall homeostasis	
Internal symbiosis	Undeveloped	Developed
Nutrient conservation	Poor	Good
Stability (resistance to external perturbations)	Poor	Good
Entropy	High	Low
Information	Low	High

Source: From Odum, E.T. 1969. The strategy of ecosystem development. *Science* 164:262–270. With permission.

Conventional tillage practices alter the distribution of organic material within the soil profile and disrupt soil aggregates, releasing and exposing sequestered organic material to microbes and invertebrates, thereby accelerating the mineralization of carbon and nitrogen. Minimum tillage and no-tillage practices disrupt soils to a lesser extent but often rely on herbicides to control weeds with nontarget effects on humans and other biota.

Irrigation enhances productivity but leads to the erosion of topsoil and the runoff of soluble nutrients and pesticides.

Proponents of sustainable agriculture have advocated that management practices be adopted that emulate the properties of mature ecosystems that Odum (1969) had envisioned. Elliott and Coleman (1988) adopted this philosophy in their article, "Let the Soil Work for Us." Management practices that promote less reliance on inorganic nutrient additions and a greater reliance on organic amendments and biotic interactions promote internal cycling of nutrients and, if adopted, would reduce the environmental degradation that conventional practices induce. What is missing in Odum's assessment and the philosophies of sustainable agricultural practices that followed is a critical mathematical link between the desired structural and process phenomena and the mathematical notions of stability and persistence. This link is important because it allows us to make the connections between ecosystem structure and function and the consequences of altering conditions on these ecosystem properties on its persistence. Food web models offer a way to integrate activities of organisms and the cycling of matter within ecological systems to make the link between observation and stability.

Food web descriptions in conjunction with population and biogeochemical cycling models have played an important role in the study of agricultural ecosystems. The descriptions and models help to frame questions, interpret empirical outcomes, and convey in numerical and visual form the complexity of interactions among species, soil organic matter, and nutrients (Parton et al., 1983; Hendrix et al., 1986; Andrén et al., 1990; Moore and de Ruiter, 1991; de Ruiter, Moore, et al., 1993; de Ruiter et al., 1994; Swift and Anderson, 1994). Three important observations that have emerged from these studies serve as the basis for this chapter. First, conventional agricultural management practices have resulted in significant losses of soil organic matter (Beare et al., 1994; Vitousek et al., 1997). Second, the patterning of nutrient flow and trophic interactions within the soil food web under conventional management practices differ from their counterparts in native or conservation management practices in ways that favor bacteria and their consumers over fungi and their consumers (Hendrix et al., 1986). Third, coincident with the losses in organic matter and changes in food web interactions and nutrient flow is a decrease in stability as measured by mathematical representations derived from these interactions (de Ruiter et al., 1995; Neutel et al., 2002).

This chapter focuses on these observations and results from models and offers propositions on how the study of trophic interactions, nutrient dynamics, and stability using food webs can assist in management decisions to optimize production, nutrient retention, and soil integrity in ways that mimic the mature ecosystems that are recommended for sustainable agriculture. We start with an overview of food web descriptions and models that have been developed to study the interactions among plants and soil organisms in natural and agricultural ecosystems. The models capture many of the structural and functional components that Odum (1969) used to contrast early development and mature ecosystems (see Table 4.1). Next, we present a set of results that link many of these structural and functional components to dynamic properties and stability. Here, we present the results of theoretical exercises that change (read disturb) the observed patterns in food web structure and assess the effects of these changes on stability. We end the chapter by connecting the theoretical results to empirical observations from studies of agricultural ecosystems demonstrating that agricultural management practices affect soil food webs in ways that parallel those used in the theoretical exercises on disturbance and stability.

4.2 Soil food web descriptions and models

A food web is a description of the trophic interactions within an ecosystem. By their nature, they are incomplete, focusing on a subset of organisms and resources and averaging over different spatial and temporal scales. Paine (1980) distinguished three types of food web descriptions—*connectedness*, *energy flux*, and *functional*—that share a common underlying structure but that utilize and convey different types of information about the system. These three descriptions provide a convenient organizational and operational framework to study and model agricultural practices and their impacts on ecosystem structure, function, and stability. Within each description, different modeling approaches and formulations are used (see Hunt et al., 1987; Moore and Hunt, 1988; Moore et al., 1993; de Ruiter et al., 1995).

The food web descriptions and models we use are based on those developed for soils of the North American shortgrass steppe developed by Hunt et al. (1987). The native shortgrass steppe system supports grazing by cattle, while cultivated systems include conventional tillage and no-tillage dryland winter wheat and irrigated corn and alfalfa. The modeling approaches presented here have been adopted by several research groups that have attempted to link the structure of soil food webs to the decomposition of organic matter and the mineralization of nutrients (Andrén et al., 1990; Brussaard et al., 1988, 1997; de Ruiter, Van Veen, et al., 1993; Hendrix et al., 1986; Hunt et al., 1987; Moore et al., 1988).

4.2.1 The connectedness food web

The connectedness food web defines the model's basic structure and foundation for the energy flux and functional descriptions (Figure 4.1). For our purposes, the web depicts the trophic interactions among functional groups of organisms and detritus within the

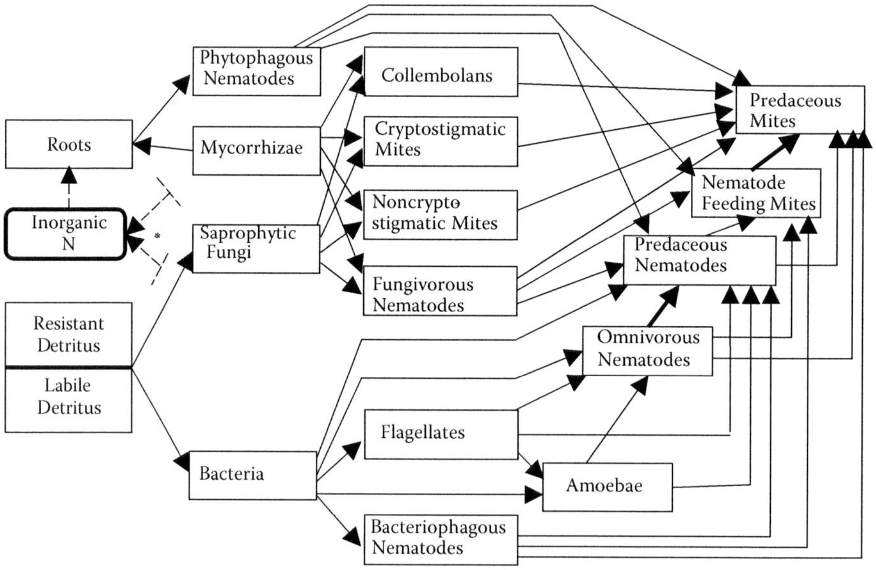

Figure 4.1 The connectedness description of the shortgrass steppe soil food web from the Central Plains Experimental Range (CPER), Nunn, Colorado. The solid vectors represent trophic interactions in terms of carbon fluxes among resources, mutualists, and consumer (prey and predator) functional groups. The dashed vectors represent nitrogen fluxes. The asterisk (*) indicates that all living organisms contribute to the inorganic nitrogen pool as explained in the text and Figure 4.3.

rhizosphere, the zone of soil affected by the growth and maintenance of plant roots (Swift et al., 1979; Coleman et al., 1983). The description condenses the high diversity and complexity within soils to a manageable level by defining the web in terms of functional groups of organisms that share similar prey and predators, feeding modes, life-history attributes, and habitat preferences (Table 4.2; Moore et al., 1988). At the base of the web are plant roots, labile (C:N ratio < 30:1) and resistant (C:N ratio > 30:1) forms of detritus, and an inorganic nitrogen source. The plant roots and their by-products and the different forms of detritus are utilized by microbes and invertebrate consumers, which are then consumed by different predators.

4.2.2 The energy flux food web

The energy flux food web description displays the amounts of biomass or population sizes of the functional groups and quantifies the rates of feeding or energy fluxes within the connectedness description (Figure 4.2). More often than not, the energy flux description depicts elemental flows, usually C or N, inferred from biomass estimates and models of trophic interactions. Table 4.3 provides the physiological and life-history parameters and biomass estimates used to estimate the elemental flows (O'Neill, 1969; Hunt et al., 1987; de Ruiter, Moore, et al., 1993).

The model starts with estimates of the biomasses of the individual functional groups. The biomass estimates are obtained from direct or indirect estimates of population sizes (e.g., morphometric conversions) obtained from field samples, preferably at several times to capture temporal dynamics or to estimate steady states or long-term averages (Moore et al., 1996).

Feeding rates are estimated using the model presented by Hunt et al. (1987). The model assumes that at the steady-state biomass any new biomass produced within a population is offset by biomass that is lost from the population (i.e., inputs equal outputs at steady state). Consumed matter is divided into a fraction that is immobilized into consumer biomass (assimilation) and a fraction that is excreted to the environment as feces and unconsumed. Of the assimilated fraction, material either is incorporated into new biomass as growth and reproduction (production) or is mineralized as inorganic material (Figure 4.3). Notice that matter is conserved and, per the second law of thermodynamics, that not all assimilated matter is converted to new biomass. Hence, feeding rates F are estimated as follows:

$$F = (dB + M)/ap \tag{4.1}$$

where F is the feeding rate (biomass time^{-1}), d is the specific death rate (time^{-1}) of the consumer, B (biomass estimated from field samples) is the population size of the consumer, M is the death rate due to predators (biomass time^{-1}), and a and p are the assimilation (percentage expressed as proportions) and production (percentage expressed as proportions) efficiencies, respectively. For consumers that feed on multiple prey types, the fluxes are weighted by the predators' feeding preferences for the respective prey.

The estimation procedure starts with top predators since the death due to predation is zero. The fluxes from prey to the top predators serve as the estimates of the prey's death due to predation. Hence, the process moves downward through the prey to the basal resources with fluxes to each prey taking into account the biomass lost to predation. To construct a dynamic version, we simply need to take into account changes in the biomasses over an interval of time t by adding $\Delta B/t$ to the numerator of Equation 4.1 (O'Neill, 1969).

Table 4.2 Criteria Used to Formulate the Functional Groups of the Shortgrass Steppe Soil Food Web for the Central Plains Experimental Range (CPER), Nunn, Colorado

Functional group	Description	Habitat	Life history	Examples (genera)
Predatory mites	Attack most soil invertebrates small enough to overcome	Air-filled pore spaces	Weeks–months	*Hypoaspis, Asca, Amblyseius, Rhodacarus, Gamasellodes, Macrocheles, Spinibdella, Cyta, Stigmaeus, Cocorhagidia*
Nematophagous mites	Attack only nematodes	Air-filled pore spaces	Weeks–months	*Alliphis, Eviphis, Alycus, Alicorhagia, Ololaelaps, Veigaia*
Predatory nematodes	Attack nematodes and bacteria (minimal)	Water film	Days–weeks	*Discolaimium, Mononchus*
Omnivorous nematodes	Consumer of bacteria and protozoa	Water film	Days–weeks	*Mesodiplogaster*
Fungivorous nematodes	Feed on fungal cytoplasm	Water film	Days–weeks	*Aphelenchus, Aphelenchoides*
Bacteriophagous nematodes	Consume bacteria	Water film	Days–weeks	*Acrobeloides, Pelodera, Rhabditis*
Collembola	Consume fungal hyphae and spores, algae, pollen	Air-filled pore spaces	Weeks–months	*Folsomia, Isotoma, Isotomdesm Hypogastura, Tullbergia, Deuterosmithurus, Sminthurus*
Mycophagous Prostigmata	Pierce fungal hyphae and consume fungal cytoplasm	Air-filled pore spaces	Weeks–months	*Tydeus, Eupodes, Tarsonemus, Bakerdania, Pediculaster, Scutacarus, Speleorchestes*
Cryptostigmata	Consume fungal hyphae and spores	Air-filled pore spaces	Weeks–months	*Haplozetes, Passalozetes, Zygoribatula, Pilogalumna, Tectocepheus, Oppiella, Ceratozetes*
Amoebas	Consume bacteria	Water-filled pore spaces Water film	Hours	*Acanthamoeba, Hartmanella, tricamoeba, Mayorella, Varella*
Flagellates	Consume bacteria	Water-filled pore spaces Water film	Hours	*Pleuromonas, Bodo, Mastigamoeba*
Phytophagous nematodes	Feed on plant roots	Water film	Days–weeks	*Helicotylenchus, Tylenchorhynchus, Xiphinema*

Source: From Moore, J.C., D.E. Walter, and H.W. Hunt. 1988. Arthropod regulation of micro- and mesobiota in below-ground detrital food webs. *Annual Review of Entomology* 33:419–439.

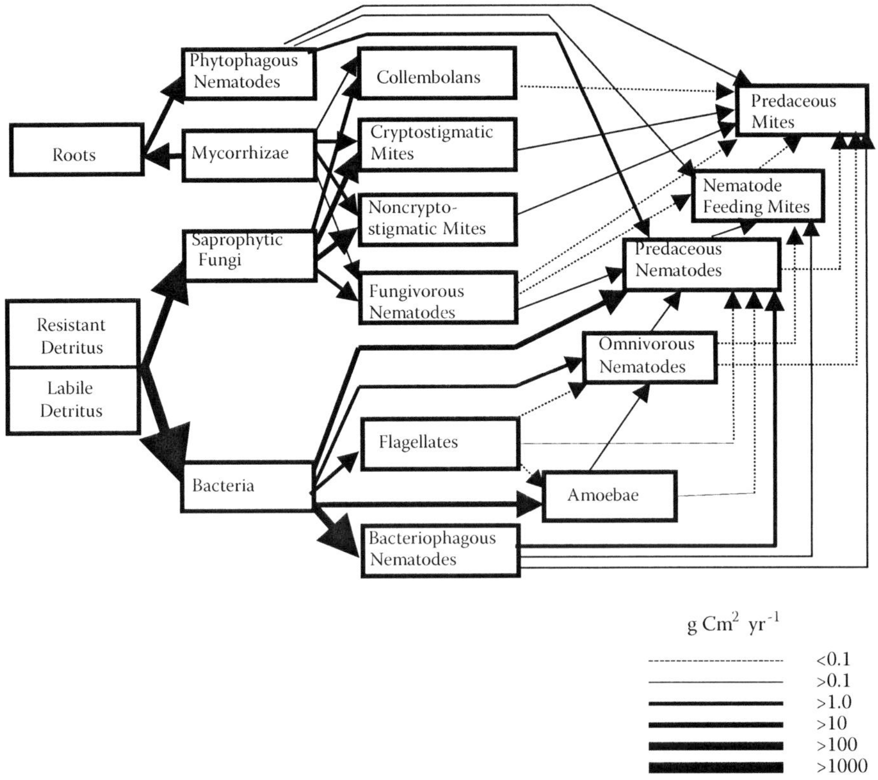

Figure 4.2 The energy flux description expressed in terms of carbon (g C m^2 y^{-1}) of the shortgrass steppe soil food web from the Central Plains Experimental Range (CPER), Nunn, Colorado. The thickness of the vectors is scaled to the estimated carbon fluxes (g C m^2 y^{-1}) among prey to predator functional groups. Although not represented in this energy flux description, the boxes for functional groups could be sized to scale to the standing biomass (g C m^2) for each functional group as well (see Moore, J.C., D.E. Walter, and H.W. Hunt. 1988. Arthropod regulation of micro- and mesobiota in below-ground detrital food webs. *Annual Review of Entomology* 33:419–439).

4.2.3 The functional food web

A functional food web can be derived from (a) the connectedness web, (b) the set of differential equations that include the trophic interactions among species to describe the population dynamics of the species, and (c) energy flux descriptions. The functional description displays impacts that each population (species or functional groups) has on the dynamics of themselves and the other species in the food web. We have adopted interaction strength as the metric to represent these impacts (Figure 4.4).

There are several different definitions of interaction strength in the literature (Paine, 1992; Berlow et al., 2004). We define interaction strength in terms of the elements of the Jacobian matrix of the system of differential equations that describe the dynamics of the food web. For example, we could use the following set of equations to describe an *n*-species community:

Table 4.3 Biomasses (kg C ha^{-1}) and Physiological and Life-History Parameters for the Different Functional Groups for the Central Plains Experimental Range (CPER) Food Web

Functional group	C/N	Turnover rate (y^{-1})	Assimilation efficiency (%)	Production efficiency (%)	Biomass (kg C ha^{-2})
Predatory mites	8	1.84	60	35	0.160
Nematophaous mites	8	1.84	90	35	0.160
Predatory nematodes	10	1.60	50	37	1.080
Omnivorous nematodes	10	4.36	60	37	0.650
Fungivorous nematodes	10	1.92	38	37	0.410
Bacteriophagous nematodes	10	2.68	60	37	5.800
Collembola	8	1.84	50	35	0.464
Mycophagous Prostigmata	8	1.84	50	35	1.360
Cryptostigmata	8	1.20	50	35	1.680
Amoebas	7	6.00	95	40	3.780
Flagellates	7	6.00	95	40	0.160
Phytophagous nematodes	10	1.08	25	37	2.900
Arbuscular mycorrhizal (AM) fungi	10	2.00	100	30	7.000
Saprobic fungi	10	1.20	100	30	63.000
Bacteria	4	1.20	100	30	304.000
Detritus	10	0.00	100	100	3,000.000
Roots	10	1.00	100	100	300.000

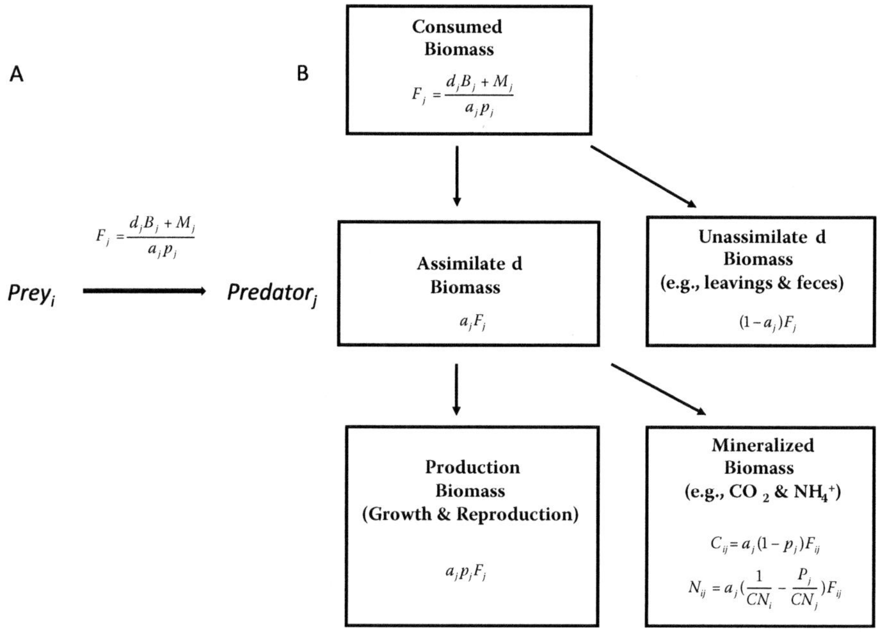

Figure 4.3 (A) Representation of a simple trophic interaction (Equation 4.1). (B) Schematic diagram partitioning the fate of consumed matter into various organic and inorganic pools.

$$\frac{dX_i}{dt} = b_i X_i - \sum_{i=1}^{n} f(X_i) X_j \tag{4.2}$$

where X_i and X_j represent the population densities of the primary producer and consumers, respectively; b_i is the specific growth rate of the primary producer; and $f(X_i)$ represents the functional response of the interaction between (species $i \neq j$) or within the populations (species $i = j$). As such, the Jacobian matrix is an $n \times n$ matrix whose elements (α_{ij}, with units of "per time") are the partial derivatives of the equations describing the growth and dynamics of the functional groups at or near equilibrium (X_i^*) (May, 1973):

$$\alpha_{ij} = [\delta(dX^*_i/dt)/\delta X_j] \tag{4.3}$$

The diagonal elements (α_{ii}) of the $n \times n$ Jacobian matrix depict the degree of self-regulation or intraspecific competition, while the paired off-diagonal elements represent the per-capita effects of predators on prey (α_{ij}) and prey on predators (α_{ji}). For our purposes, it will be important to estimate the magnitudes of the interaction strengths to understand how agricultural management practices affect structural and functional aspects of food webs and in turn their stability. The off-diagonal interaction strengths can be estimated by linking the rate equations used to model the population dynamics of the functional groups with estimates of the feeding rates among the functional groups that make up the energy flux description. To do this requires that we make two assumptions. First, we assume that

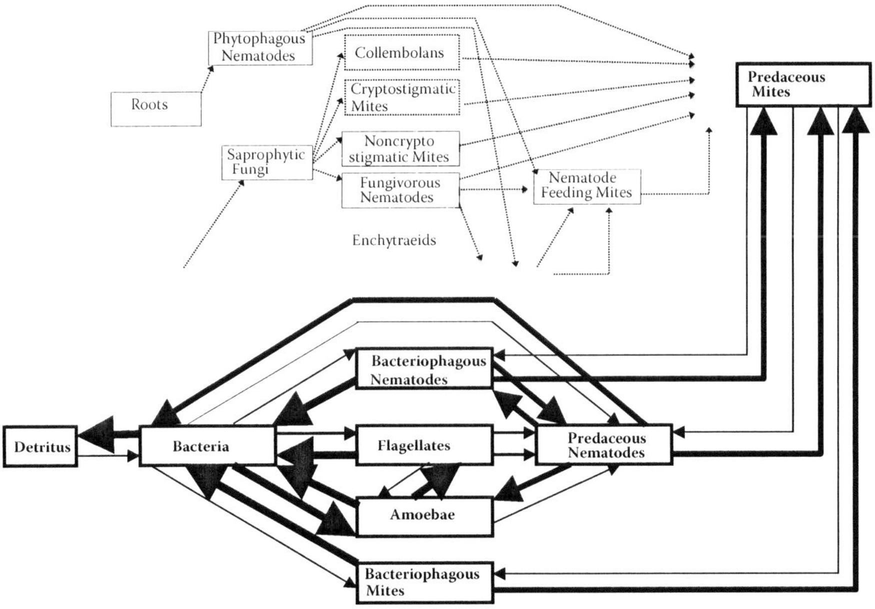

Figure 4.4 A partial functional food web description for the shortgrass steppe soil food web from the Central Plains Experimental Range (CPER), Nunn, Colorado. The description is based on the connectedness and energy flux descriptions. The arrows that depicted carbon fluxes in the energy flux description are split to reflect the negative (predator-to-prey) and positive (prey-to-predator) interaction strengths (y^{-1}) associated with each trophic interaction. A subset of the interactions is quantified here. Estimates of all interaction strengths are presented in Figure 4.5.

the steady-state biomass of functional group i (X^*_i in Equation 4.2 and Equation 4.3) in the rate equations can be approximated from long-term seasonal averages of biomass of functional group i (B_i). Second, we assume that the consumption terms in the rate equations for prey i to predator j represent the flux rates estimated as described in Equation 4.1, that is, $F_{ij} = f(X_i^*)X_j^*$. Under these assumptions, the off-diagonal interaction strengths are $\alpha_{ij} = -F_{ij}/B_j$ for the per capita effect of predator j on prey i, and $\alpha_{ji} = a_j p_j F_{ij}/B_j$ for prey i on predator j (Moore et al., 1993; de Ruiter et al., 1994, 1995).

The diagonal elements α_{ii} of the matrix depicting the degree of self-limitation for a group cannot be derived from field data or estimates of energy fluxes as described. Here, we assume that all groups possess some degree of self-limitation, and that the magnitude of the effect can be scaled s_i in a density-dependent manner to the specific death rates d_i, that is, $\alpha_{ii} = -s_i d_i$ (de Ruiter et al., 1995), or can be set at levels that ensure stability (Neutel et al., 2002).

Stability is assessed by evaluating the eigenvalues of the Jacobian matrix. If the real parts of each of the eigenvalues are negative, then the system is deemed stable. In this case, when the system is disturbed and populations deviate from their equilibrium, the populations and hence the system will return to their equilibrium. If the real part of any of the eigenvalues is zero or positive, then the system is deemed unstable. In this case, the system does not return to its original equilibrium state.

4.3 Patterns in energy flux and interaction strength

The food web descriptions of the shortgrass steppe and tillage agricultural systems reveal three important patterns in the arrangement of interactions and the distribution of biomass and energy flux within the food web that are important to their dynamic stability (May, 1972; Coleman et al., 1983; Moore and Hunt, 1988; Rooney et al., 2006). First, the biomasses of the functional groups and fluxes among the functional groups possess pyramidal arrangements with an ascending distribution of trophic position. With this arrangement, the majority of biomass and the dominant fluxes are at the lower trophic positions. Second, the food webs possess a distinct asymmetry in the pattern of interaction strength when arranged by trophic position. The effects of prey on predators are large at the base of the food web and decrease with increased trophic position, while the converse is true for the effects of predators on prey (Figure 4.5).

The connectedness structure and energy fluxes reveal a third pattern; the functional groups are compartmentalized into subsystems based on dominant flows of energy, termed *energy channels* (Moore and Hunt 1988; Moore and de Ruiter, 1997; Table 4.4). The energy channels originate from plant roots and continue through their consumers (root energy channel) and from detritus through bacteria and their consumers (bacterial energy channel) and fungi and their consumers (fungal energy channel). The boundaries of subsystems are often defined in terms of temporal or spatial scaling (Table 4.5). This follows an arrangement that Coleman et al. (1983) recognized within the rhizosphere; the dynamics of the transfer of C and N among plants and soil biota is mediated by a fast cycle dominated by bacteria and their consumers and a slow cycle dominated by fungi and their consumers. A spatial component is evident when considering the spatial extent, movement patterns, and water requirement of the organisms within each channel.

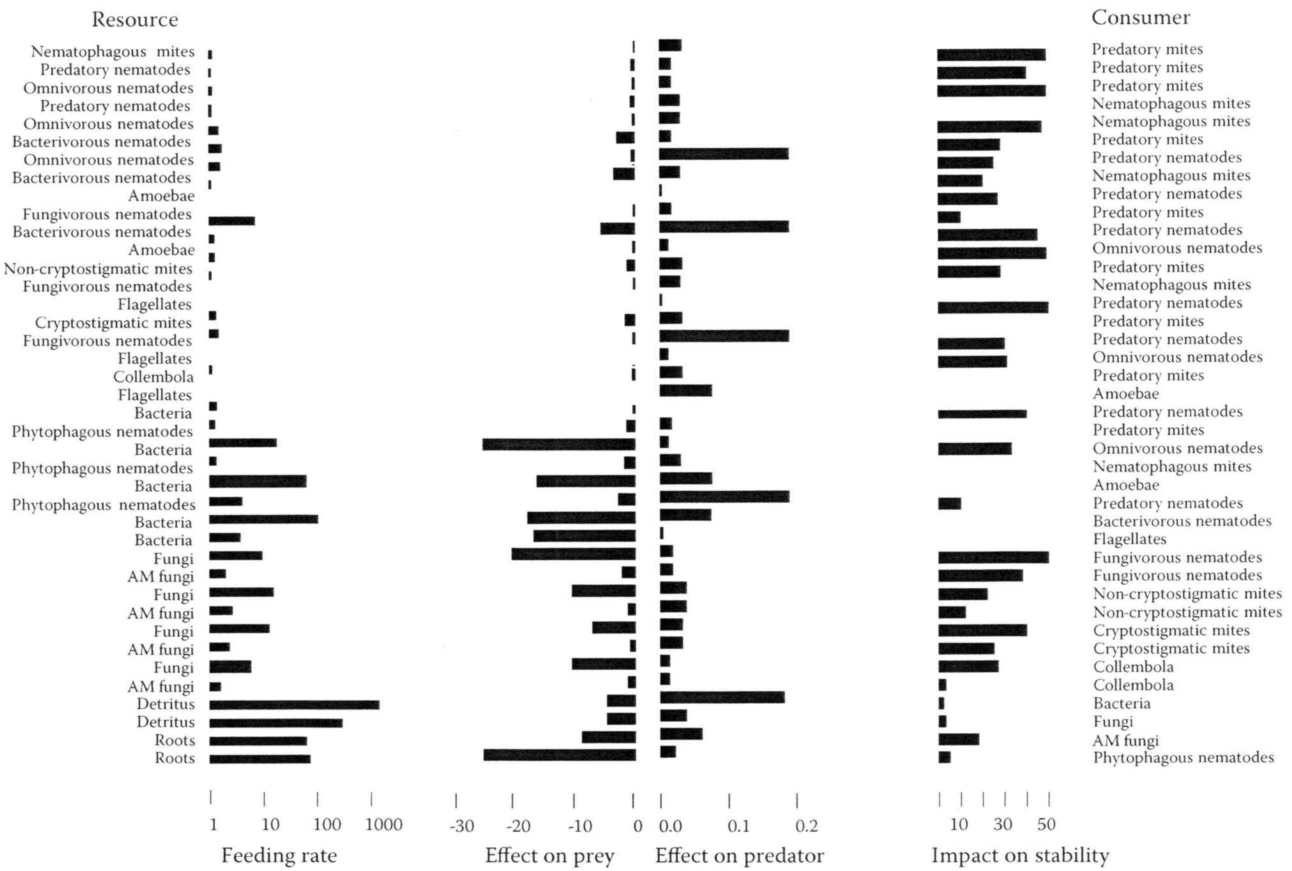

Figure 4.5 The distributions of the carbon flux rates (kg C ha^{-1} y^{-1}), interaction strengths (y^{-1}), and impact of stability of trophic interactions arranged by trophic position for functional groups of the Central Plains Experimental Range (Colorado, United States).

Table 4.4 Estimates of the Proportion of Energy (Indexed by Carbon) That the Different Functional Groups Potentially Derive from within the Bacterial, Fungal, or Plant Roots Energy Channels within the Soil Food Web of the CPER

	Energy channel		
Functional group	Bacteria	Fungi	Root
Protozoa			
Flagellates	100	0	0
Amoebae	100	0	0
Ciliates	100	0	0
Nematodes			
Phytophagous nematodes	0	0	100
Mycophagous nematodes	0	90	10
Omnivores nematodes	100	0	0
Bacteriophagous nematodes	100	0	0
Predatory nematodes	68.67	3.50	27.83
Microarthropods			
Collembola	0	90	10
Cryptostigmata	0	90	10
Mycophagous Prostigmata	0	90	10
Nematophagous mites	66.70	3.78	29.52
Predatory mites	39.54	38.56	21.91

Source: From Moore, J.C., D.E. Walter, and H.W. Hunt. 1988. Arthropod regulation of micro- and mesobiota in below-ground detrital food webs. *Annual Review of Entomology* 33:419–439.

4.4 Patterns and stability

Are the observed patterns in biomasses, energy fluxes, and interaction strengths related in any way to the stability of the systems? Odum's assessment of ecosystem development alluded to the connections between the energetic organization of the ecosystem and stability. The distribution of biomass and the magnitude of flows within the energy flow description constitute the energetic organization of the system. de Ruiter et al. (1995) assessed whether these patterns within the energetic organization of the system are linked to asymptotic stability using the eigenvalue-based analysis of the Jacobian matrix described. Two approaches were taken. In the first approach, pairs of elements within the Jacobian matrix associated with individual trophic interactions were disturbed by varying the values within plus or minus two times their mean to determine the sensitivity of the interaction to change and its impact on the stability of the web as a whole. The sensitivity of interactions and their impact on stability did not correlate to the magnitude of their fluxes (Figure 4.5). In the second approach, the connectedness structure and positioning of the nonzero elements of the Jacobian matrix were preserved, but each of the paired elements within the matrix was transposed with other pairs. The webs that possessed the observed patterning of interaction strengths that emerged from the field data were more stable than those that possessed the manipulated arrangements (Figure 4.6).

The observed compartmentalized organization of the food webs into interacting energy channels is consistent with the patterns expected for systems that are hierarchically organized as a collection of interacting subsystems. May (1972) had proposed that

Table 4.5 Habitat, Life History, and Physiological Attributes of the Dominant Taxa in the Soil Food Web

	Bacteria	Fungi	Protozoa	Nematodes	Collembola	Mites
Habitat	Water/surfaces	Free air/ surfaces	Water/surfaces	Water films/ surfaces	Free air spaces	Free air spaces
Minimum generation time (h)	0.5	4–8	2–4	120	720	720
Turnover time ($season^{-1}$)	2–3	0.75	10	2–4	2–3	2–3
Assimilation efficiency (%)	100[a]	100[a]	0.95	0.38–0.60	0.5	0.3–0.9
Production efficiency (%)	0.4–0.5	0.4–0.5	0.4	0.37	0.35	0.35–0.40
Body width	1–2 μm	3–100 μm	15–100 μm	5–120 μm	0.150–2 mm	0.08–2 mm

[a] The assimilation efficiencies of bacteria and fungi are 100% given that microbes absorb materials across their membranes as opposed to ingesting or engulfing prey or materials.

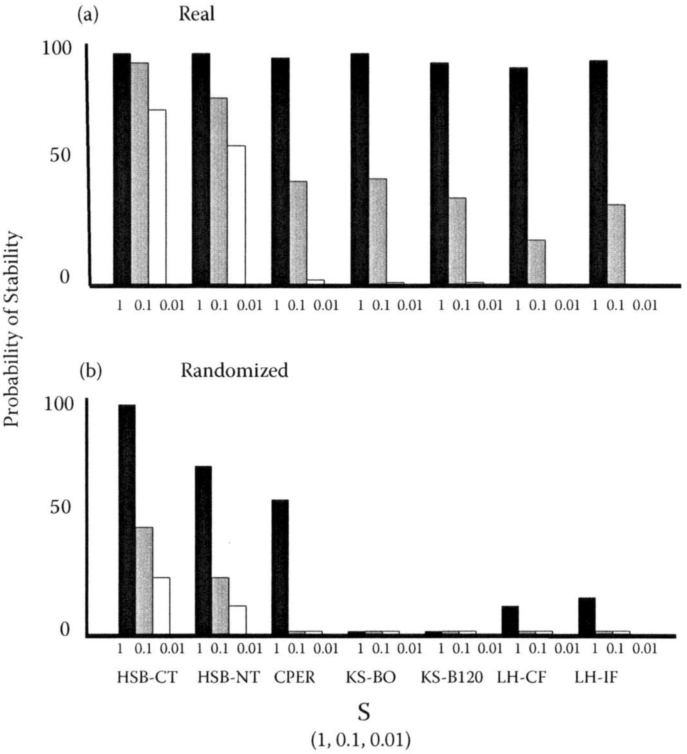

Figure 4.6 Probability of stability of (a) "real" matrices and (b) "randomized" matrices for seven different food webs whose diagonal terms were weighted by fixed values of *s* (1, 0.1, 0.01). CPER, Central Plains Experimental Range (Colorado, United States); LH, Loevinkhoeve (the Netherlands); IF, integrated farming; CF, conventional farming; HSB, Horseshoe Bend (Georgia, United States); CT, conventional tillage; NT, no tillage; KS, Kjettslinge (Sweden); B0, without fertilizer; B120, with fertilizer. (From de Ruiter, Moore et al., 1993.)

the compartmentalized arrangement of species was more likely to be stable than a random network of interactions of similar diversity and complexity. The soil food webs seem to bear this out (Moore et al., 2004; Rooney et al., 2006). Simplified representations of energy channels, one fast (e.g., a bacterial energy channel) and one slow (e.g., a fungal energy channel), coupled by a predator revealed that the coupled systems were more stable than either energy channel in isolation (Figure 4.7). Depending on the choice of parameters, the coupled system was quite tolerant to changes or shifts in the relative amount of flux through the fast or slow channel, but a tipping point or singularity in the dynamic states was evident. It appears that unimpeded flow of material through either channel has its limits, leading to instability.

The analyses presented effectively altered the energetic organization of the community by the redistribution of the biomass and directional flow of matter within the system. The adjustments to the paired interactions' strengths that de Ruiter et al. (1995) made necessarily moved biomass to higher trophic positions. The adjustment to the directional flow of material through different channels implied an alteration in the relative abundance of basal resources (bacteria or fungi). Given the steady-state assumptions, these alterations in effect implied that greater enrichment rates in the form of plant production or detritus were in play to support the biomass at the higher trophic positions.

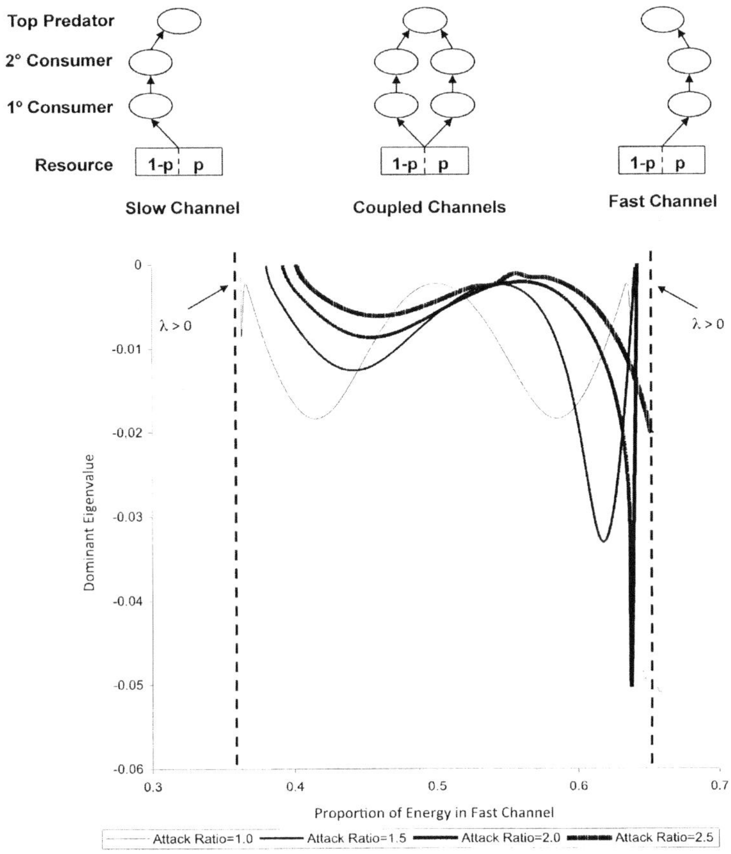

Figure 4.7 The dominant eigenvalues for models of two parallel food chains drawing energy from a basal resource (e.g., detritus) and linked by a common predator along a gradient of resource allocation. The food chains differ in the rates that transfer and turnover consumed matter and energy, that is, a fast bacterial energy channel and a slow fungal energy channel. The x-axis represents the proportion of the basal resource that is consumed by the first consumer in the fast channel (p). The y-axis represents the dominant eigenvalue (λ) for the system at each iteration of partitioning of resource. The total amount of energy passing through each system is the same for each iteration, but the attack ratio of the consumer of the fast channel differs. The dashed vertical lines represent unstable transitions. Note that the most stable configuration occurs when the two channels are coupled and unstable ($\lambda > 0$) if most of the energy passes through either the fast channel or slow channel.

The analysis also revealed that the magnitude of individual interactions does not necessarily equate to their importance to stability (Paine, 1980). Weak interactions, in terms of both energy flow and per-capita effects, can have profound effects on stability (McCann et al., 1998; de Ruiter et al., 1998). The analyses did indicate that the patterning of the interactions and their relationships among one another appear to be more important than the magnitude of the interactions per se (de Ruiter et al., 1995; Neutel et al., 2002). Observed patterning of fluxes and interaction strengths possess a compartmentalized organizational structure around energy channels (Moore et al., 1988, 2004; Rooney et al., 2006). On the whole, the energetic organization of the food web defined in terms of the distribution of biomass and fluxes and the arrangement of trophic interactions in the form of trophic loops (sensu Neutel et al., 2002) is important to the stability and persistence of the system.

4.5 Impacts of agricultural practices

Agricultural management practices affect soil food webs in ways that parallel those used in the theoretical exercises in the previous section. Tillage, organic matter amendments, irrigation, and pesticides affect the densities of soil biota, the availability of plant-limiting nutrients, and the quantity and quality of soil organic matter (Wardle, 1995; Beare, 1997; Vitousek et al., 1997; Doles et al., 2001). In this section, we connect empirical observations from field studies on conventional and minimum tillage agricultural practices to the theoretical results presented.

4.5.1 Effects of tillage on soil organic matter and nutrient dynamics

Tillage disrupts the distribution of soil organic matter within soil horizons and alters the structure of soil that develops over time in natural systems. Conventional tillage methods overturn soils, mixing standing dead plant material, surface litter, and topsoil with lower layers in the soil profile. In the process, tillage alters the aggregate structure of soil by physically disrupting the larger size fractions and creating smaller ones. This action exposes formerly occluded organic materials to biota and weathering. When coupled with the addition of inorganic fertilizers, the organic materials that were once held within the aggregates are mineralized by soil microbes, resulting in a net loss of soil organic matter and plant-limiting nutrients (Vitousek et al., 1997; Wander and Bidart 2000). The adoption of minimum tillage practices (e.g., no till and stubble mulching) that are far less disruptive to the aggregate structure, and that leave plant residues on the soil surface, or that use organic fertilizers have reversed these trends in fields that once employed conventional tillage practices (Elliott et al., 1984; Brussaard et al., 1988; Beare et al., 1994; Frey et al., 1999; Six et al., 1999; Wander and Bidart, 2000; Simpson et al., 2004).

The changes in the distribution in size classes of soil aggregates and in the concentrations of soil organic matter and plant-limiting nutrients that occur following tillage are coincident with changes in the distribution of crop residues and the relative activities of soil bacteria and fungi and their respective contributions to the decomposition of crop residues, nutrient dynamics, and soil-forming processes (Tisdall and Oades, 1982; Cambardella and Elliott, 1994). For example, Holland and Coleman (1987) studied the placement of crop residues on microbial activity in winter wheat grown on the shortgrass steppe. The study manipulated the placement of crop residues by incorporating the material into the soil profile to simulate conventional tillage or by leaving the residues on the soil surface to simulate no-tillage management. The soils with the residues incorporated into the profile promoted greater bacterial growth and rates of nitrogen mineralization, while the soils with residues placed on the soil surface promoted greater fungal growth and lower rates of mineralization.

Changes in the relative abundances and activities of bacteria and fungi induced by different tillage practices can affect the formation of soil aggregates and the size distribution of soil aggregates. Both bacteria and fungi produce amino sugars that are important to the formation of soil aggregates. The ratio of fungal-derived glucosamine (Glc) to bacterial-derived muramic acid (MurA) is a useful indicator of the relative importance of fungi to bacteria to these processes (Zhang et al., 1998). Fungi and their by-products are believed to have a dominant role in the formation of large and intermediate-size aggregates (Cambardella and Elliott, 1994). Soil aggregates from soils under no-till management tend to have higher Glc:MurA ratios than their counterparts collected from conventionally tilled plots (Guggenberger et al., 1999; Frey et al., 1999; Simpson et al., 2004).

4.5.2 Effects on soil food webs

Several large research efforts have adopted the modeling approach presented to study the impact of agricultural practices on biotic interactions and nutrient dynamics (Hendrix et al., 1986; Hunt et al., 1987; Andrén et al., 1990; Moore and de Ruiter, 1991). The studies adopted the practice of grouping biota into functional groups and in many cases used the same parameter set presented in Table 4.3 to characterize the physiologies and life histories of the groups, but with their own field biomass estimates (Table 4.6). The connectedness, energy flux, and functional food web descriptions for the sites were similar to one another. Similar functional groups were present at each site, distinct pyramids of biomass and flux rates were present, and the asymmetry in the pairwise interaction strengths apparent in Figure 4.5 appeared as well (de Ruiter et al., 1995). The compartmentalization of interactions and fluxes into the root, bacterial, and fungal energy channels were present. The models also captured the responses of soil biota and the mineralization of carbon and nitrogen with surprising accuracy (Figure 4.8), despite the fact that similar criteria were used to formulate the descriptions and parameterizations.

Numerous studies have demonstrated that agricultural practices affect the densities and activities of various taxa and the processes that they mediate, similar to the example presented in Table 4.7 comparing no-till and conventional winter wheat management practices on prairie soils in Colorado (Table 4.7). The food web approach developed in the previous section provides a way to tie these responses together and in doing so has revealed important patterns in the responses (de Ruiter et al., 1994; Schröter et al., 2003). For example, the same type of information presented in Table 4.7 but from the Lovinkhoeve Farm comparing integrated management to conventional management of winter wheat revealed that the seasonal dynamics of the energy channels were affected in ways that affected nitrogen dynamics and nitrogen availability (Figure 4.9).

Comparisons of the responses of soil food webs for systems under conventional tillage compared to no-till or reduced tillage reveal that conventional tillage practices stimulate microbial activity, increasing soil respiration and nitrogen mineralization rates (Elliott et al., 1984; Hendrix et al., 1986; Andrén et al., 1990; Moore and de Ruiter, 1991; Wardle, 1995).

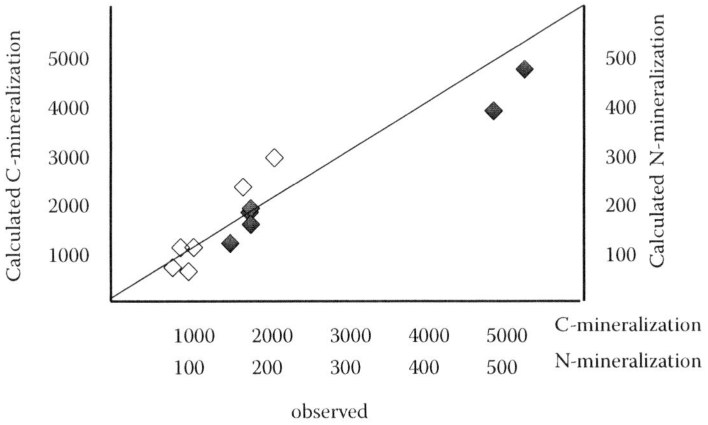

Figure 4.8 Comparison between measured and calculated mineralization rates for the food webs presented in Table 4.6. Values (kg ha^{-1} y^{-1}) are for the 0- to 25-cm depth layer, except for the Horseshoe Bend webs (0–15 cm). Open symbols, N-mineralization; closed symbols, C-mineralization.

Table 4.6 Biomass Estimates (kg C ha^{-1}) for the Functional Groups in the Different Food Webs

	CPER	LH-IF	LH-CF	HSB-NT	HSB-CT	KS-B0	KS-B120
			Microbes				
Bacteria	304	245	228	440	690	740	900
Fungi	63	3.27	2.12	160	150	1,500	2,300
AM fungi	7						
			Protozoa				
Amoebas	3.78	18.9	11.5	40[a]	50[a]	110[a]	34[a]
Flagellates	0.16	0.63	0.53				
			Nematodes				
Herbivores	2.90	0.35	0.19	0.40	0.50	0.18	0.29
Bacteriovores	5.80	0.36	0.30	0.46	1.40	0.45	0.50
Fungivores	0.41	0.13	0.08	0.12	0.08	0.20	0.12
Predators[b]	1.08	0.06	0.06	0.44	0.44		
			Arthropods				
Predatory mites	0.16	0.08	0.06	0.20[c]	0.04[c]	0.18[d]	0.28[d]
Nematophagous mites	0.16	0.006	0.004				
Cryptostigmatic mites	1.68	0.003	0.007	0.80	0.22		
Noncryptostigmatic mites	1.36	0.04	0.02	0.90	0.39		
Bacteriovorous mites	0.0003	0.001					
Fungivorous Collembola	0.46	0.38	0.47	0.30	0.09	0.17[e]	0.17[e]
Predatory Collembola	0.008	0.03					
Foliar-feeding arthropods	0.10	0.14					
Predatory herbage arthropods	0.15	0.19					
Herbivorous macroarthropods	0.19	0.19					
Microbivorous macroarthropods	0.25	0.25					
Predatory macroarthropods	0.49	0.49					
			Annelids				
Enchytraeids	0.21	0.43	0.10	0.30	4.20	3.40	
Earthworms	63.5	—	100	20	13	13	

Note: Values refer to the 0- to 25-cm depth layer, except for the Horseshoe Bend webs (0–15 cm). CPER, Central Plains Experimental Range (Colorado, United States); LH, Loevinkhoeve (Netherlands); IF, integrated farming; CF, conventional farming; HSB, Horseshoe Bend (Georgia, United States); CT, conventional tillage; NT, no tillage; KS, Kjettslinge (Sweden); B0, without fertilizer; B120, with fertilizer.

[a] Including amoebas and flagellates.
[b] Including predators and omnivores.
[c] Including all predatory arthropods.
[d] Including all predatory microarthropods.
[e] Including all microbivorous microarthropods.

Table 4.7 Microbial Biomass Carbon (kg C ha^{-1}), Soil Animal Biomass C (kg C ha^{-1}), Mineralizable Carbon (kg CO_2-C ha^{-1}), and Mineralizable Nitrogen (kg NO_3 ions ha^{-1}) in the Top 10 cm of No-Till and Stubble Mulch Fallow Winter Plots at Akron, Colorado (Elliott et al., 1984) over 5 Dates in 1982

Variable	Treatment	8 June	6 July	2 August	23 August	13 September	ANOVA*
Microbial biomass (kg C ha^{-1})	Stubble mulch	245	204	271	186	255	$P = 0.019$ (D)
	No till	329	273	299	194	256	$P = 0.054$ (T)
Collembola × 100 (kg C ha^{-1})	Stubble mulch	6.69	1.03	2.18	0.61	1.49	$P = 0.019$ (T × D)
	No till	5.58	8.29	4.01	2.60	1.49	
Acari × 10 (kg C ha^{-1})	Stubble mulch	3.77	0.47	0.91	0.40	0.92	$P = 0.020$ (T × D)
	No till	3.60	3.31	1.24	1.32	0.54	
Holophagous nematodes × 100 (kg C ha^{-1})	Stubble mulch	0.88	0.46	1.12	0.51	0.56	$P = 0.002$ (T × D)
	No till	0.48	1.69	0.96	0.57	0.32	
Protozoa (kg C ha^{-1})	Stubble mulch	1.76	0.67	0.74	1.96	1.92	$P = 0.001$ (D)
	No till	2.20	0.96	0.93	2.29	1.61	
Respired carbon (kg CO_2 ha^{-1})	Stubble mulch	49	42	56	53	24	$P = 0.075$ (D)
	No till	80	69	84	57	24	$P = 0.042$ (T)
Mineralizable nitrogen (kg NO_3 ha^{-1})	Stubble mulch	10.63	6.39	5.79	2.62	1.95	$P = 0.026$ (D × T)
	No till	12.68	5.79	8.90	−3.31	−1.38	

* Level of highest significant interaction for analysis of variance (ANOVA) of five dates (D) × two treatments (T) × three field replicates.

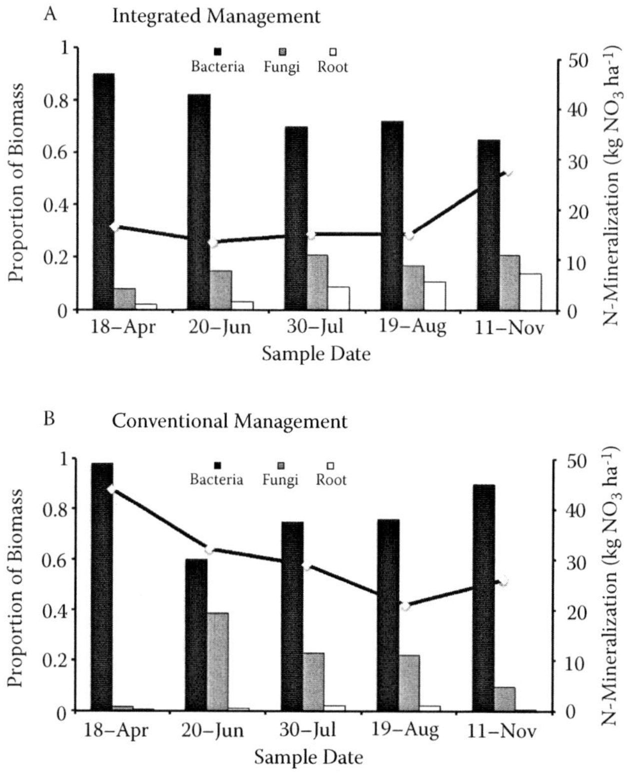

Figure 4.9 Proportion of total food web biomass for the functional groups within the bacterial, fungal, and root energy channels and potential N mineralization (kg NO_3 ha^{-1}) from soils collected over a growing season in winter wheat under integrated (A) and conventional management (B) at the Loevinkhoeve Farm, Marknesse, the Netherlands.

The increase in microbial activity is not limited to increases in microbial biomass but to a shift in the relative abundances and activities of organisms within the bacterial and fungal energy channels as well. The observed increases in microbial activity, respiration, and mineralization can be attributed to an increase in the biomass and activity of bacteria and their consumers (i.e., the bacterial energy channel) and the loss of biomass and activity of fungi and their consumers (i.e., the fungal energy channel).

Management practices affect not only the relative abundances and activities of organisms within the different energy channels but also their timing. In natural ecosystems, the timing of microbial activity and mineralization is in synchrony with plant growth and uptake. In temperate regions, this often involves a pulse of activity in both the bacterial and fungal energy channels, but with some degree of spatial separation (read compartmentalization) within the soil profile (Holland and Coleman, 1987) and temporal separation within the growing season (Moore and de Ruiter, 1991). During the fall and winter, processes proceed but at reduced rates. The seasonal cycle in some respects resets in the spring and repeats the cycle. What has emerged from comparative studies of conventional and minimum tillage practices is a pattern of biotic activity and resulting processes late in the growing season in intensively managed systems that are out of synchrony with plant growth and management. Minimum tillage practices tend to retain the seasonal rhythm in the activities of the energy channels and processes. What we see in many intensively managed systems are pulses of microbial activity during times of plant senescence or after the crop has been harvested, resulting in the decomposition of residues and the losses of plant-available nutrients through leaching (Elliott et al., 1984; Hendrix et al., 1986; Moore and de Ruiter, 1991).

The models and experimentation have offered some insight into the mechanisms at play and the long-term consequences as they relate to sustainability. Field manipulations point to mechanisms that include the placement of crop residues, the bioavailability of labile versus resistant soil carbon, the changes in aggregate structure, the relative sizes of habitable pore space, water availability, and the toxicity of different groups to pesticides, confirming this conclusion. The more intensive management practices alter the environment and placement of detritus in ways that favor the activities of bacteria and consumers of bacteria over fungi and their consumers.

The models highlight the importance of body sizes, physiologies, and life histories of the soil biota and their resistance and resilience to disturbances. Coleman (1994) noted that organisms within the bacterial pathway tend to be smaller in size and more energetically efficient and to possess shorter life histories than their counterparts in the fungal pathway (see Table 4.5). These differences affect the turnover rates of populations, and thus the mineralization rates can explain the enhanced activity of the bacterial energy channel in tilled systems. These traits can also explain differences in resistance and resilience of the pathways to disturbances (Moore et al., 1993). The fungal energy channel, dominated to a large extent by arthropods, is particularly vulnerable to the physical mixing of soils during tillage. Add to this the use of insecticides that target arthropods and we can explain in part the demise of the fungal energy channel that is comprised of many arthropod taxa.

The resilience of the channels in response to disturbances can explain the changes in the timing of the activities of the energy channels. Investigations of wetting/drying cycles and freezing/thawing cycles of soils revealed that the organisms within the bacterial channel are more resilient than those within the fungal energy channels (Allen-Morley and Coleman, 1989; Hunt et al., 1989). Models of food chains parameterized to resemble the bacterial energy channel have shorter return times compared to ones that resemble a fungal energy channel (Moore et al., 2004; Rooney et al., 2006). These results can be explained in large part by differences in the reproductive and turnover rates of organisms within the respective channels. Taken together, if we assume that both energy channels were equally affected by a tillage event with the same proportionate decline in biomass, the models would predict that the bacterial energy channel would rebound at a faster rate than the fungal pathway.

These studies demonstrated that intensive agricultural practices have disproportionately high negative effects on the organisms within the fungal pathway and conversely have high positive effects on organisms within the bacterial pathway. This coupled with the different capacities of the energy channels to respond to disturbance can explain the observed shifts in activity of the channels and the processes they mediate. Our models of coupled bacterial and fungal pathways and whole food webs provide more realistic assessment of how changes in the relative sizes of energy flows or shifts from one pathway to another might affect food stability (Figure 4.10).

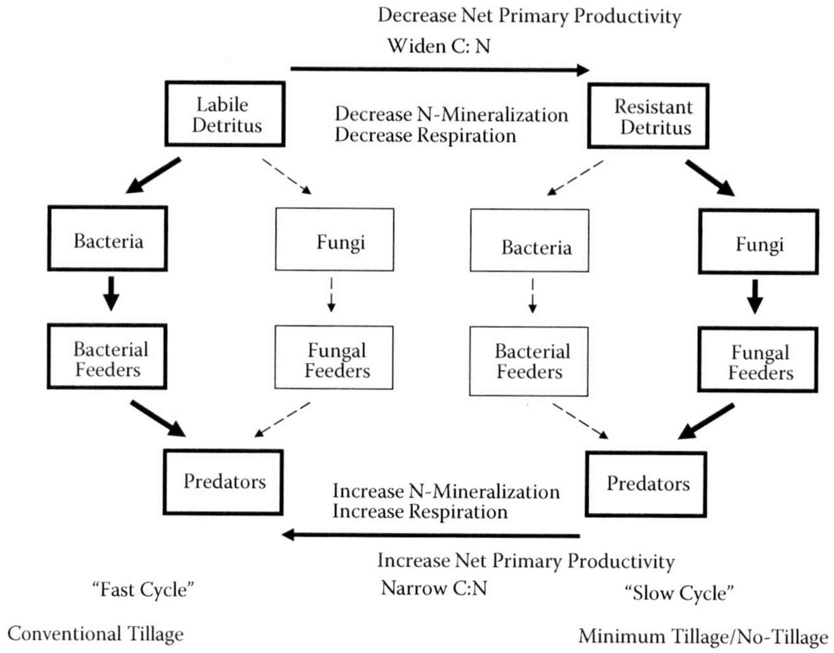

Figure 4.10 Simplified representations of the bacterial and fungal energy channels under conventional tillage and no-till or reduced tillage systems (compilation of results from Hendrix et al., 1986; Andrén et al., 1990; Moore and de Ruiter, 1991; and Beare et al., 1994).

4.6 Discussion and conclusions

The empirical studies and modeling exercises lead to a series of observations and propositions that help link soil food web structure, nutrients dynamics, and stability.

1. The utilization of resources by species is highly diverse yet compartmentalized. The compartments are organized as a series of quasi-independent interacting energy pathways. The pathways are made up of species that share similar habitat requirements and complementary life-history traits. As a result of this, organic matter is processed, and energy is transferred within the pathways at distinct rates. In soils, these pathways originate from plant roots and detritus. The detritus pathway is divided into a bacterial pathway and a fungal pathway.
2. The soil decomposer communities are tightly coupled through the quality of plant litter and soil organic matter and shared life-history characteristics. Fungi and bacteria have affinities toward substrates that differ in quality and contribute to the formation and degradation of soil organic matter in different ways given differences in the metabolic by-products they produce.
3. Disturbances associated with agricultural practices alter the bacterial and fungal pathways to differing degrees and hence the pattern of the flow of energy within ecosystems. Nutrient availability to plants and retention within the ecosystem are affected by changes in the activity of the pathways relative to plant growth. In arable soils, tillage has a disproportionately large adverse effect on the fungal pathway

compared to the bacterial pathway. Nutrient availability and retention are either tied to the activities of the fungal pathway or are a function of the combined activities of the bacterial and fungal pathways and how synchronous these activities are with plant growth.

4. Disturbances associated with agricultural practices alter the energetic organization of the soil food web in ways that affect its stability. What the theoretical exercises do is link these changes to asymptotic stability by suggesting that the observed deviations in the activities of organisms, energy channels, and soil processes induced by tillage are destabilizing, less so for the minimum tillage practices that emulate natural systems.

What the modeling, field manipulations, and laboratory experiments suggest is that it is neither the size of the interactions nor the magnitude of the nutrient flow that determines the stability of an ecosystem. Rather, it is the patterning of the interactions among species and the distribution of nutrients within the community that govern its stability. The common feature that disturbances have on ecosystems, be they natural or human induced, is to change the densities of organisms, the species composition of communities, and hence the patterning of interaction and nutrient flows within them. It is through these changes that the stability of the system is impacted. In short, models have shown us that it is the energetic organization of communities that forms the basis of ecosystem structure and asymptotic stability. Empirical studies have shown us that different management practices alter the energetic organization of communities to various degrees. Aligning management practices to promote the stable patterns that emerge from theory seems palpable.

To this point, we have used food webs as a tool to study agricultural systems under the guises of long-term asymptotic stability and steady-state assumptions. Arguably, many natural systems and agricultural systems may not possess these properties or, if they do, may be operating within the short-term transient state (i.e., a dynamic state in response to a perturbation) rather than the long-term asymptotic state. Recent discussions on transient dynamics and behavior offer an interesting twist to this discussion that affects the way in which we think about systems and management. Hastings (2001, 2004, 2010) proposed that many ecological systems may not exhibit traditional asymptotic dynamic behavior but rather are operating within a transient dynamic state in which the asymptotic behavior is uncertain. The disruption of soils through tillage and annual amendments of fertilizers, water, or herbicides may not be sustainable in a traditional dynamic systems context, but neither is the natural system. In other words, it is unclear whether the natural system would be able to persist without the regular disruption and resetting of activity afforded by the seasons.

With this in mind, the call to "Let the soil work for us" and underpinnings of sustainable agricultural practices are not centered on creating a persistent stable system. Rather, the focus is more aligned with conserving resources that are necessary to sustain the system in a perpetual transient state. Here, a sustainable practice would manage the system in a perpetual transient dynamic state, maximizing production and minimizing the loss and export of plant-limiting nutrients. In real terms, this translates to adopting practices that promote the placement of residues at the soil surface and that minimize the disruption of soils in ways that restore a balance in the activities of organisms within the bacterial and fungal energy channels found in natural systems. Soil food webs provide us a means of studying and assessing agricultural practices to optimize these aims.

Acknowledgments

This research was supported by grants from the U.S. National Science Foundation, the U.S. Department of Energy, the Dutch Ministry of Fisheries and Agriculture, and the International Agricultural Center (the Netherlands). We wish to thank the organizers of the symposium at the 2009 annual meeting of the Ecological Society of America and the editors for the opportunity to contribute to this volume.

References

Allen-Morley, C.R., and D.C. Coleman. 1989. Resilience of soil biota in various food webs to freezing perturbations. *Ecology* 70:1127–1141.

Andrén, O., T. Lindberg, U. Bostrom, M. Clarholm, A.-C. Hansson, G. Johansson, J. Lagerlof, K. Paustian, J. Persson, R. Pettersson, J. Schnurer, B. Sohlenius, and M. Wivstad. 1990. Organic carbon and nitrogen flows. *Ecological Bulletins* 40:85–126.

Beare, M.H. 1997. Fungal and bacterial pathways of organic matter decomposition and nitrogen mineralization in arable soil. In *Soil Ecology in Sustainable Agricultural Systems,* eds. L. Brussaard and R. Ferrera-Cerrato, 37–70. Lewis, CRC Press, Boca Raton, FL.

Beare, M.H., P.F. Hendrix, and D.C. Coleman. 1994. Water-stable aggregates and organic matter fractions in conventional and no-tillage soils. *Soil Science Society of America Journal* 58:777–786.

Berlow, E.L., A.M. Neutel, J.E. Cohen, P.C. de Ruiter, B. Ebenman, M. Emmerson, J.W. Fox, V.A.A. Jansen, J.I. Jones, G.D. Kokkoris, D.O. Logofet, A.J. McKane, J.M. Montoya, and O. Petchey. 2004. Interaction strengths in food webs: issues and opportunities. *Journal of Animal Ecology* 73:585–598.

Brussaard, L., V.M. Behan-Pelletier, D.E. Bignell, V.K. Brown, W. Didden, P. Flogarait, P. Fragoso, D. Wall-Freckman, V.V.S.R. Gupta, T. Hattori, D.L. Hawksworth, C. Klopatek, P. Lavelle, D.W. Mallocj, J. Rusek, B. Soderstrom, J.M. Tiedje, and R.A Virginia. 1997. Biodiversity and ecosystem functioning in soil. *Ambio* 26:563–570.

Brussaard, L., J.A. Van Veen, M.J. Kooistra, and J. Lebbink. 1988. The Dutch programme of soil ecology and arable farming systems. I. Objectives, approach, and some preliminary results. *Ecological Bulletins* 39:35–40.

Cambardella, C.A., and E.T. Elliott. 1994. Carbon and nitrogen dynamics of soil organic matter fractions from cultivated grassland soils. *Soil Science Society of America Journal* 58:123–130.

Coleman, D.C. 1994. The microbial loop concept as used in terrestrial soil ecology studies. *Microbial Ecology* 28:245–250.

Coleman, D.C., C.P.P. Reid, and C.V. Cole. 1983. Biological strategies of nutrient cycling in soil systems. In *Advances in Ecological Research,* Vol. 13, eds. A. Macfayden and E.D. Ford, 1–55. Academic Press, London.

de Ruiter, P.C., J.C. Moore, J. Bloem, K.B. Zwart, L.A. Bouwman, J. Hassink, J.A. De Vos, J.C.Y. Marinissen, W.A.M. Didden, G. Lebbink, and L. Brussaard. 1993. Simulation of nitrogen dynamics in the belowground food webs of two winter-wheat fields. *Journal of Applied Ecology* 30:95–106.

de Ruiter, P.C., A. Neutel, and J.C. Moore. 1994. Modelling food webs and nutrient cycling in agroecosystems. *Trends in Ecology and Evolution* 9:378–383.

de Ruiter, P.C., A. Neutel, and J.C. Moore. 1995. Energetics, patterns of interaction strengths, and stability in real ecosystems. *Science* 269:1257–1260.

de Ruiter, P.C., A.M. Neutel, and J.C. Moore. 1998. Biodiversity in soil ecosystems: the role of energy flow and community stability. *Applied Soil Ecology* 10:217–228.

de Ruiter, P.C., J.A. Van Veen, J.C. Moore, L. Brussaard, and H.W. Hunt. 1993. Calculation of nitrogen mineralization in soil food webs. *Plant Soil* 157:263–273.

Doles, J.L., R.J. Zimmerman, and J.C. Moore. 2001. Soil microarthropod community structure and dynamics in organic and conventionally managed apple orchards in western Colorado, USA. *Applied Soil Ecology* 18:83–96.

Elliott, E.T., and D.C. Coleman. 1988. Let the soil work for us. *Ecology Bulletins* 39:23–32.

Elliott, E.T., K. Horton, J.C. Moore, D.C. Coleman, and C.V. Cole. 1984. Mineralization dynamics in fallow dryland wheat plots, Colorado. *Plant and Soil* 79:149–155.

Frey, S.D., E.T. Elliott, and K. Paustian. 1999. Bacterial and fungal abundance and biomass in conventional and no-tillage agroecosystems along two climatic gradients. *Soil Biology and Biochemistry* 31:573–585.

Guggenberger, G., S.D. Frey, J. Six, K. Paustian, and E.T. Elliott. 1999. Bacterial and fungal cell-wall residues in conventional and no-tillage agroecosystems. *Soil Science Society of America Journal* 63:1188–1198.

Hastings, A. 2001. Transient dynamics and persistence of ecological systems. *Ecology Letters* 4:215–220.

Hastings, A. 2004. Transients: the key to long-term ecological understanding? *Trends in Ecology and Evolution* 19:39–45.

Hastings, A. 2010. Timescales, dynamics, and ecological understanding. *Ecology* 91:3471–3480.

Hendrix, P.F., R.W. Parmelee, D.A.J. Crossley, D.C. Coleman, E.P. Odum, and P.M. Groffman. 1986. Detritus food webs in conventional and no-tillage agroecosystems. *Bioscience* 36:374–380.

Holland, E.A., and D.C. Coleman. 1987. Litter placement effects on microbial communities and organic matter dynamics. *Ecology* 68:425–433.

Hunt, H.W., D.C. Coleman, E.R. Ingham, R.E. Ingham, E.T. Elliott, J.C. Moore, C.P.P. Reid, and C.R. Morley. 1987. The detrital food web in a shortgrass prairie. *Biology and Fertility of Soil* 3:57–68.

Hunt, H.W., E.T. Elliott, and D.E. Walter. 1989. Inferring trophic interactions from pulse-dynamics in detrital food webs. *Plant and Soil* 115:247–259.

May, R.M. 1972. Will a large complex system be stable? *Nature* 238:413–414.

May, R.M. 1973. *Stability and Complexity of Model Ecosystems*. Princeton University Press, Princeton, NJ.

McCann, K., A. Hastings, and G.R. Huxel. 1998. Weak trophic interactions and the balance of nature. *Nature* 395:794–798.

Moore, J.C., E.L. Berlow, D.C. Coleman, P.C. de Ruiter, Q. Dong, A. Hastings, N. Collins-Johnson, K.S. McCann, K.M. Melville, P.J. Morin, K. Nadelhoffer, A.D. Rosemond, D.M. Post, J.L. Sabo, K.M. Scow, M.J. Vanni, and D.H. Wall. 2004. Detritus, trophic dynamics and biodiversity. *Ecology Letters* 7:584–600.

Moore, J.C., and P.C. de Ruiter. 1991. Temporal and spatial heterogeneity of trophic interactions within belowground food webs. *Agriculture Ecosystems and Environment* 34:371–394.

Moore, J.C., and P.C. de Ruiter. 1997. Compartmentalization of resource utilization within soil ecosystems. In *Multitrophic Interactions in Terrestrial Systems,* eds. A. Gange and V. Brown, 375–393. Blackwell Science, Oxford, UK.

Moore, J.C., P.C. de Ruiter, and H.W. Hunt. 1993. The influence of ecosystem productivity on food web stability. *Science* 261:906–908.

Moore, J.C., P.C. de Ruiter, H.W. Hunt, D.C. Coleman, and D.W. Freckman. 1996. Microcosms and soil ecology: critical linkages between field studies and modelling food webs. *Ecology* 77:694–705.

Moore, J.C., and H.W. Hunt. 1988. Resource compartmentation and the stability of real ecosystems. *Nature* 333:261–263.

Moore, J.C., D.E. Walter, and H.W. Hunt. 1988. Arthropod regulation of micro- and mesobiota in below-ground detrital food webs. *Annual Review of Entomology* 33:419–439.

Neutel, A.M., J.A.P. Heesterbeek, and P.C. de Ruiter. 2002. Stability in real food webs: weak links in long loops. *Science* 296:1120–1123.

Odum, E.T. 1969. The strategy of ecosystem development. *Science* 164:262–270.

O'Neill, R.V. 1969. Indirect estimation of energy fluxes in animal food webs. *Journal of Theoretical Ecology* 22:284–290.

Paine, R.T. 1980. Food webs: linkage, interaction strength and community infrastructure. *Journal of Animal Ecology* 49:667–685.

Paine, R.T. 1992. Food-web analysis through field measurement of per capita interaction strength. *Nature* 355:73–75.

Parton, W.J., D.W. Anderson, C.V. Cole, and J.W.B. Stewart. 1983. Simulation of soil organic matter formation and mineralization in semiarid agroecosystems. In *Nutrient Cycling in Agricultural Ecosystems,* eds. R.R. Lowrance, R.L. Todd, L.E. Asmussen, and R.A. Leonard, 533–550. Special Publication No. 23. University of Georgia, College of Agriculture Experiment Stations, Athens, GA.

Rooney, N., K. McCann, G. Gellner, and J.C. Moore. 2006. Structural asymmetry and the stability of diverse food webs. *Nature* 442:265–269.
Schröter, D., V. Wolters, and P.C. de Ruiter. 2003. C and N mineralisation in the decomposer food webs of a European forest transect. *Oikos* 102:294–308.
Simpson, R.T., S.D. Frey, J. Six, and R.K. Thiet. 2004. Preferential accumulation of microbial carbon in aggregate structures of no-tillage soils. *Soil Science Society of America Journal* 68:1249–1255.
Six, J., E.T. Elliott, and K. Paustian. 1999. Aggregate and soil organic matter dynamics under conventional and no-tillage systems. *Soil Science Society of America Journal* 63:1350–1358.
Swift, M.J., and J.M. Anderson. 1994. Biodiversity and ecosystem function in agricultural systems. In *Biodiversity and Ecosystem Function*, eds. E.D. Schulze and H.A. Mooney, 15–41. Springer-Verlag, New York.
Swift, M.J., O.W. Heal, and J.M. Anderson. 1979. *Decomposition in Terrestrial Ecosystems*. University of California Press, Berkeley.
Tisdall, J.M., and J.M. Oades. 1982. Organic matter and water-stable aggregates in soils. *Journal of Soil Science* 3:141–161.
Vitousek, P.M., H.A. Mooney, J. Lubchenco, and J.M. Mellilo. 1997. Human domination of the earth's ecosystems. *Science* 277:494–499.
Wall, D.W., and J.C. Moore. 1999. Interactions underground: soil biodiversity, mutualism and ecosystem processes. *BioScience* 49:109–117.
Wander, M.M., and G.M. Bidart. 2000. Tillage practice influences physical protection, bioavailability and composition of particulate organic matter. *Biology and Fertility of Soils* 32:360–367.
Wardle, D.A. 1995. Impacts of disturbance on detritus food webs in agro-ecosystems of contrasting tillage and weed management practices. *Advances in Ecological Research* 26:105–185.
Zhang, X., W. Amelung, Y. Yuan, and W. Zech. 1998. Amino sugar signature of particle-size fractions in soils of the native prairie as affected by climate. *Journal of Soil Science* 163:220–229.

chapter 5

Community composition of soil organisms under different wheat-farming systems

Klaus Birkhofer
Lund University

T. Martijn Bezemer
Netherlands Institute of Ecology (NIOO-KNAW)

Katarina Hedlund
Lund University

Heikki Setälä
University of Helsinki

Contents

5.1 Wheat-farming systems and soil organisms

5.1.1 Organic farming

Agricultural land under organic management globally constitutes around 30.4 million hectares out of 4676.9 million hectares total agricultural land (0.64%; Willer et al. 2008), and wheat was fourth among crops in world production in 2008 (FAOSTAT 2011). The term *organic farming* dates back prior to 1940, when it was defined as a way to improve soil fertility through the application of animal feces (Heckman 2006). Today, organic farming refers to management strategies that avoid the use of inorganic fertilizers, pesticides, and growth regulators to minimize environmental impact. These management strategies have therefore been promoted as ecosystem-oriented agricultural production systems and may contribute to a sustainable agriculture in the future (Häni et al. 1998). A long-term study by Mäder et al. (2002) in Switzerland suggested that the profit under organic farming is comparable to conventional farming for a number of row crops, while associated environmental impact and soil degradation are reduced. Yields were on average 20% higher in conventionally managed crops, but costs for pesticides and inorganic fertilizers adjusted farmers' profits to the same level. The extrapolation of results of such small-scale studies and related reviews to the potential of organic farming for sustaining a global food supply (Badgley et al. 2007; Badgley and Perfecto 2007) has been challenged (Avery 2007; Cassman 2007). However, consumer awareness is increasing in several countries, and there is a growing demand for environmentally sound agricultural products (Tilman et al. 2002; Willer and Kilcher 2010). One of the mechanisms that have been put forward to explain the relatively high yield in organic farms without making use of pesticides and inorganic fertilizers is that organic crops are associated with more complex soil food webs (Haubert et al. 2009; Crowder et al. 2010). These food webs can ensure sufficient nutrient availability through decomposition of organic matter by soil organisms and may maintain pest densities at low levels (Zehnder et al. 2007; Birkhofer, Bezemer, et al. 2008).

In this chapter, we describe, for a long-term agricultural experiment, the effects of different wheat-farming systems on the community composition of selected soil organisms. We further provide information about the taxonomic resolution that is operational to study the impact of management decisions (taxonomic sufficiency) and relate responses of different soil organism groups to each other, to soil properties, and to ecosystem functions.

5.1.2 *Taxonomic sufficiency*

Recent approaches in biodiversity assessment are based on the definition of functional groups or the use of higher taxonomic orders to serve the challenge of making biodiversity research functional (Letourneau and Bothwell 2008) and economically feasible (Mandelik et al. 2007). Detailed taxonomic information (e.g., species level) is often excluded from these studies, and functional groups or traits are defined at higher taxonomic levels (e.g., Ferris et al. 2004; Franklin et al. 2005). These approaches are motivated by the fact that species identification is laborious and that the scientific community suffers from an ongoing loss of taxonomic experts. However, our understanding of how community composition based on different taxonomic resolutions in soil organisms relates to management practices or to functional characteristics in different farming systems is still scarce. The concept of taxonomic sufficiency is based on the assessment of how much taxonomic information can be given up, without losing characteristic response patterns of communities to anthropogenic disturbance.

While there has been a lively debate and intense research on this topic over the last decade in aquatic sciences (Maurer 2000; Terlizzi et al. 2003; Bevilacqua et al. 2009), the subject is not as prominent in terrestrial systems or particularly in soil ecology. Three studies that focused on taxonomic sufficiency in comparing patterns of ant communities in different habitats (Pik et al. 1999; Andersen et al. 2002; Schnell et al. 2003) suggested that a coarser taxonomic resolution is sufficient as community composition differed between habitats at all resolution levels. Báldi (2003) found a close correlation between diversity measures at different taxonomic levels in Diptera and Acari but did not analyze responses of communities to different land-use or management strategies. For other invertebrate groups, such as Carabidae or Lepidoptera, results were not convincing and failed to produce clear correlations between diversity measures at different resolutions (Mandelik et al. 2007).

Bertrand et al. (2006) argued that the aggregation of data from species to higher taxonomic levels is generally flawed by a misconception about Linnaean classification. Allocations in current taxonomic classifications are based on a heterogeneous mixture of historical and contemporary views, and even if all taxa would be monophyletic, those pooled into the same rank would simply denote separate clades without further equivalence. Bertrand et al. (2006) concluded that these ranks are no more comparable than any other, nonnested taxa ("such as, for example, the genus *Rattus* and the phylum Arthropoda"), and that taxonomic surrogacy lacks justification. Consequently, the concept of taxonomic sufficiency has been repeatedly questioned in aquatic studies (Maurer 2000; Terlizzi et al. 2003; Bevilacqua et al. 2009; Terlizzi et al. 2009) but rarely been challenged in terrestrial ecology, partly because of the often-inconsistent taxonomy in soil organisms (Wall et al. 2008). Swift et al. (2004) further stated that "the abstraction 'diversity' has often not been separated from the specific attributes of a community of organisms that is under study in any particular location or system."

In light of that comment and the limited knowledge about the response of community composition in soil organisms to farming systems, we study response patterns from five soil-living/-dwelling taxa to four wheat-farming systems in an agricultural long-term experiment. We use data from our case study to compare patterns based on multivariate community data at different resolution levels (taxonomic and functional) and then directly relate the observed patterns at different taxonomic and functional levels to data on soil properties and associated ecosystem functions. The following section provides a brief overview of the analyzed taxonomic groups, primarily focusing on studies from row crops and methods that are directly comparable to results from our case study.

5.1.3 *Effects on soil organisms*

Soil organisms play an important role in agricultural systems as they directly affect nutrient dynamics, soil structure, and plant growth (Barrios 2007; Power 2010). The ongoing intensification of agricultural management significantly reduces the complexity of soil communities and could result in a reduced biological capacity for self-regulation in response to anthropogenic disturbances (Birkhofer, Bezemer, et al. 2008; Postma-Blaauw et al. 2010). Soil properties change slowly under transition from conventional to organic farming and may require several years or even decades to reach a new equilibrium (Clark et al. 1998; Fließbach et al. 2007). Therefore, it is crucial to study the impact of organic farming in agroecosystems with a long-term farming system history (Raupp et al. 2006). In the following sections, we discuss published results that compare effects of different farming systems on soil organisms. We limit the discussion to bacteria, fungi, nematodes, enchytraeids, and generalist predators since these groups were also studied in our case study from a long-term field experiment (for general textbooks, see Benckiser 1997; Collins and Qualset 1999). This is followed by a case study relating community responses of soil organisms from four wheat-farming systems to each other, to soil properties, and to functional characteristics.

5.1.3.1 *Bacteria and Fungi*

Soil bacteria and fungi are the most important primary decomposers in agricultural soils (up to 10^9 cells g^{-1} soil; Kennedy 1999) and contribute significantly to soil nutrient availability (Adl 2003). The biomass of these organisms is negatively affected by conventional compared to integrated or organic management (Bloem et al. 1994; Kong et al. 2011; Ullrich et al. 2011), but we only have a limited understanding of how the community composition changes under different farming systems. The driver of microbial communities is the carbon content of the soil, which can vary with management practice (Alvarez 2005). Methodological advances (e.g., PLFA [phospholipid fatty acid] analysis, Frostegård and Bååth 1996; or fatty acid methyl ester [FAME] profiling, e.g., Acosta-Martinez et al. 2007) allow for an analysis of management impacts on the biomass of different functional groups of bacteria and fungi (Grayston et al. 2004; Joergensen and Emmerling 2006).

PLFA techniques utilize the fact that after extraction specific fatty acids (FAs) can be ascribed to membranes of particular groups of bacteria or fungi and therefore provide insight into the microbial community structure and biomass (FA biomarkers; e.g., Zelles 1999). Moeskops et al. (2010) found that all analyzed groups of bacteria and fungi (gram-positive and gram-negative bacteria, actinomycetes, total bacteria, AMF [arbuscular mycorrhizal fungi], and fungi) showed higher absolute concentrations in organically managed fields that also had higher soil organic carbon content. Comparison of relative proportions of individual FAs revealed that actinomycetes were relatively more abundant in conventionally managed fields, and AMF contributed more to the overall FA composition in organically managed fields. Zhang et al. (2005) found a decreasing contribution of the fungal biomarker FA with land-use intensification that ranged from plantation woodlots to conventionally managed cereal systems. In contrast, Yeates et al. (1997) reported that grassland soils had higher amounts of fungal PLFA with organic management, but there were no such differences for bacterial PLFAs. Bossio et al. (1998) found that different PLFAs were more abundant in organic or conventional systems, ultimately leading to a clear differentiation between microbial communities in both management systems. Again, soil organic carbon content was consistently higher in the organically managed fields in this study. Several PLFAs were more abundant in conventionally managed systems, with a lower organic matter content in a study by Lundquist et al. (1999), including biomarker PLFAs for actinomycetes, resembling results by Moeskops et al. (2010).

Crop rotation and tillage practice affect soil organic carbon content (Ibekwe et al. 2002; Larkin 2003; Larkin and Honeycutt 2006) and may contribute to differences in microbial community composition between organic and conventional systems. More subtle management differences (e.g., organic farming with or without biodynamic compost preparations) have not resulted in similar patterns (Carpenter-Boggs et al. 2000). Organic and conventional farming systems differ markedly in their PLFA profiles, and soil organic carbon content is among the most important drivers of microbial community changes.

5.1.3.2 *Nematoda*

Nematodes comprise a diverse and abundant group of soil organisms (up to 9×10^6 individuals m^{-2}; Neher & Barbercheck 1999) that, depending on their mode of feeding, can be classified into various trophic groups within the soil food web. They are commonly categorized as plant feeders, fungal feeders, bacterial feeders, and omnivores or carnivores (Bongers 1988). Plant-feeding nematodes are among the major soilborne pests in agricultural crops. However, the other groups of nematodes fulfill important roles in soil food webs through their involvement in making nutrients available for the plant, while predacious nematodes can play a role in controlling pest organisms, including plant parasitic nematodes (Wardle et al. 2005). Impacts of agricultural practices not only can directly affect nematodes but also can indirectly alter nematode community composition through changes in resource or food availability or quality. Therefore, the community composition of nematodes may be particularly sensitive to agricultural intensification, and several indices based on nematode community composition have been developed to provide reliable indicators of anthropogenic disturbance (e.g., Bongers 1990). Briar et al. (2007) reported that bacterivorous nematodes were more abundant in organically managed fields at five of nine sampling dates over a 4-year field study. However, the other trophic groups showed a much weaker response to farming system. Ferris et al. (2004) also found consistently higher numbers of bacterivorous nematodes, but only during the midseason in organically grown tomatoes. No such difference was observed at the time of tomato planting or at the end of the growing season. All other trophic groups did not show a consistent response or varied too much to draw further conclusions. Sanchez-Moreno et al. (2009) also showed that bacterial-feeding nematodes were significantly more abundant in organically managed fields. The relative contribution of plant- and fungal-feeding nematodes to the whole assemblage was greater in conventional fields.

In contrast to these studies, other publications only reported minor responses of nematode communities to different farming systems. For example, van Diepeningen et al. (2006) found no significant differences between organic and conventional farming systems for any of the nematode trophic groups. The proportion of bacterivorous nematodes also did not respond to organic management but was positively related to nitrogen status in pasture soils in New Zealand (Parfitt et al. 2005). Neher (1999) observed a greater abundance of individuals from five families (Criconematidae, Heteroderidae, Plectidae, Prismatolaimidae, and Tylencholaimidae) in organically compared to conventionally managed soils. However, among all trophic groups and three different community indices, only the plant parasitic nematode maturity index differed significantly between farming systems. Yeates et al. (1997) did not find pronounced differences for bacterivorous nematodes, but fungivorous and plant-feeding nematodes were more abundant in organically managed grassland soils.

5.1.3.3 *Enchytraeidae*

Enchytraeid worms belong to the soil mesofauna and are abundantly present (numbers sometimes exceeding 30,000 individuals m^{-2}; Didden et al. 1997) in various terrestrial ecosystems. These worms with omnivorous feeding habits (Didden 1993) have been shown to strongly regulate ecosystem processes of organic matter decomposition (Standen 1978) and nutrient (C and N) mineralization (Briones et al. 1998; Cole et al. 2000; van Vliet et al. 2004) through a positive effect of their grazing on microbial activity (Hedlund and Augustsson 1995). It has been shown that the activities of enchytraeid worms lead to increased plant growth (Setälä 1995) and soil porosity (van Vliet et al. 1993), and that their role in the decomposer food web can be proportionally greater than that of other groups (Laakso and Setälä 1999). Enchytraeids depend on dead organic matter and therefore may benefit from organic amendments, while the application of inorganic fertilizers may affect population size and diversity negatively (Didden et al. 1997). Insecticide applications lead to negative, neutral, or positive responses by enchytraeid populations (Didden 1991), with particularly negative effects of fungicide application (Moser et al. 2004). Recent comparisons between organically and conventionally managed systems have not documented abundance or diversity differences for enchytraeids (Parfitt et al. 2005; Birkhofer, Bezemer, et al. 2008). Interestingly, however, in a study by Bezemer et al. (2010) enchytraeid abundance and richness responded characteristically to the presence of specific plant species, indicating that enchytraeid worms have specialized plant preferences. The authors also observed that enchytraeid biomass depended on the surrounding plant community. As organically managed cereal fields generally have richer plant communities in the presence of arable weeds (Gabriel et al. 2006), enchytraeid communities may differ in species composition from patterns observed in conventionally managed wheat fields.

5.1.3.4 Generalist predators

Studies about organic and conventional farming effects on the abundance and species richness of generalist predators were comprehensively summarized in three review articles (Bengtsson et al. 2005; Fuller et al. 2005; Hole et al. 2005). All major groups of generalist predators contribute to biological control of agricultural pests as an abundant part of the natural enemy assemblage (Symondson et al. 2002; spiders up to 80 individuals m^{-2}, carabids up to 50 individuals m^{-2}; Ekschmitt et al. 1997). Several of the reviewed studies suggested a negative impact of conventional farming practices on generalist predator populations. The majority of publications focused on the consequences of insecticide application (e.g., Wang et al. 2001; Thomas and Jepson 1997) or mechanical management practices (e.g., Holland and Reynolds 2003; Thorbek and Bilde 2004). Side effects of herbicide application on surface-dwelling generalist predators are primarily indirect (e.g., Navntoft et al. 2006), with weak or no direct lethal effects on spiders (Baines et al. 1998; Haughton et al. 2001) or ground beetles (Brust 1990; Taylor 2006). Most authors reported delayed indirect effects on generalist predators that only occurred several months after herbicides were applied (but see Bell et al. 2002). This negative impact was attributed to habitat changes as herbicides reduce plant diversity and structural complexity in agroecosystems. Such characteristics can affect predators through their effects on prey availability, altered microclimate, or the availability of refuge from intraguild predation (e.g., Birkhofer, Wise, et al. 2008).

In general, most studies identified positive effects of organic management on generalist predator activity density (Mäder et al. 2002; Shah et al. 2003), with a few studies in cereals showing contrasting negative effects on ground beetles (Moreby et al. 1994). Effects of organic farming on generalist predator diversity are less predictable. Species and functional-group diversity of carabids and spiders can be enhanced (Glück and Ingrisch 1990; Basedow 1998; Kromp 1999; Diekötter et al. 2010), reduced (Weibull et al. 2003), or remain unaffected (Purtauf et al. 2005; Schmidt et al. 2005; Birkhofer, Wise, et al. 2008) by organic farming. Most studies focusing on effects of agricultural intensification on predator populations analyzed abundance or diversity indices, but the importance of community composition for associated ecosystem services has recently been emphasized (e.g., pest control, Crowder et al. 2010; von Berg et al. 2010; Birkhofer et al. 2011), as it has been shown that generalist predator species or functional groups differ in their potential to suppress pests (e.g., Lang et al. 1999; Birkhofer, Gavish-Regev, et al. 2008).

5.2 A case study

Here, we present a case study from an agricultural field trial in which farming systems primarily differed in terms of fertilization (identical tillage, crop rotation, and soil type, Table 5.1), which allows us to directly relate differences in microbial community composition to different long-term fertilization strategies and study responses of soil fauna to microbial community change. We relate patterns to each other and to multivariate responses of microbes, other invertebrate groups, soil properties, and ecosystem functions to quantify the amount of information that is lost by aggregation of data in higher-order taxonomic or functional categories.

The DOK (bio**d**ynamic, bio**o**rganic and conventional [**k**onventionell] treatments) trial in Therwil, Switzerland, is a long-term agricultural experiment that was established in 1978 by the Agroscope Reckenholz-Tänikon (ART) research station and the Research

Table 5.1 Description of the Study Sites in the DOK (Biodynamic, Organic, and Conventional Farming Systems) Trial in Therwil, Switzerland

	Biodynamic	Bioorganic	Conventional	Mineral
(a) Farming system				
Abbreviation	BIODYN	BIOORG	CONFYM	CONMIN
Organic fertilizer	Composted FYM	Rotted FYM	Stacked FYM	—
Mineral fertilizer	—	—	NPK	NPK
Weed control	Mechanical	Mechanical and herbicides		
Disease control	Indirect methods	Chemical (thresholds)		
Pest control	Plant extracts, biocontrol	Chemical (thresholds)		
Special treatments	Biodynamic preparation	—	Plant growth regulators	
(b) Soil properties				
Soil pH	5.8	5.4	5.2	4.5
Soil water content	0.24	0.24	0.25	0.24
C_{org} (% dw soil)	1.6	1.4	1.3	1.2
N_{tot} (% dw soil)	0.19	0.17	0.16	0.15
(c) Ecosystem functions				
N mineralization (mg kg^{-1} wk^{-1})	4.8	4.9	5.4	3.7
Grain yield (dt ha^{-1})	42	43	55	54
Grain protein content (N g kg^{-1} dw)	23	24	28	28
(d) Nutrients (kg ha^{-1} y^{-1})				
$C_{organic}$	1,818	2,272	2,272	0
N_{total}	99	102	157	122
$N_{mineral}$	31	35	101	122
P	23	27	41	40
K	165	157	258	250

Note: (a) Farming system, (b) measured soil properties, (c) estimated ecosystem functions, and (d) fertilizer (farmyard manure, FYM; mineral, NPK) and mean annual nutrient amendments between 1978 and 2005 to all systems during four crop rotation periods. For more details, see Birkhofer et al. (2008a).

Institute of Organic Agriculture (FiBL). It has been established to compare selected conventional agricultural practices to organic agriculture over relevant temporal scales and under identical soil type and climatic conditions. The soil is a haplic luvisol on deep deposits of alluvial loess. Mean precipitation is 785 mm per year, with an annual average temperature of 9.5°C. We sampled wheat plots of two organic farming systems (BIODYN, BIOORG); one conventional (CONFYM) system, receiving farmyard manure and mineral fertilizer; and a second conventional system mimicking stockless farming (CONMIN, receiving mineral fertilizers only) in May 2005 after all plots were managed according to the respective system for 27 years. The BIODYN system differs from the BIOORG system by the application of composted farmyard manure instead of rotted farmyard manure and by the use of biodynamic preparations (horn-manure and horn-silica; see Mäder et al. 2000 for further details). Crop rotation and soil tillage regime were identical for all 5 × 20 m plots. Synthetic insecticides were last applied in 2001 in the conventional systems, and soils of the CONFYM and CONMIN system were treated with a molluscicide in 2003 and 2004. A single herbicide containing the active ingredients carfentrazone and isoproturon was applied to winter wheat in both conventional systems on 1 April 2005, and the growth regulator trinexapac-ethyl was applied on 22 April 2005. The two conventional systems received a total of 90 kg N ha^{-1} as calcium ammonium nitrate in March and April 2005; BIOORG received a total of 44 m^3 slurry ha^{-1} in two rates (N_{tot} 77.9, N_{min} 33.2, organic matter 1,335 kg ha^{-1}); and BIODYN received 30 m^3 ha^{-1} (N_{tot} 46.2, N_{min} 28.5, organic matter 651 kg ha^{-1}) only once (for further details on the DOK trial, see Mäder et al. 2002 and Fließbach et al. 2007).

All four replicated wheat plots of each farming system were sampled on 9 May 2005 by collecting one small soil core (8-cm diameter, 5-cm deep) and one sample of bulk soil (1 kg) from the upper 5-cm layer of randomly chosen locations at the northern and southern end of each of the 16 plots (minimum edge distance 1.5 m, $N = 32$). Bulk soil was used to analyze chemical soil properties, FA profiles, and nematode communities; soil cores were sampled to analyze enchytraeid communities. Soil pH was measured in 0.1 *M* KCl solution in a soil-to-liquid ratio of 1:3 (w/v). Total soil nitrogen and carbon were measured using an elemental analyzer (Carlo Erba, Milan, Italy) and were taken to represent total soil nitrogen content (N_{tot}) and soil organic carbon content (C_{org}) as concentrations of carbonates are low at the study site. To measure soil water content (SWC), fresh soil samples were dried at 105°C for 72 h (for mean values of soil properties, see Table 5.1; for more details, see Birkhofer, Bezemer, et al. 2008).

Lipid extractions for analysis of PLFAs and NLFAs (neutral lipid fatty acids) were made on 3 g of fresh soil according to the work of Frostegård et al. (1993) and Hedlund (2002). The resulting FAMEs were separated on a Hewlett-Packard 6890 gas chromatograph. Relative retention times of the FAMEs were compared to those of standards (for more details, see Birkhofer, Bezemer, et al. 2008). The classification of FAs into functional groups was based on published references (Frostegård and Bååth 1996; White et al. 1996; Olsson 1999; D'Angelo et al. 2005): gram-positive bacteria at i14:0, i15:0, a15:0, i16:0, i17:0, and a17:0; gram-negative bacteria at 16:1ω9, 16:1ω7c, 16:1ω7t, and 18:1ω7; sulfate-reducing bacteria at 10Me16:0, cy17:0, 10Me17:0, 10Me18:0, and cy19:0; NLFA 16:1ω5 at AMF; 18:2ω6 at nonmycorrhizal fungi.

The total number of nematodes was counted, and a minimum of 150 nematodes was further identified to family or genus level according to Bongers (1988) and classified into different feeding groups according to Yeates et al. (1993). Enchytraeids were extracted from soil using a combination of cold and hot wet funnel extraction methods to maximize the extraction efficiency. Soil samples were first submerged in cold water for 24 h in a plastic sieve (Dunger and Fiedler 1997) and later heated for 3–4 h in a wet funnel (O'Connor 1955). All individuals were identified to species or genus level.

To estimate densities of soil-dwelling generalist predators (Araneae and Carabidae), an area of 2.0 m^2 was fenced (PVC [polyvinyl chloride] barriers; 50 cm high, 10 cm sunk into the soil), and four pitfall traps containing a water detergent mix were placed along the inner barrier and left open for 14 days (3 to 17 May). Pitfall samples were transferred to 70% ethanol, sieved, and later hand sorted. Vegetation-living spiders were sampled on 16 May under dry conditions using an Eco-Vac insect suction sampler (EcoTech, Bonn, Germany). An area of 0.7 m^2 was fenced on each of the two sides of each plot (PVC barriers 50 cm high) and immediately suction sampled for 90 s. Samples were transferred to glass vessels containing a small amount of ethyl acetate. Suction samples were stored at −10°C and later sieved and hand sorted using a dissecting microscope to separate animals from debris. Functional grouping according to nematode feeding modes (six groups; Yeates et al. 1993), spider-hunting strategies (nine groups; Uetz et al. 1999), and carabid-feeding mode (four groups) or wing morphology (three groups) (Lindroth 1985, 1986; Luff 1998; Ribera 2001) was based on the most comprehensive publications for each taxon.

Yield was estimated on 23 July from a center 1.7 × 10 m plot in each of the analyzed fields. To measure the nitrogen content of wheat plants in each field, four grain samples were bulked and dried at 80°C for 10 h, followed by a measurement of nitrogen content by X-ray fluorescence (XRF) spectroscopy (for details, see Mäder et al. 2007). Potential nitrogen mineralization rate in soil was determined by incubating 200-g soil samples at 20°C and about 50% water-holding capacity for 6 weeks. Nitrogen mineralization rate was calculated from the increase in mineral N concentration between week 6 and week 2; data from week 1 were excluded to reduce bias from sample handling (for mean values of ecosystem functions, see Table 5.1; for more details, see Birkhofer, Bezemer, et al. 2008). Figure 5.1 exemplifies the differences in soil surface structure and wheat plant development between CONMIN and BIOORG plots at the time of sampling.

The effect of different farming systems on the community composition of soil organisms at different taxonomic resolutions was tested by analysis of similarities (ANOSIM; Anderson et al. 2008). Prior to ANOSIM, all biotic data were log(n + 1) transformed to weigh down the importance of abundant species. Bray-Curtis distances were used to

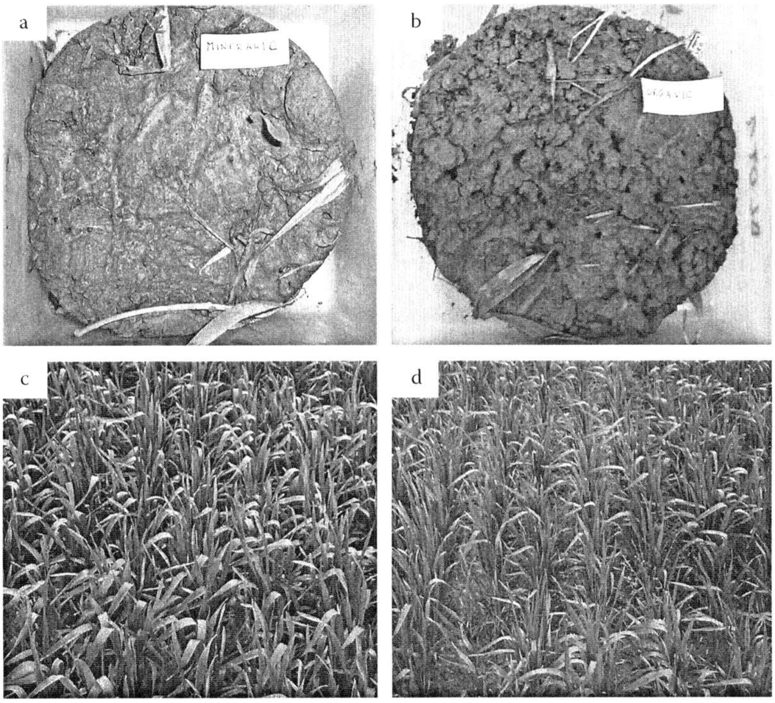

Figure 5.1 Structural differences between (a) and (c) conventionally (CONMIN) and (b) and (d) organically (BIOORG) managed plots at the time of sampling in the DOK trial, Switzerland, with (a) and (b) showing differences in soil surface structure and (c) and (d) showing differences in wheat plant growth in May 2005.

create resemblance matrices for all biotic variables. Statistical significance in ANOSIM was tested based on 9,999 permutation runs. After a statistically significant main test, significant pairwise comparisons between farming systems were further tested by similarity percentage (SIMPER; Clarke and Gorley 2006) tests to identify individual variables that were responsible for the observed differences. Two-dimensional nonmetric multidimensional scaling (NMDS) was used as an ordination method to display significant results. All NMDS ordinations were calculated with the Kruskal fit scheme 1, 150 restarts, and a stability criterion of 0.001. Variables on soil properties and ecosystem functions were normalized by subtracting the variable's mean from all values and then dividing by the standard deviation of the variable. Euclidean distances were used for these subsets of normalized abiotic variables to create NMDS plots. To relate resemblance matrices of soil organisms based on different taxonomic and functional group resolutions, soil properties, and ecosystem function data with each other, we calculated a new resemblance matrix based on ρ (Spearman rank correlation) values of all pairwise comparisons between those matrices and show this data in a second-stage NMDS plot (Clarke and Gorley 2006).

Table 5.2 provides an overview of the abundance and diversity of the studied soil and surface-dwelling fauna in the DOK trial. None of the diversity indices was significantly affected by farming system; for results on univariate abundance data, soil properties,

Table 5.2 Mean Abundance/Amount, Species/Genus Richness, Inverse Simpson Index, and Pielou's Evenness for PLFAs, Nematodes, Enchytraeids, Spiders, and Carabids per Plot in Four Replicated Wheat-Farming Systems

	Farming system	Mean abundance/ amount[a]	Mean species/ genus[b] richness	Inverse Simpson diversity $(1-\lambda)$	Pielou's evenness (J')
Fatty acids (nmol g soil dw^{-1})	BIODYN	62[a]	N/A	0.91	0.91
	BIOORG	59[a]	N/A	0.91	0.92
	CONFYM	56[a]	N/A	0.91	0.92
	CONMIN	43[a]	N/A	0.91	0.92
Nematoda (ind 100 g soil dw^{-1})	BIODYN	4966	27[b]	0.87	0.72
	BIOORG	5020	28[b]	0.86	0.70
	CONFYM	5008	30[b]	0.89	0.73
	CONMIN	2955	28[b]	0.87	0.72
Enchytraeidae (ind m^{-2})	BIODYN	22,000	10	0.83	0.76
	BIOORG	28,000	10	0.84	0.81
	CONFYM	15,000	10	0.86	0.78
	CONMIN	14,000	8	0.81	0.74
Araneae (ind m^{-2})	BIODYN	10	11	0.82	0.69
	BIOORG	9	12	0.80	0.69
	CONFYM	5	9	0.85	0.77
	CONMIN	6	11	0.85	0.76
Carabidae (ind m^{-2})	BIODYN	8	10	0.85	0.78
	BIOORG	7	11	0.86	0.80
	CONFYM	5	9	0.79	0.77
	CONMIN	6	9	0.81	0.79

Note: For abbreviations and details, see Table 5.1. (a) Amount for FAs, abundances for animals; (b) Nematodes identified to genus level, all other species level.

or ecosystem functions, see the work of Birkhofer, Bezemer, et al. (2008). FA profiles differed significantly between farming systems if based on the concentration of individual FAs in the soil (Figure 5.2a; ANOSIM, $R = 0.454$; $P < 0.001$). The profiles differed significantly between CONMIN and BIODYN, CONMIN and BIOORG, and CONMIN and CONFYM. The FA that contributed most to these differences was 18:1ω7 as the concentration of this biomarker FA for gram-negative bacteria was always lower in the CONMIN system (Figure 5.3a). A biomarker NLFA for AMF (16:1ω5) had lower concentrations in CONMIN soils compared to the other systems (Figure 5.3b). FA profiles still differed significantly between farming systems if individual FAs were pooled into biomarker groups (Figure 5.3c; $R = 0.358$; $P = 0.003$). At this organizational level, pairwise comparisons only detected significant differences between CONMIN and BIODYN ($R = 0.667$; $P = 0.029$) and CONMIN and BIOORG ($R = 0.802$; $P = 0.029$). The major subset of FAs responsible for this difference was the biomarkers for gram-negative bacteria, which had higher concentrations in the two organically managed systems.

The community composition of nematodes differed significantly between farming systems at the genus level (Figure 5.2b; $R = 0.738$; $P < 0.001$), with significant differences

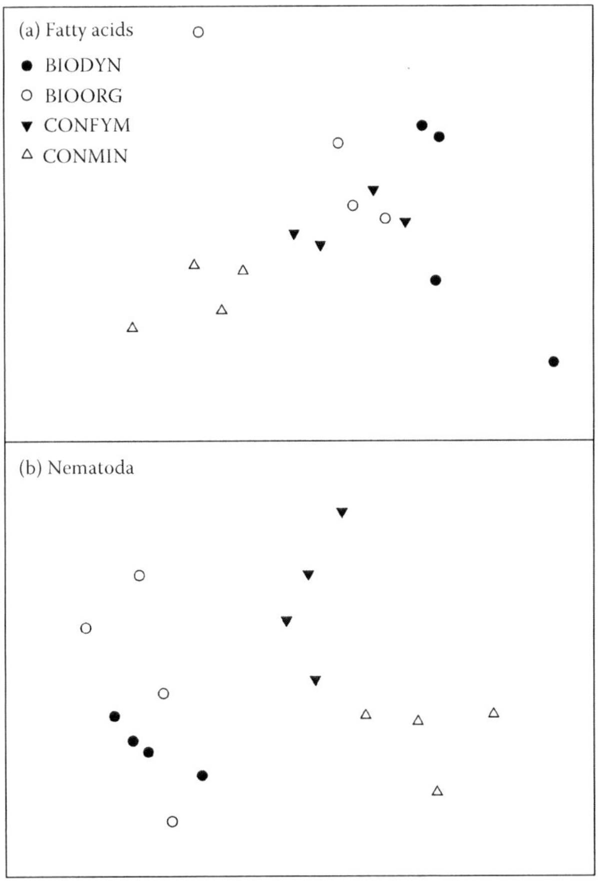

Figure 5.2 Two-dimensional NMDS ordination of (a) profiles based on concentrations of 17 fatty acids (2-D [two-dimensional] stress = 0.084) and (b) nematode genus composition (2-D stress = 0.185) in 16 plots of four different farming systems (see Table 5.1).

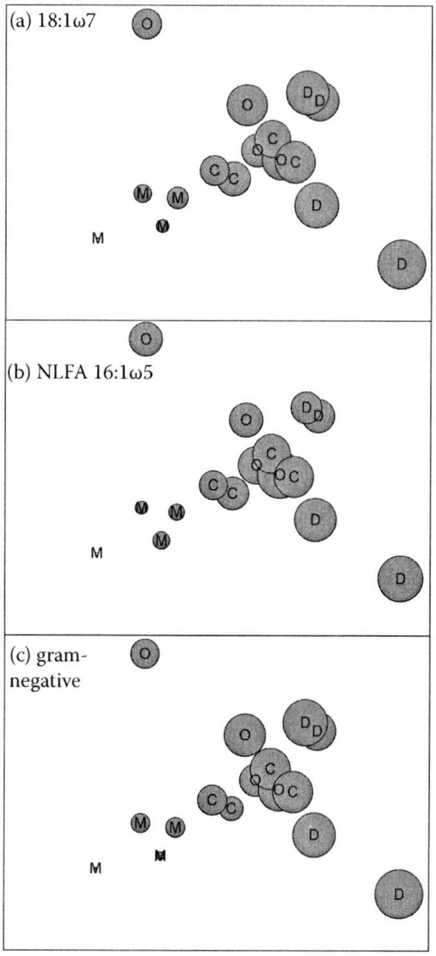

Figure 5.3 Two-dimensional NMDS ordination based on concentrations of 17 fatty acids in 16 plots of biodynamic, bioorganic, conventional with farmyard manure or conventional only with mineral fertilizer (stockless farming) farming systems (see Table 5.1; 2-D stress = 0.084). Bubble sizes are scaled according to the data range and superimposed based on concentrations of (a) 18:1ω7 (biomarker for gram-negative bacteria, data range = 0.091–0.118 mol%), (b) NLFA 16:1ω5 (biomarker for AMF, 0.035–0.052 mol%), and (c) combined biomarker FAs for gram-negative bacteria (0.191–0.233 mol%).

between communities of all farming systems with the exception of BIODYN versus BIOORG. The genus that contributed most to the observed differences between CONMIN and BIODYN or BIOORG was the plant and hyphal feeder *Tylenchus,* which was almost absent from CONMIN soils (Figure 5.4a). The genus *Rhabditis,* which encompasses bacterivores, further discriminated CONFYM from BIODYN, BIOORG, and CONMIN nematode communities as individuals from this genus were common in CONFYM but absent from the other systems (Figure 5.4b). At the family level, nematode communities also differed between farming systems ($R = 0.508$; $P = 0.001$), but at this taxonomic resolution, family composition in BIOORG did not differ from BIODYN or from CONFYM. The fungal feeding Aphelenchidae were absent from BIODYN but abundant in the other systems (Figure 5.4c). In contrast, Anguinidae, which are considered plant or fungal feeders, occurred at much

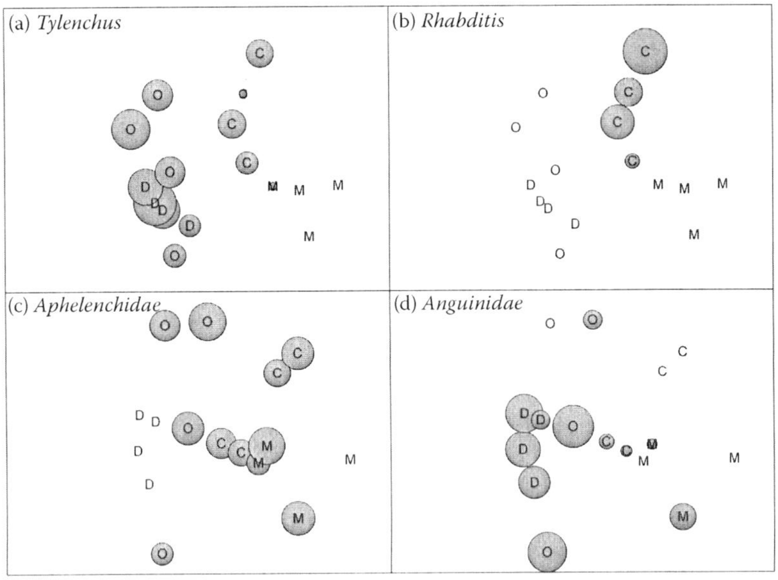

Figure 5.4 Two-dimensional NMDS ordination based on (a) and (b) nematode genus composition (44 genera, 2-D stress = 0.185) or (c) and (d) family composition (29 families, 2-D stress = 0.155) in 16 plots of biodynamic, bioorganic, conventional with farmyard manure and conventional only with mineral fertilizer (stockless farming) farming systems (see Table 5.1). Bubble sizes are scaled according to the data range and superimposed based on densities of (a) *Tylenchus* sp. (data range = 0–3 ind g soil dw^{-1}), (b) *Rhabditis* sp. (0–3 ind g soil dw^{-1}), (c) Aphelenchidae (0–0.5 ind g soil dw^{-1}), and (d) Anguinidae (0–0.8 ind g soil dw^{-1}).

lower abundances in conventionally managed soils (Figure 5.4d). At the level of trophic groups, differences between farming systems were still significant ($R = 0.291$; $P = 0.019$), but only the trophic group composition in BIODYN differed significantly from both conventionally managed systems. Trophic group composition in BIOORG soils only tended to differ from CONMIN. Hyphal and plant feeders primarily contributed to these differences as this functional group was always more abundant in organically managed systems.

Enchytraeid communities did not differ significantly between farming systems, independent of whether analyzed on species ($R = 0.047$; $P = 0.322$) or genus ($R = 0.102$; $P = 0.204$) level. The community composition of spiders also did not differ significantly between farming systems at the species ($R = 0.177$; $P = 0.113$), family ($R = 0.052$; $P = 0.232$), or functional group ($R = 0.081$; $P = 0.182$) level. The same result was observed for carabid beetles, not showing a significant difference between farming systems if analyzed at species ($R = 0.001$; $P = 0.478$), family ($R = -0.055$; $P = 0.678$), or functional group level based on either wing morphology ($R = -0.102$; $P = 0.844$) or trophic group ($R = 0.017$; $P = 0.388$).

Multivariate patterns in community composition of some soil organisms were closely related to patterns in soil properties (pH, SWC, C_{org}, and N_{tot}) as the community composition of nematodes based on genus or family levels resembled plot-specific soil properties (Figure 5.5). Nematode functional group composition and microbial community structure based on FA profiles or biomarker groups still resembled multivariate patterns for soil properties. The observed patterns in community composition of enchytraeids, carabids, and spiders were less closely related to soil properties. In carabids, community composition based

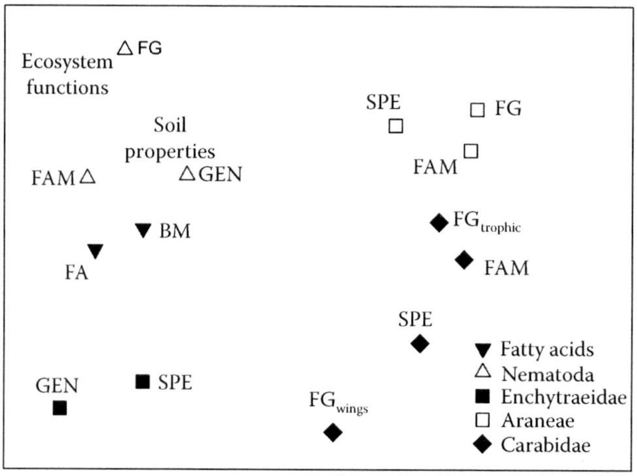

Figure 5.5 Two-dimensional second-stage NMDS based on pairwise rank correlates between individual NMDS plots. Patterns in multivariate community composition are related to each other at different levels of resolution for fatty acids (profiles based on individual FAs vs. biomarker groups); Nematoda (functional groups vs. genus vs. family); Enchytraeidae (genus vs. species); Araneae (functional groups vs. species vs. family); and Carabidae (functional groups based on feeding guild or wing morphology vs. species vs. family) and to multivariate patterns observed for soil properties (pH, SWC, C_{org}, and N_{tot}) and ecosystem functions (N mineralization, yield, and grain protein content). The closer two symbols are located to each other, the more their observed multivariate patterns are related to each other. BM, biomarker; FAM, family; FG, functional groups; GEN, genus; SPE, species.

on feeding preferences related more closely to soil properties than functional grouping according to wing morphology.

Relating patterns in community composition of soil organisms to multivariate patterns in ecosystem functions (N mineralization, yield, and grain protein content) suggest that nematode community composition was closely related to functions at all three organizational levels (Figure 5.5). A weaker relationship was observed between microbial community composition based on FA profiles or functional groups and multivariate patterns of functional parameters. Arthropod or enchytraeid community composition did not relate well to the analyzed ecosystem functions, independent of the taxonomic or functional resolution.

Multivariate patterns at both taxonomic levels (family and genus) for nematode community composition related closely to each other (Figure 5.5). However, aggregation of individuals to the functional group level changed the information content of the multivariate data. While multivariate patterns based on taxonomic levels resembled the observed patterns in soil properties (pH, SWC, C_{org}, and N_{tot}), functional group composition of nematodes rather reflected the multivariate patterns based on ecosystem functions (N mineralization, yield, and grain protein content). The differences between the information used to relate microbial community composition based on FA profiles or biomarker grouping to soil properties and ecosystem functions were almost redundant, and the same holds true for Enchytraeid communities analyzed at the species or genus level and spider community composition analyzed at any taxonomic or functional level. Carabid beetles showed the most pronounced difference in information content if analyzed at different taxonomic or functional resolutions. Multivariate analyses of carabid communities at different taxonomic levels or functional grouping showed that family and trophic group composition were closely related to each other (Figure 5.5). However, community composition patterns observed at the species level or by wing morphology type did not resemble family or trophic group composition.

5.3 *Discussion*

Based on diversity indices, we did not observe significant differences between the four farming systems; however, community composition differed significantly for two of the five analyzed groups of soil organisms. FA profiles of microorganisms and nematode community composition showed distinct patterns separating communities in the farming system that only received synthetic fertilizer (CONMIN) from communities of the three systems that received manure.

Biomarker FAs for gram-negative bacteria (18:1ω7) and AMF (NLFA 16:1ω5) both had higher concentrations in organically managed fields. Moeskops et al. (2010) observed similar responses by biomarker FAs for AMF comparing organic versus conventional horticultures and concluded that the high sensitivity of this group was related to a reduced colonization potential of AMF (see also Bending et al. 2004) and a lower inoculum (see also Mäder et al. 2002) in conventionally managed fields. According to Peacock et al. (2001), gram-negative bacteria suffer from the application of inorganic fertilizers and increase in concentration under application of manure. Organic substrates promote the growth of gram-negative bacteria, and the pronounced differences in the availability of organic carbon in our study system most likely contributed to the observed pattern (see also Bossio and Scow 1998). The patterns of FA profiles in the four farming systems supported the general trends for a dominance of gram-positive bacteria and a reduction of fungal biomass in conventionally managed systems (see also Gryndler et al. 2006; Esperschütz et al. 2007).

Nematode communities showed a significant, but more complex, response to farming systems compared to FA profiles. The abundance of individuals from the genus *Tylenchus* contributed to a separation of the NPK-only (CONMIN) treatment from other farming systems. *Tylenchus* is known to feed on root hairs or hyphae (Thorne 1961), and the application of herbicide prior to our sampling followed by reduced weed presence may have contributed to the lower abundance of this species in conventionally managed systems. Population sizes of this species were higher under the application of manure compared to NPK fertilizer (Liang et al. 2009), and our results suggest that the addition of organic substrates promotes *Tylenchus* densities. Individuals from the genus *Rhabditis* were only found in the system that received synthetic fertilizer and manure (CONFYM) but were absent from the two organically managed systems and the system that only received synthetic fertilizer. Bacterivorous nematodes generally benefit from the application of compost or manure and have lower abundances in fields that are only treated with synthetic fertilizers (Hu and Qi 2010), but it remains unclear why this genus was absent from organically managed fields. At the family level, the single genus within the family Aphelenchidae (*Aphelenchus* sp.), which are fungal feeders, was absent from the biodynamically managed fields, and individuals from the family Anguinidae had higher abundances in fields that received manure. Populations of *Aphelenchus* sp. have been shown to benefit from cultivation (mechanical weed management) but to suffer from organic mulching (Wardle et al. 1995). Addition of organic substrate may have contributed to the observed lower abundance of this genus in biodynamically managed plots, but it remains unclear why individuals from this genus did not show a similar response in bioorganic fields. Individuals from the genus *Ditylenchus* (Anguinidae) most likely belong to the fungal-feeding guild. As fungal biomass is highest in the organically managed systems of the DOK trial (see Birkhofer, Bezemer, et al. 2008), the increased abundance of this genus probably reflects the higher availability of resources. It further remains unresolved why one family of fungal-feeding nematodes (Anguinidae) benefits from higher fungal biomasses, whereas another family from the same trophic group does not show such response to high-fungal biomasses in the biodynamically managed plots (Aphelenchidae).

It has been suggested that agricultural intensification will have a stronger impact on the abundance and diversity of larger (enchytraeids and nematodes) versus smaller taxa (bacteria and fungi; Postma-Blaauw et al. 2010); however, this was not present if community composition was considered in our study. Our results suggest that community composition of taxa with larger body sizes (Enchytraeidae, Araneae, and Carabidae) did not respond significantly to differences in farming systems. The abundance and species richness of smaller soil organisms may suffer more from anthropogenic disturbances, compared to larger, surface-active arthropod taxa (Dauber et al. 2005; Birkhofer, Bezemer, et al. 2008). In soils, however, the suggested stronger impact of agricultural intensification on the abundance and diversity of larger (enchytraeids and nematodes) versus smaller taxa (bacteria and fungi; Postma-Blaauw et al. 2010) was not present if community composition was considered in our study. In contrast, our results suggest that bacterial, fungal, and nematode community composition responded stronger to intensified farming practices than larger endogaeic (enchytraeids) or epigeic (spider or ground beetle) communities. In summary, we therefore suggest that less-mobile, soil-living species either may be affected more seriously by intense management or may fail to recolonize disturbed fields as successfully as larger, more mobile taxa.

Relating patterns of soil organism communities to the observed resemblance of plots in terms of soil properties or ecosystem functions suggests that less-mobile, endogaeic organisms (bacteria, fungi, and nematodes) closely resemble multivariate patterns observed for soil properties and ecosystem functions. In contrast, the community composition of enchytraeids, carabids, and spiders did not resemble the similarity between plots based on soil and functional characteristics. Endogeic taxa are probably more seriously affected by agricultural management (see previous section), and such agricultural intensification also directly alters soil properties. Indirect effects of agricultural management on soil organisms include the alteration of habitat structure (e.g., Blackshaw et al. 2003) or soil biotic conditions (e.g., Haynes and Naidu 1998). It is therefore likely that multivariate patterns in soil properties and community composition of soil organisms are linked to each other and to management practices.

The observed close relationship between field-specific functional parameters, community composition in nematodes, and FA profiles at least partly reflects the assumed linkage between N content in wheat plants, N mineralization rates, abundance of bacterivorous nematodes, and bacterial activity. Bouwman and Zwart (1994), for example, demonstrated that grazing through nematodes promotes bacterial activity and therefore alters rates of N mineralization (see also Bloem et al. 1994). Populations of some nematode families can even be limited by the availability of bacteria in agricultural soils (Bouwman and Zwart 1994). Our observation that patterns in nematode community composition resemble patterns for the composition of bacterial and fungal communities supports the assumption of close interactions between bacterial and fungal resources and consumers at higher trophic levels. We further demonstrated that not only the abundance of nematodes or bacteria relate to the quantified ecosystem functions but also observed close relationships between management practices, community composition of nematodes, FA profiles, and multivariate patterns in ecosystem functions.

The second-stage NMDS relates several multivariate patterns to each other to help identify if taxonomic sufficiency is a valid concept in studying community-level responses to agricultural management. Figure 5.5 suggests that only limited additional information is gained from higher taxonomic resolution comparing patterns within taxa but between different taxonomic or functional resolutions. Nevertheless, compositional data on individual FA concentrations and nematode genera provided a significant separation of organic systems from systems that received either manure and NPK or only NPK fertilizers. Such detailed contrasts were lost if community composition was analyzed at coarser resolutions. From such results, it remains unclear if the critical comment about the information loss caused by aggregating data into more or less arbitrary formed higher taxonomic orders (Bertrand et al. 2006) is supported by our analyses. However, it seems plausible that information could be lost after aggregation as individuals in higher-order categories have less-well-defined ecological niches and therefore may not show a pronounced response to anthropogenic disturbance. Functional groups are defined at different taxonomic resolutions (e.g., nematode at genus level, spiders at family level, and carabids at species level) and include different information about community-level responses to disturbance. In our case study, nematode and carabid patterns in functional group composition contained different information compared to patterns observed at the highest possible taxonomic resolution. Báldi (2003) suggested that richness patterns at higher taxonomic orders are well correlated with species richness patterns for some terrestrial arthropod groups. However, this approach to taxonomic sufficiency only holds true if research solely focuses on richness measures. As soon as multivariate community responses are within the focus of a study, aggregation into higher taxonomic levels or functional groups almost certainly changes the information content (see also Maurer 2000). In such cases, different species/genera fall within the same functional groups, and the resulting multivariate patterns in functional group composition may therefore not resemble taxonomic composition.

We conclude that a low-level taxonomic resolution in soil organisms often provides more detailed and ecologically meaningful information about response patterns of communities to anthropogenic disturbance. Our case study particularly highlighted the added value of genus-level nematode determination and individual FA data for identifying differences between organic and conventional wheat-farming systems. However, additional multivariate analyses of functional group composition are important if functional groups can be built without resembling taxonomic groups too closely.

References

Acosta-Martinez, V., M.M. Mikha, and M.F. Vigil. 2007. Microbial communities and enzyme activities in soils under alternative crop rotations compared to wheat-fallow for the Central Great Plains. *Applied Soil Ecology* 37:41–52.

Adl, S.M. 2003. *The Ecology of Soil Decomposition*. CABI, Wallingford, UK.

Alvarez, R. 2005. A review of nitrogen fertilizer and conservation tillage effects on soil organic carbon storage. *Soil Use and Management* 21:38–52.

Andersen, A.N., B.D. Hoffmann, W.J. Muller, and A.D. Griffiths. 2002. Using ants as bioindicators in land management: simplifying assessment of ant community responses. *Journal of Applied Ecology* 39:8–17.

Anderson, M.J., R.N. Gorley, and K.R. Clarke. 2008. *PERMANOVA+ for PRIMER: Guide to Software and Statistical Methods*. PRIMER-E, Plymouth, UK.

Avery, A. 2007. "Organic abundance" report: fatally flawed. *Renewable Agriculture and Food Systems* 22:321–323.

Badgley, C., J. Moghtader, E. Quintero, E. Zakem, M.J. Chappell, K. Aviles-Vazquez, A. Samulon, and I. Perfecto. 2007. Organic agriculture and the global food supply. *Renewable Agriculture and Food Systems* 22:86–108.

Badgley, C., and I. Perfecto. 2007. Can organic agriculture feed the world? *Renewable Agriculture and Food Systems* 22:80–82.

Baines, M., C. Hambler, P.J. Johnson, D.W. Macdonald, and H. Smith. 1998. The effects of arable field margin management on the abundance and species richness of Araneae (spiders). *Ecography* 21:74–86.

Báldi, A. 2003. Using higher taxa as surrogates of species richness: a study based on 3700 Coleoptera, Diptera, and Acari species in Central-Hungarian reserves. *Basic and Applied Ecology* 4:589–593.

Barrios, E. 2007. Soil biota, ecosystem services and land productivity. *Ecological Economics* 64:269–285.

Basedow, T. 1998. The species composition and frequency of spiders (Araneae) in fields of winter wheat grown under different conditions in Germany. *Journal of Applied Entomology* 122:585–590.

Bell, J.R., A.J. Haughton, N.D. Boatman, and A. Wilcox. 2002. Do incremental increases of the herbicide glyphosate have indirect consequences for spider communities? *Journal of Arachnology* 30:288–297.

Benckiser, G. 1997. *Fauna in Soil Ecosystems*. Dekker, New York.

Bending, G.D., M.K. Turner, F. Rayns, M.C. Marx, and M. Wood. 2004. Microbial and biochemical soil quality indicators and their potential for differentiating areas under contrasting agricultural management regimes. *Soil Biology & Biochemistry* 36:1785–1792.

Bengtsson, J., J. Ahnstrom, and A.C. Weibull. 2005. The effects of organic agriculture on biodiversity and abundance: a meta-analysis. *Journal of Applied Ecology* 42:261–269.

Bertrand, Y., F. Pteijel, and G.W. Rouse. 2006. Taxonomic surrogacy in biodiversity assessments, and the meaning of Linnaean ranks. *Systematics and Biodiversity* 4:149–159.

Bevilacqua, S., S. Fraschetti, L. Musco, and A. Terlizzi. 2009. Taxonomic sufficiency in the detection of natural and human-induced changes in marine assemblages: a comparison of habitats and taxonomic groups. *Marine Pollution Bulletin* 58:1850–1859.

Bezemer, T.M., M.T. Fountain, J.M. Barea, S. Christensen, S.C. Dekker, H. Duyts, R. van Hal, J.A. Harvey, K. Hedlund, M. Maraun, J. Mikola, A.G. Mladenov, C. Robin, P.C. de Ruiter, S. Scheu, H. Setälä, P. Šmilauer, and W.H. van der Putten. 2010. Divergent composition but similar function of soil food webs of individual plants: plant species and community effects. *Ecology* 91:3027–3036.

Birkhofer, K., T.M. Bezemer, J. Bloem, M. Bonkowski, S. Christensen, D. Dubois, F. Ekelund, A. Fliessbach, L. Gunst, K. Hedlund, P. Mader, J. Mikola, C. Robin, H. Setälä, F. Tatin-Froux, W.H. Van Der Putten, and S. Scheu. 2008. Long-term organic farming fosters below and aboveground biota: implications for soil quality, biological control and productivity. *Soil Biology & Biochemistry* 40:2297–2308.

Birkhofer, K., A. Fließbach, D.H. Wise, and S. Scheu. 2011. Arthropod food webs in organic and conventional wheat farming systems: a stable isotope approach. *Agricultural and Forest Entomology* 13:197–204.

Birkhofer, K., E. Gavish-Regev, K. Endlweber, Y.D. Lubin, K. Von Berg, D.H. Wise, and S. Scheu. 2008. Cursorial spiders retard initial aphid population growth at low densities in winter wheat. *Bulletin of Entomological Research* 98:249–255.

Birkhofer, K., D.H. Wise, and S. Scheu. 2008. Subsidy from the detrital food web, but not microhabitat complexity, affects the role of generalist predators in an aboveground herbivore food web. *Oikos* 117:494–500.

Blackshaw, R.E., R.N. Brandt, H.H. Janzen, T. Entz, C.A. Grant, and D.A. Derksen. 2003. Differential response of weed species to added nitrogen. *Weed Science* 51:532–539.

Bloem, J., G. Lebbink, K.B. Zwart, L.A. Bouwman, S.L.G.E. Burgers, J.A. de Vos, and P.C. de Ruiter. 1994. Dynamics of microorganisms, microbivores and nitrogen mineralization in winter-wheat fields under conventional and integrated management. *Agriculture Ecosystems & Environment* 51:129–143.

Bongers, T. 1988. *De Nematoden van Nederland*. Pirola, the Netherlands.

Bongers, T. 1990. The maturity index—an ecological measure of environmental disturbance based on nematode species composition. *Oecologia* 83:14–19.

Bossio, D.A., and K.M. Scow. 1998. Impacts of carbon and flooding on soil microbial communities: phospholipid fatty acid profiles and substrate utilization patterns. *Microbial Ecology* 35:265–278.

Bossio, D.A., K.M. Scow, N. Gunapala, and K.J. Graham. 1998. Determinants of soil microbial communities: effects of agricultural management, season, and soil type on phospholipid fatty acid profiles. *Microbial Ecology* 36:1–12.

Bouwman, L.A., and K.B. Zwart. 1994. The ecology of bacterivorous protozoans and nematodes in arable soils. *Agriculture Ecosystems & Environment* 51:145–160.

Briar, S.S., P.S. Grewal, N. Somasekhar, D. Stinner, and S.A. Miller. 2007. Soil nematode community, organic matter, microbial biomass and nitrogen dynamics in field plots transitioning from conventional to organic management. *Applied Soil Ecology* 37:256–266.

Briones, M.J.I., J. Carreira, and P. Ineson. 1998. *Cognettia sphagnetorum* (Enchytraeidae) and nutrient cycling in organic soils: a microcosm experiment. *Applied Soil Ecology* 9:289–294.

Brust, G.E. 1990. Direct and indirect effects of 4 herbicides on the activity of carabid beetles (Coleoptera, Carabidae). *Pesticide Science* 30:309–320.

Carpenter-Boggs, L., A.C. Kennedy, and J.P. Reganold. 2000. Organic and biodynamic management: effects on soil biology. *Soil Science Society of America Journal* 64:1651–1659.

Cassman, K. 2007. Editorial response by Kenneth Cassman: can organic agriculture feed the world—science to the rescue? *Renewable Agriculture and Food Systems* 22:83–84.

Clark, M.S., W.R. Horwath, C. Shennan, and K.M. Scow. 1998. Changes in soil chemical properties resulting from organic and low-input farming practices. *Agronomy Journal* 90:662–671.

Clarke, K.R., and R.N. Gorley. 2006. *PRIMER v6: User Manual*. PRIMER-E, Plymouth, UK.

Cole, L., R.D. Bardgett, and P. Ineson. 2000. Enchytraeid worms (Oligochaeta) enhance mineralization of carbon in organic upland soils. *European Journal of Soil Science* 51:185–192.

Collins, W.W., and C.O. Qualset. 1999. *Biodiversity in Agroecosystems*. CRC Press, New York.

Crowder, D.W., T.D. Northfield, M.R. Strand, and W.E. Snyder. 2010. Organic agriculture promotes evenness and natural pest control. *Nature* 466:109–123.

D'Angelo, E.M., A.D. Karathanasis, E.J. Sparks, S.A. Ritchey, and S.A. Wehr-McChesney. 2005. Soil carbon and microbial communities at mitigated and late successional bottomland forest wetlands. *Wetlands* 25:162–175.

Dauber, J., T. Purtauf, A. Allspach, J. Frisch, K. Voigtlander, and V. Wolters. 2005. Local vs. landscape controls on diversity: a test using surface-dwelling soil macroinvertebrates of differing mobility. *Global Ecology and Biogeography* 14:213–221.

Didden, W.A.M. 1991. Population ecology and functioning of Enchytraeidae in some arable farming systems. Dissertation, Agricultural University, Wageningen, the Netherlands.

Didden, W.A.M. 1993. Ecology of terrestrial Enchytraeidae. *Pedobiologia* 37:2–29.

Didden, W.A.M., H.-C. Fründ, and U. Graefe. 1997. Enchytraeids. In *Fauna in Soil Ecosystems*, ed. G. Benckiser, 135–172. Dekker, New York.

Diekötter, T., S. Wamser, V. Wolters, and K. Birkhofer. 2010. Landscape and management effects on structure and function of soil arthropod communities in winter wheat. *Agriculture Ecosystems & Environment* 137:108–112.

Dunger, W., and H.J. Fiedler. 1997. *Methoden der Bodenbiologie*. Gustav Fischer Verlag, Jena, Germany.

Ekschmitt, K., M. Weber, and V. Wolters. 1997. Spiders, carabids and staphylinids: the ecological potential of predatory macroarthropods. In *Fauna in Soil Ecosystems*, ed. G. Benckiser, 307–362. Dekker, New York.

Esperschütz, J., A. Gattinger, P. Mader, M. Schloter, and A. Fliessbach. 2007. Response of soil microbial biomass and community structures to conventional and organic farming systems under identical crop rotations. *Fems Microbiology Ecology* 61:26–37.

FAOSTAT. 2011. Food and Agriculture Organization of the United Nations. http://faostat.fao.org/default.aspx (accessed 3 January 2011).

Ferris, H., R.C. Venette, and K.M. Scow. 2004. Soil management to enhance bacterivore and fungivore nematode populations and their nitrogen mineralisation function. *Applied Soil Ecology* 25:19–35.

Fließbach, A., H.R. Oberholzer, L. Gunst, and P. Mäder. 2007. Soil organic matter and biological soil quality indicators after 21 years of organic and conventional farming. *Agriculture Ecosystems & Environment* 118:273–284.

Franklin, E., W.E. Magnusson, and F.J. Luizao. 2005. Relative effects of biotic and abiotic factors on the composition of soil invertebrate communities in an Amazonian savanna. *Applied Soil Ecology* 29:259–273.
Frostegård, A., and E. Bååth. 1996. The use of phospholipid fatty acid analysis to estimate bacterial and fungal biomass in soil. *Biology and Fertility of Soils* 22:59–65.
Frostegård, A., E. Bååth, and A. Tunlid. 1993. Shifts in the structure of soil microbial communities in limed forests as revealed by phospholipid fatty-acid analysis. *Soil Biology & Biochemistry* 25:723–730.
Fuller, R.J., L.R. Norton, R.E. Feber, P.J. Johnson, D.E. Chamberlain, A.C. Joys, F. Mathews, R.C. Stuart, M.C. Townsend, W.J. Manley, M.S. Wolfe, D.W. Macdonald, and L.G. Firbank. 2005. Benefits of organic farming to biodiversity vary among taxa. *Biology Letters* 1:431–434.
Gabriel, D., I. Roschewitz, T. Tscharntke, and C. Thies. 2006. Beta diversity at different spatial scales: plant communities in organic and conventional agriculture. *Ecological Applications* 16:2011–2021.
Glück, E., and S. Ingrisch. 1990. The effect of biodynamic and conventional agriculture management on Erigoninae and Lycosidae spiders. *Journal of Applied Entomology* 110:136–148.
Grayston, S.J., C.D. Campbell, R.D. Bardgett, J.L. Mawdsley, C.D. Clegg, K. Ritz, B.S. Griffiths, J.S. Rodwell, S.J. Edwards, W.J. Davies, D.J. Elston, and P. Millard. 2004. Assessing shifts in microbial community structure across a range of grasslands of differing management intensity using CLPP, PLFA and community DNA techniques. *Applied Soil Ecology* 25:63–84.
Gryndler, M., J. Larsen, H. Hrselova, V. Rezacova, H. Gryndlerova, and J. Kubat. 2006. Organic and mineral fertilization, respectively, increase and decrease the development of external mycelium of arbuscular mycorrhizal fungi in a long-term field experiment. *Mycorrhiza* 16:159–166.
Häni F., E. Boller, and S. Keller. 1998. Natural regulation at the farm level. In *Enhancing Biological Control: Habitat Management to Promote Natural Enemies of Agricultural Pests*, eds. C.H. Pickett and R.L. Bugg, 161–210. University of California Press, Berkeley.
Haubert, D., K. Birkhofer, A. Fliessbach, M. Gehre, S. Scheu, and L. Ruess. 2009. Trophic structure and major trophic links in conventional versus organic farming systems as indicated by carbon stable isotope ratios of fatty acids. *Oikos* 118:1579–1589.
Haughton, A.J., J.R. Bell, A. Wilcox, and N.D. Boatman. 2001. The effect of the herbicide glyphosate on non-target spiders: Part I. Direct effects on *Lepthyphantes tenuis* under laboratory conditions. *Pest Management Science* 57:1033–1036.
Haynes, R.J., and R. Naidu. 1998. Influence of lime, fertilizer and manure applications on soil organic matter content and soil physical conditions: a review. *Nutrient Cycling in Agroecosystems* 51:123–137.
Heckman, J. 2006. A history of organic farming: transitions from Sir Albert Howard's war in the soil to USDA National Organic Program. *Renewable Agriculture and Food Systems* 21:143–150.
Hedlund, K. 2002. Soil microbial community structure in relation to vegetation management on former agricultural land. *Soil Biology & Biochemistry* 34:1299–1307.
Hedlund, K., and A. Augustsson. 1995. Effects of enchytraeid grazing on fungal growth and respiration. *Soil Biology & Biochemistry* 27:905–909.
Hole, D.G., A.J. Perkins, J.D. Wilson, I.H. Alexander, F. Grice, and A.D. Evans. 2005. Does organic farming benefit biodiversity? *Biological Conservation* 122:113–130.
Holland, J.M., and C.J.M. Reynolds. 2003. The impact of soil cultivation on arthropod (Coleoptera and Araneae) emergence on arable land. *Pedobiologia* 47:181–191.
Hu, C., and Y.C. Qi. 2010. Effect of compost and chemical fertilizer on soil nematode community in a Chinese maize field. *European Journal of Soil Biology* 46:230–236.
Ibekwe, A.M., A.C. Kennedy, P.S. Frohne, S.K. Papiernik, C.H. Yang, and D.E. Crowley. 2002. Microbial diversity along a transect of agronomic zones. *FEMS Microbiology Ecology* 39:183–191.
Joergensen, R.G., and C. Emmerling. 2006. Methods for evaluating human impact on soil microorganisms based on their activity, biomass, and diversity in agricultural soils. *Journal of Plant Nutrition and Soil Science* 169:295–309.
Kennedy, A.C. 1999. Microbial diversity in agroecosystem quality. In *Biodiversity in Agroecosystems*, eds. W.W. Collins and C.O. Qualset, 1–18. CRC Press, Boca Raton, FL.
Kong, A.Y.Y., K.M. Scow, A.L. Cordova-Kreylos, W.E. Holmes, and J. Six. 2011. Microbial community composition and carbon cycling within soil microenvironments of conventional, low-input, and organic cropping systems. *Soil Biology & Biochemistry* 43:20–30.

Kromp, B. 1999. Carabid beetles in sustainable agriculture: a review on pest control efficacy, cultivation impacts and enhancement. *Agriculture, Ecosystem & Environment* 74:187–228
Laakso, J., and H. Setälä. 1999. Sensitivity of primary production to changes in the architecture of belowground food webs. *Oikos* 87:57–64.
Lang, A., J. Filser, and J.R. Henschel. 1999. Predation by ground beetles and wolf spiders on herbivorous insects in a maize crop. *Agriculture Ecosystems & Environment* 72:189–199.
Larkin, R.P. 2003. Characterization of soil microbial communities under different potato cropping systems by microbial dynamics, substrate utilization, and fatty acid profiles. *Soil Biology & Biochemistry* 35:1451–1466.
Larkin, R.P., and C.W. Honeycutt. 2006. Effects of different 3-year cropping systems on soil microbial communities and *Rhizoctonia* diseases of potato. *Phytopathology* 96:68–79.
Letourneau, D.K., and S.G. Bothwell. 2008. Comparison of organic and conventional farms: challenging ecologists to make biodiversity functional. *Frontiers in Ecology and the Environment* 6:430–438.
Liang, W.J., Y.L. Lou, Q. Li, S. Zhong, X.K. Zhang, and J.K. Wang. 2009. Nematode faunal response to long-term application of nitrogen fertilizer and organic manure in Northeast China. *Soil Biology & Biochemistry* 41:883–890.
Lindroth, C.H. 1985. *The Carabidae of Fennoscandia and Denmark,* Volume 15, Part 1. Scandinavian Science Press, Vinderup, Denmark.
Lindroth, C.H. 1986. *The Carabidae of Fennoscandia and Denmark,* Volume 15, Part 2. Scandinavian Science Press, Vinderup, Denmark.
Luff, M.L. 1998. *Provisional Atlas of the Ground Beetles (Coleoptera, Carabidae) of Britain.* Biological Records Centre, Huntingdon, UK.
Lundquist, E.J., K.M. Scow, L.E. Jackson, S.L. Uesugi, and C.R. Johnson. 1999. Rapid response of soil microbial communities from conventional, low input, and organic farming systems to a wet/dry cycle. *Soil Biology & Biochemistry* 31:1661–1675.
Mäder, P., S. Edenhofer, T. Boller, A. Wiemken, and U. Niggli. 2000. Arbuscular mycorrhizae in a long-term field trial comparing low-input (organic, biological) and high-input (conventional) farming systems in a crop rotation. *Biology and Fertility of Soils* 31:150–156.
Mäder, P., A. Fließbach, D. Dubois, L. Gunst, P. Fried, and U. Niggli. 2002. Soil fertility and biodiversity in organic farming. *Science* 296:1694–1697.
Mäder, P., D. Hahn, D. Dubois, L. Gunst, T. Alföldi, H. Bergmann, M. Oehme, T. Amadò, H. Schneider, U. Graf, A. Velimirov, A. Fließbach, and U. Niggli. 2007. Wheat quality in organic and conventional farming: results of a 21 year field experiment. *Journal of the Science of Food and Agriculture* 87:1826–1837.
Mandelik, Y., T. Dayan, V. Chikatunov, and V. Kravchenko. 2007. Reliability of a higher-taxon approach to richness, rarity, and composition assessments at the local scale. *Conservation Biology* 21:1506–1515.
Maurer, D. 2000. The dark side of taxonomic sufficiency (TS). *Marine Pollution Bulletin* 40:98–101.
Moeskops, B., Sukristiyonubowo, D. Buchan, S. Sleutel, L. Herawaty, E. Husen, R. Saraswati, D. Setyorini, and S. De Neve. 2010. Soil microbial communities and activities under intensive organic and conventional vegetable farming in West Java, Indonesia. *Applied Soil Ecology* 45:112–120.
Moreby, S.J., N.J. Aebischer, S.E. Southway, and N.W. Sotherton. 1994. A comparison of the flora and arthropod fauna of organically and conventionally grown winter-wheat in Southern England. *Annals of Applied Biology* 125:13–27.
Moser, T., C.A.M. Van Gestel, S.E. Jones, J.E. Koolhaas, J.M.L. Rodrigues, and J. Rombke. 2004. Ring-testing and field-validation of a Terrestrial Model Ecosystem (TME)—an instrument for testing potentially harmful substances: effects of carbendazim on enchytraeids. *Ecotoxicology* 13:89–103.
Navntoft, S., P. Esbjerg, and W. Riedel. 2006. Effects of reduced pesticide dosages on carabids (Coleoptera:Carabidae) in winter wheat. *Agricultural and Forest Entomology* 8:57–62.
Neher, D.A. 1999. Nematode communities in organically and conventionally managed agricultural soils. *Journal of Nematology* 31:142–154.
Neher, D.A., and Barbercheck, M.E. 1999. Diversity and function of soil Mesofauna. In *Biodiversity in Agroecosystems,* eds. W.W. Collins and C.O. Qualset, 27–48. CRC Press, Boca Raton, FL.

O'Connor, F.B. 1955. Extraction of enchytraeid worms from a coniferous forest soil. *Nature* 175: 815–816.

Olsson, P.A. 1999. Signature fatty acids provide tools for determination of the distribution and interactions of mycorrhizal fungi in soil. *Fems Microbiology Ecology* 29:303–310.

Parfitt, R.L., G.W. Yeates, D.J. Ross, A.D. Mackay, and P.J. Budding. 2005. Relationships between soil biota, nitrogen and phosphorus availability, and pasture growth under organic and conventional management. *Applied Soil Ecology* 28:1–13.

Peacock, A.D., M.D. Mullen, D.B. Ringelberg, D.D. Tyler, D.B. Hedrick, P.M. Gale, and D.C. White. 2001. Soil microbial community responses to dairy manure or ammonium nitrate applications. *Soil Biology & Biochemistry* 33:1011–1019.

Pik, A.J., I. Oliver, and A.J. Beattie. 1999. Taxonomic sufficiency in ecological studies of terrestrial invertebrates. *Australian Journal of Ecology* 24:555–562.

Postma-Blaauw, M.B., R.G.M. De Goede, J. Bloem, J.H. Faber, and L. Brussaard. 2010. Soil biota community structure and abundance under agricultural intensification and extensification. *Ecology* 91:460–473.

Power, A.G. 2010. Ecosystem services and agriculture: tradeoffs and synergies. *Philosophical Transactions of the Royal Society B–Biological Sciences* 365:2959–2971.

Purtauf, T., I. Roschewitz, J. Dauber, C. Thies, T. Tscharntke, and V. Wolters. 2005. Landscape context of organic and conventional farms: influences on carabid beetle diversity. *Agriculture Ecosystems & Environment* 108:165–174.

Raupp, J., C. Pekrun, M. Oltmanns, and U. Köpke. 2006. *Long Term Field Experiments in Organic Farming*. ISOFAR Scientific Series No. 1. Verlag Dr. Köstery, Berlin.

Ribera, I., S. Doledec, I.S. Downie, and G.N. Foster. 2001. Effect of land disturbance and stress on species traits of ground beetle assemblages. *Ecology* 82:1112–1129.

Sanchez-Moreno, S., N.L. Nicola, H. Ferris, and F.G. Zalom. 2009. Effects of agricultural management on nematode-mite assemblages: soil food web indices as predictors of mite community composition. *Applied Soil Ecology* 41:107–117.

Schmidt, M.H., I. Roschewitz, C. Thies, and T. Tscharntke. 2005. Differential effects of landscape and management on diversity and density of ground-dwelling farmland spiders. *Journal of Applied Ecology* 42:281–287.

Schnell, M.R., A.J. Pik, and J.M. Dangerfield. 2003. Ant community succession within eucalypt plantations on used pasture and implications for taxonomic sufficiency in biomonitoring. *Austral Ecology* 28:553–565.

Setälä, H. 1995. Growth of birch and pine-seedlings in relation to grazing by soil fauna on ectomycorrhizal fungi. *Ecology* 76:1844–1851.

Shah, P.A., D.R. Brooks, J.E. Ashby, J.N. Perry, and I.P. Woiwod. 2003. Diversity and abundance of the coleopteran fauna from organic and conventional management systems in southern England. *Agricultural and Forest Entomology* 5:51–60.

Standen, V. 1978. Influence of soil fauna on decomposition by microorganisms in blanket bog litter. *Journal of Animal Ecology* 47:25–38.

Swift, M.J., A.M.N. Izac, and M. Van Noordwijk. 2004. Biodiversity and ecosystem services in agricultural landscapes—are we asking the right questions? *Agriculture Ecosystems & Environment* 104:113–134.

Symondson, W.O.C., Sunderland, K.D., and Greenstone, M.H. 2002. Can generalist predators be effective biocontrol agents? *Annual Review of Entomology* 47:561–594.

Taylor, R.L., B.D. Maxwell, and R.J. Boik. 2006. Indirect effects of herbicides on bird food resources and beneficial arthropods. *Agriculture Ecosystems & Environment* 116:157–164.

Terlizzi, A., M.J. Anderson, S. Bevilacqua, S. Fraschetti, M. Wlodarska-Kowalczuk, and K.E. Ellingsen. 2009. Beta diversity and taxonomic sufficiency: do higher-level taxa reflect heterogeneity in species composition? *Diversity and Distributions* 15:450–458.

Terlizzi, A., S. Bevilacqua, S. Fraschetti, and F. Boero. 2003. Taxonomic sufficiency and the increasing insufficiency of taxonomic expertise. *Marine Pollution Bulletin* 46:556–561.

Thomas, C.F.G., and P.C. Jepson. 1997. Field-scale effects of farming practices on linyphiid spider populations in grass and cereals. *Entomologia Experimentalis Et Applicata* 84:59–69.

Thorbek, P., and T. Bilde. 2004. Reduced numbers of generalist arthropod predators after crop management. *Journal of Applied Ecology* 41:526–538.
Thorne, G. 1961. *Principles of Nematology*. McGraw-Hill, New York.
Tilman, D., K.G. Cassman, P.A. Matson, R. Naylor, and S. Polasky. 2002. Agricultural sustainability and intensive production practices. *Nature* 418:671–677.
Uetz, G.W., J. Halaj, and A.B. Cady. 1999. Guild structure of spiders in major crops. *Journal of Arachnology* 27:270–280.
Ullrich, S.D., J.S. Buyer, M.A. Cavigelli, R. Seidel, and J.R. Teasdale. 2011. Weed seed persistence and microbial abundance in long-term organic and conventional cropping systems. *Weed Science* 59:202–209.
Van Diepeningen, A.D., O.J. De Vos, G.W. Korthals, and A.H.C. Van Bruggen. 2006. Effects of organic versus conventional management on chemical and biological parameters in agricultural soils. *Applied Soil Ecology* 31:120–135.
Van Vliet, P.C.J., M.H. Beare, D.C. Coleman, and P.F. Hendrix. 2004. Effects of enchytraeids (Annelida:Oligochaeta) on soil carbon and nitrogen dynamics in laboratory incubations. *Applied Soil Ecology* 25:147–160.
Van Vliet, P.C.J., L.T. West, P.F. Hendrix, and D.C. Coleman. 1993. The influence of Enchytraeidae (Oligochaeta) on the soil porosity of small microcosms. *Geoderma* 56:287–299.
Von Berg, K., C. Thies, T. Tscharntke, and S. Scheu. 2010. Changes in herbivore control in arable fields by detrital subsidies depend on predator species and vary in space. *Oecologia* 163:1033–1042.
Wall, D.H., M.A. Bradford, M.G. St John, J.A. Trofymow, V. Behan-Pelletier, D.D.E. Bignell, J.M. Dangerfield, W.J. Parton, J. Rusek, W. Voigt, V. Wolters, H.Z. Gardel, F.O. Ayuke, R. Bashford, O.I. Beljakova, P.J. Bohlen, A. Brauman, S. Flemming, J.R. Henschel, D.L. Johnson, T.H. Jones, M. Kovarova, J.M. Kranabetter, L. Kutny, K.C. Lin, M. Maryati, D. Masse, A. Pokarzhevskii, H. Rahman, M.G. Sabara, J.A. Salamon, M.J. Swift, A. Varela, H.L. Vasconcelos, D. White, and X.M. Zou. 2008. Global decomposition experiment shows soil animal impacts on decomposition are climate-dependent. *Global Change Biology* 14:2661–2677.
Wang, Y., R.L. Crocker, L.T. Wilson, G. Smart, X. Wei, W.T. Nailon, and P.P. Cobb. 2001. Effect of nematode and fungal treatments on nontarget turfgrass-inhabiting arthropod and nematode populations. *Environmental Entomology* 30:196–203.
Wardle, D.A., W.M. Williamson, G.W. Yeates, and K.I. Bonner. 2005. Trickle-down effects of aboveground trophic cascades on the soil food web. *Oikos* 111:348–358.
Wardle, D.A., G.W. Yeates, R.N. Watson, and K.S. Nicholson. 1995. The detritus food-web and the diversity of soil fauna as indicators of disturbance regimes in agroecosystems. *Plant and Soil* 170:35–43.
Weibull, A.C., O. Ostman, and A. Granqvist. 2003. Species richness in agroecosystems: the effect of landscape, habitat and farm management. *Biodiversity and Conservation* 12:1335–1355.
White, D.C., J.O. Stair, and D.B. Ringelberg. 1996. Quantitative comparisons of in situ microbial biodiversity by signature biomarker analysis. *Journal of Industrial Microbiology* 17:185–196.
Willer, H., and L. Kilcher. 2010. *The World of Organic Agriculture—Statistics and Emerging Trends 2010*. IFOAM, Bonn, Germany.
Willer, H., Yussefi-Menzler, M., and N. Sorensen. 2008. *The World of Organic Agriculture—Statistics and Emerging Trends 2008*. Earthscan, Sterling, VA.
Yeates, G.W., R.D. Bardgett, R. Cook, P.J. Hobbs, P.J. Bowling, and J.F. Potter. 1997. Faunal and microbial diversity in three Welsh grassland soils under conventional and organic management regimes. *Journal of Applied Ecology* 34:453–470.
Yeates, G.W., T. Bongers, R.G.M. De Goede, D.W. Freckman, and S.S. Georgieva. 1993. Feeding-habits in soil nematode families and genera: an outline for soil ecologists. *Journal of Nematology* 25:315–331.
Zehnder, G., G.M. Gurr, S. Kuhne, M.R. Wade, S.D. Wratten, and E. Wyss. 2007. Arthropod pest management in organic crops. *Annual Review of Entomology* 52:57–80.
Zelles, L. 1999. Fatty acid patterns of phospholipids and lipopolysaccharides in the characterisation of microbial communities in soil: a review. *Biology and Fertility of Soils* 29:111–129.
Zhang, W.-J., W.-Y. Rui, C. Tu, H.G. Diab, F.J. Louws, J.P. Mueller, N. Creamer, M. Bell, M.G. Wagger, and S. Hu. 2005. Responses of soil microbial community structure and diversity to agricultural deintensification. *Pedosphere* 15:440–447.

chapter 6

The biological basis for nitrogen management in agroecosystems

A. Stuart Grandy
University of New Hampshire

Cynthia Kallenbach
University of New Hampshire

Terry D. Loecke
University of Nebraska–Lincoln

Sieglinde S. Snapp
Michigan State University

Richard G. Smith
University of New Hampshire

Contents

6.1 Current challenges with nitrogen management

Nitrogen (N) inputs into agricultural systems are estimated to be in the range of 130–170 Tg N y^{-1}, and almost 90% of these inputs originate from inorganic fertilizers and other anthropogenic sources (Smil, 1999). In general, crop recovery of fertilizer N is typically less than 50% and often no more than 30% (Galloway and Cowling, 2002; Cui et al., 2010). The high rates of global N fertilizer inputs, coupled with the inefficient recovery of N by crops, have increased biologically reactive soil N at farm, regional, and global scales (Hoang and Alauddin, 2010), with far-reaching environmental consequences, including declines in water quality and biological diversity and increases in atmospheric trace gas concentrations. Developing integrated agricultural systems that efficiently cycle N is essential to maintaining and increasing agricultural productivity and minimizing declines in environmental quality and ecosystem services. To do this, we need to improve our management of inorganic fertilizer inputs and our understanding of the soil biological processes that regulate N transformations.

6.2 On-farm consequences of nitrogen management

Increases in both biological and inorganic nitrogen use in agricultural systems have substantially improved crop yields but not without unintended environmental trade-offs. Excess N in agricultural soils can lead to increases in soil acidification, accelerated soil organic matter (SOM) turnover, and altered microbial community structure and function. Soil acidification is a potentially negative side effect of nitrification and other N transformations and can be exacerbated by the coupling of base cations to NO_3^- leaching. Soil acidification is a widespread problem in areas that frequently experience high fertilizer application rates, including the U.S. Great Plains, parts of Eastern Europe, and interior China (Tarkalson et al., 2006; Guo et al., 2010).

The soil biota, which promote soil aggregation and regulate nutrient mineralization and trace gas emissions, are also strongly influenced by N fertilizer applications, although the direction of change for particular taxa is often difficult to predict. For example, increases, decreases, and no change in microbial biomass following fertilizer applications have been detected (O'Donnell et al., 2001; Ramirez et al., 2010), and the duration of reported changes varies (Ryan et al., 2009; Singh and Ghoshal, 2010). Similarly, synthetic N application in agricultural systems can increase soil CO_2 emissions and soil C loss (Al-Kaisi et al., 2008; Russell et al., 2009; Kwon and Hudson, 2010), but responses remain difficult to predict (Taylor and Townsend, 2010), and reports showed accumulation (Al-Kaisi et al., 2008; Reay et al., 2008; Poirier et al., 2009), loss (Hofmann et al., 2009; Khan et al., 2007; Mulvaney et al., 2009), or no change in soil C (Halvorson et al., 2002).

6.3 Off-farm environmental impacts of nitrogen management

Agricultural systems are typically managed independently of their surrounding ecosystems, yet N originating in agricultural soils can have far-reaching effects on environmental quality. Excess N that escapes agricultural fields can have a negative impact on drinking water and air quality; increase atmospheric greenhouse gas concentrations; alter coastal, marine, and forest ecosystem productivity and biodiversity; and change SOM decomposition rates. Galloway et al. (2003) described this as the "cascading effects" of N: One N atom can be transformed multiple times into a variety of molecular forms (NH_3, NH_4^+, NO_3^-, NO_x, N_2, N_2O) as it travels through the environment, creating numerous opportunities for environmental interactions.

The predominant pathways of N loss are through leaching, denitrification, and under some circumstances, ammonia volatilization (Bouwman et al., 2009). Many of the major losses of N involve NO_3^- and are thus dependent on nitrification, the microbially mediated process that oxidizes NH_4^+ to NO_3^-. Excessive NO_3^- concentrations in freshwater and coastal marine ecosystems can trigger rapid growth of heterotrophic and phototrophic organisms, accelerating microbial respiration and depleting oxygen through eutrophication. Although eutrophication in freshwater lakes has declined slightly since the 1990s (Conley et al., 2009), it remains a major global threat to coastal and some freshwater ecosystems. For example, the Gulf of Mexico contains one of the world's largest areas affected by eutrophication and has a resulting "dead zone" that can encompass as much as 18,000 km^2 (Turner and Rabalais, 2003). This problem is related to the release of large quantities of U.S. Midwest-derived NO_3^- into the gulf via the Mississippi River basin, which drains the heavily fertilized U.S. corn belt (David et al., 2010). NO_3^--N fluxes from the Republic of Korea, northeastern United States, and Mississippi River Basin are estimated at roughly 1,700, 1,200, and 600 kg N km^{-2} y^{-1}, respectively (Howarth et al., 2008), compared to 100 kg N km^{-2} y^{-1} in ecosystems with fewer anthropogenic N inputs (Howarth et al., 2002).

The denitrification process releases N_2O, NO_x, and N_2 into the atmosphere. N_2 is the unreactive end product of denitrification that completes the global N cycle. NO_x contributes to the formation of tropospheric ozone pollution and the production of acid rain, while N_2O depletes the stratospheric ozone layer and is a potent greenhouse gas with a global warming potential (GWP) nearly 300 times that of carbon dioxide. Atmospheric N_2O concentrations have increased from 270 ppb prior to industrialization to 320 ppb today and continue to rise at an estimated rate of 0.3–0.6% y^{-1}. Nearly 70% of current N_2O emissions are linked to fertilizer use (Smil, 1999; Halvorson et al., 2008), and fourfold increases in emissions due to expanded synthetic fertilizer use are projected to occur by 2100 (Intergovernmental Panel on Climate Change [IPCC], 2001). Currently, annual N_2O emissions from U.S. croplands are estimated to be upward of 500 Gg (U.S. Environmental Protection Agency [EPA], 2009). Nitrogen losses to the atmosphere can also occur through volatilization of NH_3. Volatilization rates are dependent on NH_4^+ and NH_3 concentrations in the soil, which are a function of pH and can be high where fertilizers are broadcast on the surface and soil pH is greater than 8. NH_3 emissions from animal wastes and fertilizers increased from 6.6 Tg N y^{-1} in 1860 to 32 Tg N y^{-1} in 1993 and are projected to rise exponentially to 77 Tg N y^{-1} by 2050 (Beusen et al., 2008; Bouwman et al., 2009).

Agricultural N that enters the atmosphere and freshwater and marine environments can also have strong indirect effects on ecosystem structure and function. Chronic N deposition is pervasive in many forest ecosystems and has led to excess N in these previously N-limited ecosystems. Northeastern U.S. forests and parts of Eastern Europe have experienced some of the highest levels of N deposition and in some cases have begun to show declines in forest productivity (Aber et al., 1998; Lamersdorf and Borken, 2004) and plant diversity (Bobbink et al., 2010). Several studies have shown that N deposition increases NO_3^- leaching and denitrification (Aber et al., 1998; Fang et al., 2011) and can have varying effects on decomposition dynamics (Berg and Matzner, 1997; Waldrop et al., 2004; Grandy et al., 2008).

Nitrogen fertilizer use has greatly increased agricultural productivity and played a major part in sustaining current population levels. Synthetic N fertilizers will become even more important in the decades ahead as the human population approaches 9 billion people. Sustaining agricultural productivity while maintaining or even enhancing other ecosystem services depends on improving nitrogen use efficiency (NUE). NUE is a function of complex interactions between plant-, soil-, and management-related factors, but at the core of NUE is the concept of synchrony. *Synchrony* refers to whether N availability coincides with plant N demand: Most agricultural systems are highly susceptible to environmental N losses because N availability in space and time is not closely aligned with plant N needs. This is due not only to fertilizer mismanagement and variable environmental conditions but also to our lack of understanding of the biological processes that regulate N availability. Many studies have focused on ways to improve fertilizer management, but we also need to understand better the biological processes that account for much of the N that is used by crops. In the sections that follow, we explore synchrony and how it is influenced by spatial and temporal biological processes as well as different forms of N. The interactions between these three factors—spatial and temporal N dynamics and the form or chemical complexity of soil N—determine when and where N is available and the degree to which plants will be able to access that N when it is most needed.

6.4 Nitrogen use efficiency

Nitrogen use efficiency broadly refers to the amount of N taken up by a crop relative to the amount applied or available in the soil. While in the broadest sense there is widespread agreement with this definition, there are disagreements over how NUE should be calculated (Dawson et al., 2008). Ultimately, the correct approach depends on context, what biomass components are targeted, and how a unit of N is defined. For example, Moll et al. (1982) defined NUE as the ratio of grain yield to plant-available N in the soil, but many studies have used applied fertilizer N uptake. Nitrogen fertilizer inputs are much easier to quantify than total available soil N during the growing season. An additional advantage of this approach is that N fertilizer uptake efficiency can be used to compare plant traits and genetics that influence NUE, as well as plant responses to different rates and types of fertilizer. The disadvantage of an approach that focuses exclusively on grain yield per unit of N fertilizer input is that it largely ignores the substantial internal provision of N that occurs in many soils and thus provides little information about N retention and conservation in cropping systems. A better understanding of long-term N mass balance, the fundamental controls over N mineralization-immobilization dynamics, and the regulators of plant-soil-microbe interactions is needed to develop more resource-efficient systems that conserve N and minimize environmental losses (Dawson et al., 2008). Further, measurements of grain N alone—while useful for providing insights into yield components—do not account for N accumulation in other biomass components. Total plant N uptake needs to be figured into estimates of N balance and retention.

Although the concepts underlying NUE are relatively straightforward, the plant and soil components related to NUE can be complex and vary across ecosystems. NUE is controlled by multivariate interactions between plant traits, genetics, and environmental factors. From the plant perspective, there is a strong physiological and genetic component to NUE (Dawson et al., 2008) that includes plant N utilization efficiency, N carryover between seasons in perennials, and N uptake, retention, and conversion efficiency (Cassman et al., 2002). Additional plant traits such as root architecture, responses to nutrient limitation, and temporal patterns of demand also strongly influence NUE and point to the need for including plant breeding in integrated efforts to increase NUE. Environmental conditions and soil processes also play a key role in NUE. For example, the yield and N demand of a crop are closely related to solar radiation, temperature, and moisture and their interaction with other environment and management-related factors, including pest populations and weed competition (Cassman et al., 2002).

6.5 Synchrony

At the core of NUE is the concept of synchrony. Synchrony refers to whether soil N availability coincides with plant N demand (Campbell et al., 1992); most agricultural systems are highly susceptible to environmental N losses because N availability in space and time

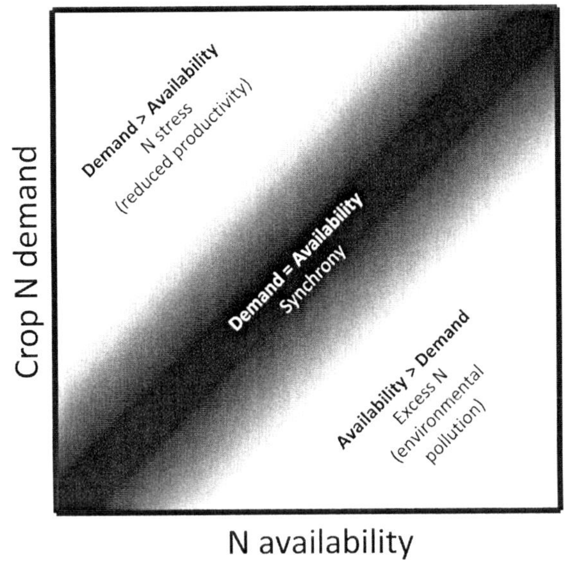

Figure 6.1 Conceptual model of nitrogen use efficiency. Inequalities in either crop demand or nutrient availability lead to reduced resource use efficiency, with concomitant negative impacts on crop performance or environmental quality, respectively.

is not closely aligned with plant N needs (Figure 6.1). Synchrony is influenced by plant traits, environmental factors, and their interactions. Crews and Peoples (2005) defined two types of asynchrony that can occur in cropping systems. One type of asynchrony is characterized by nutrient availability being insufficient to meet plant demand. Under these conditions, environmental losses are often low, but yields are limited by N availability. Most temperate, terrestrial ecosystems are N limited, and plant productivity is typically increased by increasing the supply of available N. However, the widespread use of N fertilizers has led to another type of asynchrony in which N availability outpaces plant demand. In these conditions, yield may be maximized but at the consequence of environmental losses through leaching and denitrification.

For decades, scientists have experimented with ways to improve synchrony to enhance N fertilizer use efficiency (Cassman et al., 1992; Crews and Peoples, 2005; Ross et al., 2008). One line of research has focused on the development of fertilizers or chemical inhibitors that slow the release of N, but their deployment has been limited because of cost and variable yield responses. Another line of research has been focused on spatially and temporally explicit fertilization practices. Fertilizer banding and split applications have both shown promise in some studies, but their effects on yields and environmental N losses are often inconsistent because of the overriding importance of interactions between soil type, temperature, and moisture in regulating N availability (Fan and Li, 2010). Thus, despite years of intensive research, N fertilizer use efficiency in modern, industrialized cropping systems continues often to remain below 50% and in developing countries may be significantly less (Cui et al., 2010).

While the timing of N fertilizer uptake is a key component of synchrony, we also have to consider the additional challenge that soil-derived N presents to improving synchrony. Nitrogen mineralization resulting from SOM turnover can generate large amounts of inorganic N that is also highly susceptible to environmental losses. For example, in productive Midwestern prairie soils, N reserves in the top 20 cm of soil are in the range of 4,000 kg N ha^{-1}, and the annual internal supply of N through mineralization may be more than 200 kg N ha^{-1} (Cassman et al., 2002).

This soil-derived N can provide a significant amount of N to crops, but we have not yet figured out how to synchronize soil N mineralization and plant N demand because of the diversity and complexity of processes controlling both. For example, the temporal drivers of decomposition dynamics may be related to but not directly aligned with the drivers of annual crop productivity. Similarly, the microscale processes that influence the spatial distribution of N in heterogeneous soils are difficult to manage and are strongly influenced by factors other than plant growth. Further, limitations to studying soil microbial community structure, size, and function have hampered our ability to understand N mineralization-immobilization dynamics, and many studies in agricultural systems have focused on the fate of fertilizers, but far fewer have focused on those fundamental plant-soil-microbe interactions that influence synchrony (Fierer et al., 2009). Given these challenges, improving N synchrony in cropping systems is a difficult task; however, it is arguably the most important objective of agricultural research, if not all of soil ecology, to better understand, and ultimately manage, N synchrony in agricultural soils. Recent advances in understanding the spatial and temporal processes that regulate internal N cycling and availability, including soil biological processes and their interactions with physical and chemical soil properties, are opening up new ways to think about and study synchrony. With this information, we need to construct a better understanding of soil biological communities and associated ecosystem processes that control N mineralization-immobilization dynamics, including soil aggregation, sorption, and other physical processes. In the following sections, we address N mineralization and immobilization dynamics with particular attention given to the spatial and temporal processes that regulate N availability in agroecosystems.

6.6 *Nitrogen mineralization-immobilization*

Even when N fertilizer applications are high, a significant proportion of the N taken up by crops is derived from N mineralization. Nitrogen mineralization is the conversion of complex organic forms of N to more simple molecules that plants can utilize, including NO_3^-, NH_4^+, and amino acids (Robertson and Groffman, 2007). As such, the process of N mineralization is inextricably tied to decomposition and the activities of all the soil organisms involved in that process. In the past, decomposition was often attributed primarily to bacteria, but we now know that other organisms, including archaea and fungi, as well as the higher soil organisms such as microarthropods, nematodes, and earthworms, are also important (Ayres et al., 2009). Indeed, the interactions between trophic levels, rather than the activity of a single group or organism per se, have a pivotal role in regulating decomposition processes and nutrient turnover. For example, grazing by microarthropods can increase microbial diversity and activity, while earthworm activities can dramatically change the spatial and temporal distribution of microbial activity, decomposition, and nutrient mineralization (Cragg and Bardgett, 2001; Hättenschwiler and Gasser, 2005; Milcu et al., 2008).

The grazing of microbial communities by microarthropods and other taxa releases N in a process that has been described as a "microbial loop" (Clarholm, 1985). In this process, microarthropods consume organic matter primarily to access the microorganisms colonizing it. When protozoans and microarthropods graze on microbes, they ingest more N than they need for cellular maintenance and growth and release this excess N into the environment. This can be understood through simple stoichiometric relationships between organismal C and N. Microbes have a C:N ratio of about 9:1 (Cleveland and Liptzin, 2007), which is lower than the C:N ratio of many microarthropods and protozoans. To maintain their C:N ratios, grazers need to release some of the excess N that results from ingesting microbes (Osler and Sommerkorn, 2007). Wickings and Grandy (2011) carried out an experiment showing the strong effects of microarthropods on decomposition and nutrient cycling. They compared litter decomposition dynamics in the lab with and without the oribatid mite *Scheloribates moestus*. Litter decomposing with this microarthropod had 19% greater respiration rates, almost fivefold greater NH_4^+-N concentrations, and substantially more NO_3^--N (7 vs. 1 mg g^{-1} dry litter). Further, it was found that mites altered the chemistry of litter during decomposition by changing the relative abundance of litter polysaccharides and by producing frass with unique structural chemistry relative to the original plant litter. Although the broader significance of these changes in chemistry could not be derived from a short-term study, feedbacks from changes in litter chemistry may influence decomposer communities and N mineralization.

The microbial loop is an example of N availability exceeding decomposer demand. Often, however, microorganisms need additional N to decompose organic matter with a C:N ratio that is greater than 20–25:1. When the N content of organic matter does not meet the requirements dictated by cellular stoichiometry, organisms need to acquire N from the surrounding environment. This process is termed N immobilization and results in declines in soil inorganic N as decomposer communities utilize it to break down organic matter. Given that N mineralization-immobilization is key to soil N availability, we need to understand better the factors tipping the balance toward one or the other process. Again, we can look to the microbial requirements for C and N for a general answer. Microbial growth efficiency (the amount of substrate converted to biomass relative to the total amount of the resource consumed) varies depending on substrate quality, microbial community structure, and other factors, but 45% is a typical approximation (Six et al., 2006; Thiet et al., 2006). If we assume that the average microbial C:N ratio is about 9:1 (Cleveland and Liptzin, 2007; Kallenbach and Grandy, 2011), then a substrate C:N ratio of about 20:1 represents the threshold for N mineralization and immobilization. Above this threshold, microbes will utilize soil inorganic N to decompose organic matter, resulting in immobilization. Below this threshold, there is adequate N in the litter to support decomposition, and any excess N will be released to the soil environment (i.e., mineralization). Indeed, C:N ratios provide reasonable broad-scale predictions of N mineralization-immobilization dynamics, but the specific chemical structure of organic matter (i.e., the concentrations of lignin, lipids, and polysaccharides) can be just as important as C:N ratios (Melillo et al., 1982; Berg, 2000; Grandy and Neff, 2008).

Nitrogen mineralization is thus strongly dependent on substrate C:N ratios and organic matter chemistry, but like other biological processes, it is also sensitive to soil moisture and temperature (Cook et al., 2010). Further, microbial community structure can account for as much as 15–20% of the variation in decomposition rates, and changes in trophic-level diversity can strongly influence decomposition and N mineralization (Ayres et al., 2006; Strickland et al., 2009). Given the range of controls over N mineralization and the different temporal and spatial scales at which they operate, it should come as no surprise that predicting N release from organic sources is difficult. It is even harder to try to synchronize N mineralization with plant N needs over space and time.

6.7 Temporal synchrony

In annual cropping systems, soil nitrogen uptake primarily occurs during a narrow window of peak crop growth that lasts only weeks (Figure 6.2). The period of maximum N demand is thus short and intense, but seasonal N availability in temperate cropping systems can fluctuate dramatically depending on management and environmental conditions. Herein lies the problem: The rates of internal inorganic N production, as well as inputs of fertilizer N, are often highest when plant demand is low. Both exogenous inputs and N mineralization-immobilization dynamics play key roles in soil N dynamics, but both may be strongly influenced by factors unrelated to those driving plant N demand. The problem of temporal synchrony is thus a complex one that involves the supply of inorganic and organic nutrient sources and their relationships to environmental factors and plant and soil management (Robertson and Groffman, 2007).

Depending on the quality of available organic matter, N mineralization rates are often highest in the spring before, or in the summer after, crop N demand crests. Legume residues and other organic matter with low C:N ratios and low lignin concentrations decompose quickly in warm soils. The release of N from these materials typically occurs before plants are able to use it. Organic materials with high C:N ratios and high lignin concentrations break down more slowly and often immobilize inorganic N before there is a net mineralization of N. The decomposition of these residues can deplete soil N when it is most needed by crops only to release it once crop N demand wanes. Asynchrony can also be a problem in the fall, particularly in temperate cropping systems. The incorporation of legume cover crops into warm soils followed by a winter fallow can increase soil N concentrations when there is no crop N demand (Crews and Peoples, 2005). A second wave of N mineralization is common in the fall when crop residues have accumulated in the soil and temperatures still promote microbial activity. Further, residual N fertilizer that has not been utilized by plants or microbes may be lost in the fall. These losses may be exacerbated by fall tillage, which remains a common practice in many regions where weather or soil type make tillage in the spring difficult.

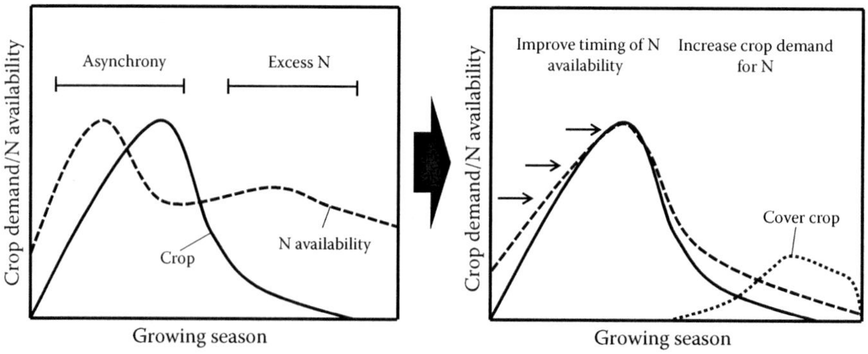

Figure 6.2 Temporal asynchrony between N availability and plant demand in temperate cropping systems. Nitrogen availability often exceeds crop demand early in the growing season and following crop senescence or harvest. Management practices that affect the timing of N release from fertilizer applications or N uptake by the crop, and that maximize N uptake during periods when cash crop demand is low, will improve the synchrony of N and reduce N losses to the environment.

Inorganic fertilizers are also managed in ways that do not optimize the recovery of N. Spring applications before or at planting are highly susceptible to environmental losses through leaching, and maximum N_2O emissions typically occur shortly after fertilizer applications (McSwiney and Robertson, 2005). The use of two or more fertilizer applications, fertigation, and slow-release fertilizers have all shown potential to improve synchrony in some systems but have not been widely adopted because of concerns about cost and yield responses. Thus, whether N is derived from organic sources via mineralization or from inorganic fertilizers, matching N availability with plant needs remains a major challenge. Further, the problem of temporal synchrony is often intertwined with that of spatial heterogeneity.

6.8 *Spatial dynamics*

There is increasing interest in the role of soil spatial heterogeneity in controlling crop nutrient use efficiency, productivity, and off-site ecosystem degradation (Ettema and Wardle, 2002; Patzold et al., 2008; Loecke and Robertson, 2009b). Spatial heterogeneity of labile organic amendments such as crop residues, cover crops, manure, and compost can influence temporal N dynamics in agroecosystems by several mechanisms and at various spatial and temporal scales (Figure 6.3). At the spatial scales at which organic amendments tend to be aggregated or clumped (1 mm to 1 m), the degree of this aggregation (i.e., patch size) influences the prevalence and rates of microbial N transformations (see Loecke's Chapter 3 in this volume for further discussion). Overlapping spatial scales of organic amendment patch sizes and plant root zones allows for interactions between spatially and temporally heterogeneous microbial N transformations and plant root foraging behavior. The potential role of these interactions in modifying soil inorganic N supply and crop N uptake synchrony are severalfold.

Competition between soil microorganisms and plants for soil inorganic N has been a topic of considerable debate (Kaye and Hart, 1997; Hodge et al., 2000; Schimel and Bennett, 2004) but is pivotal to understanding N synchrony (Korsaeth et al., 2002). Theory and a small body of laboratory-based empirical work suggest that the interactions of spatially heterogeneous soil resources, especially ephemeral patches of labile organic matter, and spatially selective root proliferation (i.e., root foraging) may alter plant-soil microbe competition for soil N (Hodge, 2006; Loecke and Robertson, 2009a). Wang and Bakken (1997) and Korsaeth et al. (2001) tested this idea by varying the distance separating microsites with N-rich clover leaves and N-poor oat straw to determine whether N derived from the N-rich sites was immobilized in the microbial biomass of the N-poor sites, the timing of this immobilization, and the quantity of soil N taken up by barley plants (*Hordeum vulgare*). As the distance between N-rich and N-poor microsites increased from 0 to 9 mm, the plant intercepted a greater quantity of N derived from the N-rich microsites; at the same time, microbial biomass N in the N-poor microsite decreased. When the N-rich and N-poor sites were in close proximity to each other (0 to 3 mm), plant N uptake was delayed. These studies were carried out under controlled environments but may nonetheless reflect plant-microbe interactions in agroecosystems.

We have taken the available data and developed a conceptual model highlighting the relationships between patch size and distribution (referred to here as grain size) and plant-microbe competition for resources. Our model uses a continuous grain size distribution representing differences in patch size and distribution and breaks this continuum into three zones. In zone A, corresponding to Figure 6.3a, the N-rich and N-poor microsites are too small and tightly packed together for the roots to proliferate selectively into N-rich

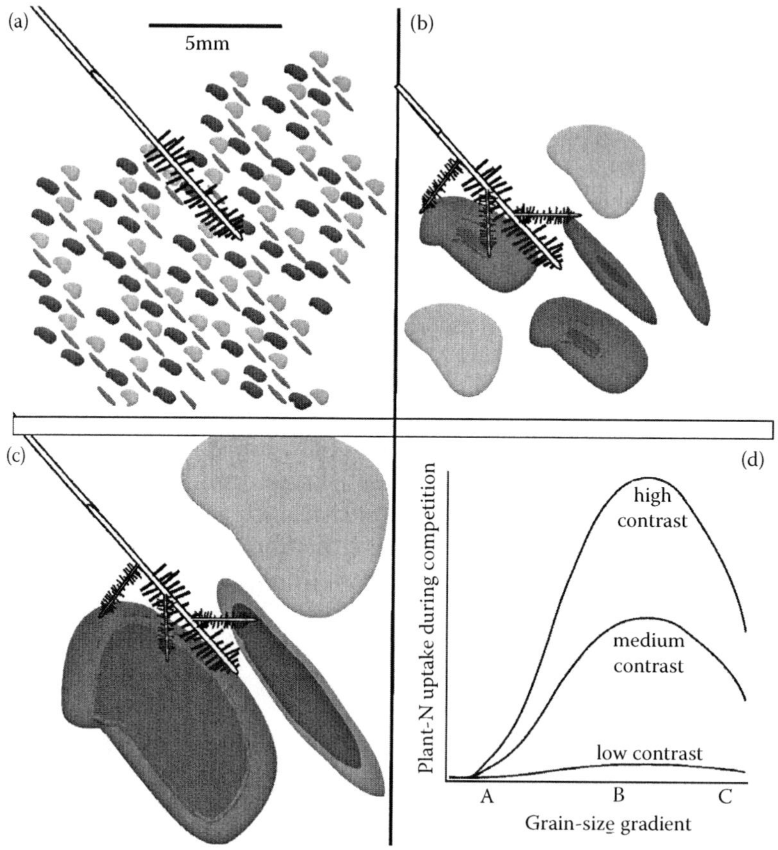

Figure 6.3 A conceptual diagram of how grain size distribution of N-poor (light gray) and N-rich (darker gray) microsites influences competition between plant roots and microorganisms in N-poor microsites for inorganic N derived from N-rich microsites. Panels (a), (b), and (c) illustrate a relative contrast in the size of N-rich and N-poor microsites compared to a maize root. Panel (d) illustrates hypothesized relationships between grain size distribution and the outcome of plant-microbe competition for N with zones corresponding to panels (a)–(c). The dark microsite interiors in panel (c) indicate the potential for development of anaerobic centers in larger microsites. See text for further details.

microsites; thus, the plants are ineffective at intercepting inorganic N flowing (by diffusion and hydrologic transport) from N-rich to N-poor microsites. In zone B, the microsites are separated at a sufficient scale for the plant roots to respond to the inorganic N gradients and selectively proliferate into and around the N-rich microsites, thus increasing the probability that inorganic N will flow toward the root surface and not be intercepted by the microbes in the N-poor microsites. In contrast, in zone C the plants are able to proliferate into the N-rich microsites, but the size of the microsite promotes anaerobic processes in the microsite centers and the conversion of NO_3^- to N gas (NOx, N_2O, or N_2) via denitrification. Alternatively, when moisture availability is limited, root proliferation into the large N-rich microsites may limit plant water acquisition and microbial activity preferentially in the large N-rich microsites. The influence of microsite grain size distribution on plant-microbe competition for inorganic N will interact with the relative quality differences between the microsites. As illustrated with the three separate lines (low, medium, and high) in Figure 6.3d, the contrast between microsite quality (or N richness) is predicted to interact with microsite grain size such that the larger the contrast in quality, the larger the potential for plant roots to compete with microbes in the N-poor microsites is. We see in Figure 6.3 that the contrast in microsite quality can alter root proliferation, a critical component of this conceptual hypothesis.

An example of a cropping system with both N-rich and N-poor residues is one with small-grain cereals grown in conjunction with legume cover crops, which is a common component of biologically managed crop rotations in the U.S. corn belt (e.g., Liebman et al., 2008). The small-grain crops grow and senesce while the legume remains under the grass canopy. Following cereal harvest, a mixture of N-poor straw and N-rich legume shoots is incorporated into the soil prior to planting the subsequent crop. As of yet, the degree to which the spatial heterogeneity of this mixture of straw and legume residue influences the temporal dynamics of soil inorganic N availability during the growing season remains unknown. What is known is that many crops are capable of displaying optimal root foraging behavior (Drew, 1975). Crop roots have the potential to proliferate in N-rich microsites and avoid N-poor microsites when plant growth is N limited (Figure 6.4). Furthermore, roots of maize and other crops are capable of foraging across neighboring plants' root zones to acquire inorganic N from N-rich patches (Hodgen et al., 2009). However, we do not have a good handle on the occurrence of root proliferation into N-rich microsites in the field, and its importance in terms of plant N acquisition has not been well established. The few field data reported on the location of crop plant roots relative to organic matter-rich microsites indicate a strong spatial association (Van Noordwijk et al., 1993), but the available data do not lend themselves to broad predictions about root foraging. We need to understand better the factors that regulate root foraging given that it may be an important mechanism crops use to acquire N.

6.10 *Managing synchrony*

It can be argued that managing nitrogen mineralization-immobilization dynamics to improve synchrony is one of the great challenges in agroecosystem science. However, given the complex array of processes on which mineralization-immobilization dynamics depend, is it realistic to think that we will ever be able to predict the release of N from organic sources? Soil biological communities, temperature and moisture, soil texture, organic matter quality, and crop management all vie for their place in the hierarchy of controls over nitrogen mineralization. Predicting the effects of these factors in isolation is difficult; it seems almost Sisyphean to predict their interactive effects on N mineralization-immobilization dynamics and synchrony.

However, despite the many challenges, there have been exciting new advances in understanding decomposition and N dynamics that may ultimately improve agroecosystem management. For example, there has been a major push to understand better the priming phenomenon in soils. For decades, the priming phenomenon—the stimulation of microbial activity and soil organic C or N turnover via the addition of relatively small resource inputs—was considered an experimental artifact. However, it is now well established that this is a real process caused by resource stimulation of microbial communities (Kuzyakov et al., 2000; Fontaine et al., 2004; Garcia-Pausas and Paterson, 2011), and that it may influence C and N cycling in agricultural soils (Neill, 2011). We are still in the earliest stages of understanding priming, but future advances in crop breeding and genetics may lead to the development of new crop varieties that are capable of stimulating

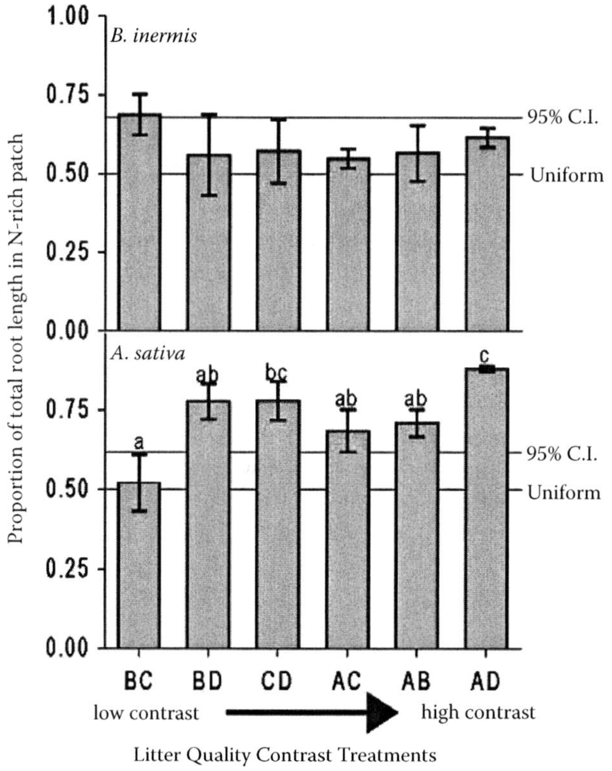

Figure 6.4 Proportion of the total root length of *Bromus inermis* (top) and *Avena sativa* (bottom) located in patches of higher quality (i.e., lower C:N) litter when roots had free access to two patch types that varied in quality. Capital letter treatment codes below each bar indicate the types of litter patches (red clover shoots, red clover roots, green oat leaves, or senesced oat straw) available to roots in each treatment, where BC equals red clover roots versus green oat leaves, BD equals red clover roots versus oat straw, CD equals green oat leaves versus oat straw, AC equals red clover shoots versus green oat leaves, AB equals red clover shoots versus red clover roots, and AD equals red clover shoots versus oat straw. The uniform line at 0.5 represents an even distribution of roots into each patch type. The 95% C.I. (confidence interval) line represents the upper confidence interval around the uniform distribution line. Treatment means falling on or below this 95% C.I. are statistically indistinguishable from the uniform root length distribution (i.e., no selective foraging). Different letters above the vertical bars indicate significant differences (Tukey-Kramer adjustment for multiple comparisons). Note that *A. sativa* plant biomass was on average about 10-fold that of *B. inermis* at harvest (55 days after planting). This likely indicates that plant growth rate (i.e., N demand) relative to patch quality (i.e., N supply rate) controls root selective foraging.

N mineralization by releasing labile root-derived compounds or specific signaling molecules into the rhizosphere (Drinkwater and Snapp, 2007). The management of crop residue diversity and quality, as well as the spatial distribution of crop residue inputs, may also provide opportunities for using priming to improve NUE. To get there, advances are needed in crop variety development and in our understanding of litter decomposition, root foraging behavior, and SOM dynamics, particularly interactions between SOM, mineral surfaces, and the formation and turnover of soil aggregates (Grandy and Neff, 2008; Behrens et al., 2008; Schmidt et al., 2011). With these advances, we will build a foundation of knowledge that will serve as a compass in our efforts to manage the priming phenomenon to improve the synchrony between N mineralization and plant N uptake.

Developing integrated soil-crop management systems that diversify the forms and sources of soil N may be another way to improve N synchrony (Chen et al., 2011). Mineral fertilizer N and urea, which is rapidly converted to mineral forms in most cropping systems, are highly susceptible to environmental losses. These fertilizers are often applied at times when crop N demand is low, and they are much more mobile and reactive than organic N. In contrast, N that is derived from organic fertilizers or from the turnover of native SOM generally becomes available more slowly than N derived from mineral fertilizers. These fertilizers may limit environmental N losses by reducing the pool size of soil inorganic N at times when plants do not need it and also provide opportunities for plants to prime the release of N in the rhizosphere when they most need it. However, N mineralization, like other biological processes, is also strongly dependent on soil moisture and temperature, microbial communities, and trophic-level interactions (Robertson and Groffman, 2007; Wickings and Grandy, 2011). Given the range of controls over N mineralization and the different temporal and spatial scales at which they operate, predicting N release from organic sources—and synchronizing N mineralization with plant N needs over space and time—is difficult.

An alternative approach to relying on either mineral or organic forms of N is to use both and to do so with specific knowledge of plant N requirements and soil N availability determined through soil tests. Recent reports have shown that mixing organic and inorganic forms of N can have positive interactive effects on yields and NUE (Vanlauwe et al., 2001; Chivenge et al., 2011). Further, organic forms of N can be diversified to include legumes, animal manures, or a range of different crop residues. There are a number of reasons why a mixed fertility strategy may increase yields and N uptake. First, it may be a way to supply adequate N throughout the season without ever having to saturate the soil with inorganic N. Inorganic N fertilizers are used to meet crop N demands early in the season and during peak crop growth; organic fertilizers provide a longer-term, slower release throughout the season and may immobilize excess inorganic N. Second, the use of organic fertilizers increases SOM N pools, which increases the potential long-term supply of N. Third, a diverse pool of organic and inorganic forms of N may provide opportunities for crops to partition N, thereby reducing competition for N among crops and weeds or intercrops (Smith et al., 2010). Finally, micronutrients may be made available through the use of organic fertilizers (Zingore et al., 2008; Chivenge et al., 2011).

One of the most effective practices to diversifying and building soil N pools is to integrate legumes into a rotation sequence (Carlsson and Huss-Danell, 2003; Snapp et al., 2005; Herridge et al., 2008; Mazzoncini et al., 2011). However, the net biological nitrogen fixation benefits from a given species will depend on the extent of growth and N fixation, as well as on the extent of N removal in harvested product. Grain harvest in a crop with a high harvest index such as soybean will remove over half of the aboveground N, which in some cases may be more than the N that was fixed. A review of elite soybean lines showed that 70% of the genotypes removed as much N as they fixed (Sanginga, 2003). Although there are clear economic reasons why farmers choose legumes with a high harvest index, there is a trade-off in that little N remains in the soil system after harvest.

Crop diversification can address multiple objectives related to improving soil quality and increasing the diversity of N inputs. For example, Smith et al. (2008) examined the effects of rotational diversity on soil inorganic N pools. Cropping system diversity ranged from one crop species grown over a 3-y period up to six species over the same period. The most diverse treatments included legume/small-grain cover crops and increased corn yields by more than 100% relative to the corn monoculture. Corn yields in the highly diverse system were not different from county averages during the experimental period, despite the fact that no chemical fertilizer inputs were used, and were closely correlated with soil inorganic N. Experimentation is under way to investigate intercrop mixtures of species with different growth habits, but it is challenging to optimize growth habit combinations for complimentary resource use as one species or the other will often dominate in a given year or environmental condition. Ecological insight into the factors that mediate species coexistence may provide clues about which suites of species may make the best intercrop partners or suggest strategies for minimizing competition for nutrient resources between crop species (Booth and Swanton, 2002; Smith et al., 2010). Although there are successful examples of diverse mixtures, both in forages and in grain production systems (Snapp et al., 2010), these examples are the exception, and minimizing competition between crops may require temporal separation through different planting times, spatial separation through markedly different growth habits (Sogbedji et al., 2006), or fertility management practices aimed at promoting resource partitioning among cooccurring crop species (Smith et al., 2010).

Crop diversification can thus increase N fixation as well as the complexity of residue inputs. Another advantage of increasing crop diversity is that it can increase the intensity and temporal pattern of plant N demand. Management practices that increase the magnitude or duration of N uptake will reduce the accumulation of plant-available N in the soil. For example, in northern temperate region annual cropping systems, crop N demand is low following planting in the spring and again in the fall prior to harvest (Robertson and Vitousek, 2009). During these periods, cropping systems are vulnerable to environmental losses, but these losses can be minimized by extending the duration of living plant cover in the field. This can be achieved by growing cover or catch crops to take up excess N during periods when cash crops are not actively utilizing soil N (Dinnes et al., 2002; Constantin et al., 2011) as well as growing intercrops that maximize N uptake during the growing season and beyond (Whitmore and Schroder, 2007). In addition, management practices that improve crop growth and the capacity of the crop to take up N, such as suppression of pests, will reduce the potential for N to accumulate in the soil (Crews and Peoples, 2005).

6.11 Conclusions

With global population expected to reach 9 billion people in the decades ahead, agricultural scientists and ecologists are looking for new ways to increase crop productivity that minimize environmental impacts. Already we have indications that the continued improvement of crop genetics coupled with strategic fertilizer use can increase NUE and reduce environmental N losses (Bundy et al., 2011). We have also seen a widespread push among land managers, scientists, and policy makers to understand better the role of cropping system diversity and cover crop use in enhancing yields and tightening nutrient cycles. At the same time, scientists continue to engineer improved slow-release fertilizers (Ni et al., 2011); advance our understanding of the spatial heterogeneity of nutrient availability and yields and their regulating mechanisms at scales ranging from plants to fields (Senthilkumar et al., 2009); and gain new insights into the relationships between nitrogen fertilizer use, trace gas emissions, and SOM dynamics under variable climatic and management conditions (Robertson et al., 2000; Grandy et al., 2006). Thus, we have good reason to be optimistic that we can, in fact, develop high-yielding cropping systems that provide adequate food for the growing human population without mortgaging our ecosystems and the key services that they provide.

To get there, however, we are going to have to understand better the biological and ecological processes regulating N cycling in agricultural systems; we also need to advance our understanding of how soil communities and processes respond to differences in edaphic soil properties, climatic conditions, and management. With the widespread application of molecular methods to soils, we are beginning to fully characterize and quantify the diversity of soil microbes. We are also beginning to gain insights into the factors driving the distribution of microorganisms across soils, link patterns in soil diversity to soil functions through targeting functional genes and metabolic pathways, and understand the broader-scale ecological relevance of soil biological processes occurring at very small scales. These are promising advances, and as our knowledge of soil biogeochemical processes moves forward, we need to look for every opportunity to apply emerging scientific concepts to managing N in agroecosystems.

References

Aber, J., W. McDowell, K. Nadelhoffer, A. Magill, G. Berntson, M. Kamakea, S. McNulty, W. Currie, L. Rustad, and I. Fernandez. 1998. Nitrogen saturation in temperate forest ecosystems—hypotheses revisited. *Bioscience* 48:921–934.

Al-Kaisi, M. M., M. L. Kruse, and J. E. Sawyer. 2008. Effect of nitrogen fertilizer application on growing season soil carbon dioxide emission in a corn-soybean rotation. *Journal of Environmental Quality* 37:325–332.

Ayres, E., K. M. Dromph, and R. D. Bardgett. 2006. Do plant species encourage soil biota that specialise in the rapid decomposition of their litter? *Soil Biology and Biochemistry* 38:183–186.

Ayres, E., H. Steltzer, B. L. Simmons, R. T. Simpson, J. M. Steinweg, M. D. Wallenstein, N. Mellor, W. J. Parton, J. C. Moore, and D. H. Wall. 2009. Home-field advantage accelerates leaf litter decomposition in forests. *Soil Biology and Biochemistry* 41:606–610.

Behrens, S., T. Losekann, J. Pett-Ridge, P. K. Weber, W. O. Ng, B. S. Stevenson, I. D. Hutcheon, D. A. Relman, and A. M. Spormann. 2008. Linking microbial phylogeny to metabolic activity at the single-cell level by using enhanced element labeling-catalyzed reporter deposition fluorescence in situ hybridization (EL-FISH) and NanoSIMS. *Applied and Environmental Microbiology* 74:3143–3150.

Berg, B. 2000. Litter decomposition and organic matter turnover in northern forest soils. *Forest Ecology and Management* 133:13–22.

Berg, B., and E. Matzner. 1997. Effect of N deposition on decomposition of plant litter and soil organic matter in forest systems. *Environmental Reviews* 5:1–25.

Beusen, A. H. W., A. F. Bouwman, P. S. C. Heuberger, G. Van Drecht, and K. W. Van Der Hoek. 2008. Bottom-up uncertainty estimates of global ammonia emissions from global agricultural production systems. *Atmospheric Environment* 42:6067–6077.

Bobbink, R., K. Hicks, J. Galloway, T. Spranger, R. Alkemade, M. Ashmore, M. Bustamante, S. Cinderby, E. Davidson, F. Dentener, B. Emmett, J. W. Erisman, M. Fenn, F. Gilliam, A. Nordin, L. Pardo, and W. De Vries. 2010. Global assessment of nitrogen deposition effects on terrestrial plant diversity: a synthesis. *Ecological Applications* 20:30–59.

Booth, B. D., and C. J. Swanton. 2002. Assembly theory applied to weed communities. *Weed Science* 50:2–13.

Bouwman, A. F., A. H. W. Beusen, and G. Billen. 2009. Human alteration of the global nitrogen and phosphorus soil balances for the period 1970–2050. *Global Biogeochemical Cycles* 23:GB0A04, DOI: 10.1029/2009GB003576.

Bundy, L. G., T. W. Andraski, M. D. Ruark, and A. E. Peterson. 2011. Long-term continuous corn and nitrogen fertilizer effects on productivity and soil properties. *Agronomy Journal* 103:1346–1351.

Campbell, C. A., R. P. Zentner, F. Selles, V. O. Biederbeck, and A. J. Leyshon. 1992. Comparative effects of grain lentil wheat and monoculture wheat on crop production, N-economy and N-fertility in a brown chernozem. *Canadian Journal of Plant Science* 72:1091–1107.

Carlsson, G., and K. Huss-Danell. 2003. Nitrogen fixation in perennial forage legumes in the field. *Plant and Soil* 253:353–372.

Cassman, K. G., D. C. Bryant, A. E. Fulton, and L. F. Jackson. 1992. Nitrogen supply effects on partitioning of dry-matter and nitrogen to grain of irrigated wheat. *Crop Science* 32:1251–1258.

Cassman, K. G., A. Dobermann, and D. T. Walters. 2002. Agroecosystems, nitrogen-use efficiency, and nitrogen management. *Ambio* 31:132–140.

Chen, X. P., Z. L. Cui, P. M. Vitousek, K. G. Cassman, P. A. Matson, J. S. Bai, Q. F. Meng, P. Hou, S. C. Yue, V. Romheld, and F. S. Zhang. 2011. Integrated soil-crop system management for food security. *Proceedings of the National Academy of Sciences of the United States of America* 108:6399–6404.

Chivenge, P., B. Vanlauwe, and J. Six. 2011. Does the combined application of organic and mineral nutrient sources influence maize productivity? A meta-analysis. *Plant and Soil* 342:1–30.

Clarholm, M. 1985. Interactions of bacteria, protozoa and plants leading to mineralization of soil nitrogen. *Soil Biology and Biochemistry* 17:181–187.

Cleveland, C. C., and D. Liptzin. 2007. C: N : P stoichiometry in soil: is there a "Redfield ratio" for the microbial biomass? *Biogeochemistry* 85:235–252.

Conley, D. J., H. W. Paerl, R. W. Howarth, D. F. Boesch, S. P. Seitzinger, K. E. Havens, C. Lancelot, and G. E. Likens. 2009. Controlling eutrophication: nitrogen and phosphorus. *Science* 323:1014–1015.

Constantin, J., N. Beaudoin, F. Laurent, J. P. Cohan, F. Duyme, and B. Mary. 2011. Cumulative effects of catch crops on nitrogen uptake, leaching and net mineralization. *Plant and Soil* 341:137–154.

Cook, J. C., R. S. Gallagher, J. P. Kaye, J. Lynch, and B. Bradley. 2010. Optimizing vetch nitrogen production and corn nitrogen accumulation under no-till management. *Agronomy Journal* 102:1491–1499.

Cragg, R. G., and R. D. Bardgett. 2001. How changes in soil faunal diversity and composition within a trophic group influence decomposition processes. *Soil Biology and Biochemistry* 33:2073–2081.

Crews, T. E., and M. B. Peoples. 2005. Can the synchrony of nitrogen supply and crop demand be improved in legume and fertilizer-based agroecosystems? A review. *Nutrient Cycling in Agroecosystems* 72:101–120.

Cui, Z. L., X. P. Chen, and F. S. Zhang. 2010. Current nitrogen management status and measures to improve the intensive wheat-maize system in China. *Ambio* 39:376–384.

David, M. B., L. E. Drinkwater, and G. F. McIsaac. 2010. Sources of nitrate yields in the Mississippi River basin. *Journal of Environmental Quality* 39:1657–1667.

Dawson, J. C., D. R. Huggins, and S. S. Jones. 2008. Characterizing nitrogen use efficiency in natural and agricultural ecosystems to improve the performance of cereal crops in low-input and organic agricultural systems. *Field Crops Research* 107:89–101.

Dinnes, D. L., D. L. Karlen, D. B. Jaynes, T. C. Kaspar, J. L. Hatfield, T. S. Colvin, and C. A. Cambardella. 2002. Nitrogen management strategies to reduce nitrate leaching in tile-drained Midwestern soils. *Agronomy Journal* 94:153–171.

Drew, M. C. 1975. Comparison of effects of a localized supply of phosphate, nitrate, ammonium and potassium on growth of seminal root system, and shoot, in barley. *New Phytologist* 75:479–490.

Drinkwater, L. E., and S. S. Snapp. 2007. Nutrients in agroecosystems: rethinking the management paradigm. *Advances in Agronomy* 92:163–186.

Ettema, C. H., and D. A. Wardle. 2002. Spatial soil ecology. *Trends in Ecology and Evolution* 17:177–183.

Fan, X. H., and Y. C. Li. 2010. Nitrogen release from slow-release fertilizers as affected by soil type and temperature. *Soil Science Society of America Journal* 74:1635–1641.

Fang, Y. T., M. Yoh, K. Koba, W. X. Zhu, Y. Takebayashi, Y. H. Xiao, C. Y. Lei, J. M. Mo, W. Zhang, and X. K. Lu. 2011. Nitrogen deposition and forest nitrogen cycling along an urban-rural transect in southern China. *Global Change Biology* 17:872–885.

Fierer, N., A. S. Grandy, J. Six, and E. A. Paul. 2009. Searching for unifying principles in soil ecology. *Soil Biology and Biochemistry* 41:2249–2256.

Fontaine, S., G. Bardoux, L. Abbadie, and A. Mariotti. 2004. Carbon input to soil may decrease soil carbon content. *Ecology Letters* 7:314–320.

Galloway, J. N., J. D. Aber, J. W. Erisman, S. P. Seitzinger, R. W. Howarth, E. B. Cowling, and B. J. Cosby. 2003. The nitrogen cascade. *Bioscience* 53:341–356.

Galloway, J. N., and E. B. Cowling. 2002. Reactive nitrogen and the world: 200 years of change. *Ambio* 31:64–71.

Garcia-Pausas, J., and E. Paterson. 2011. Microbial community abundance and structure are determinants of soil organic matter mineralisation in the presence of labile carbon. *Soil Biology and Biochemistry* 43:1705–1713.

Grandy, A. S., T. D. Loecke, S. Parr, and G. P. Robertson. 2006. Long-term trends in nitrous oxide emissions, soil nitrogen, and crop yields of till and no-till cropping systems. *Journal of Environmental Quality* 35:1487–1495.

Grandy, A. S., and J. C. Neff. 2008. Molecular C dynamics downstream: the biochemical decomposition sequence and its impact on on soil organic matter structure and function. *Science of the Total Environment* 404:221–446.

Grandy, A., R. Sinsabaugh, J. Neff, M. Stursova, and D. Zak. 2008. Nitrogen deposition effects on soil organic matter chemistry are linked to variation in enzymes, ecosystems and size fractions. *Biogeochemistry* 91:37–49.

Guo, J. H., X. J. Liu, Y. Zhang, J. L. Shen, W. X. Han, W. F. Zhang, P. Christie, K. W. T. Goulding, P. M. Vitousek, and F. S. Zhang. 2010. Significant acidification in major Chinese croplands. *Science* 327:1008–1010.

Halvorson, A. D., S. J. Del Grosso, and C. A. Reule. 2008. Nitrogen, tillage and crop rotation effects on nitrous oxide emissions from irrigated cropping systems. *Journal of Environmental Quality* 37:1337–1344.

Halvorson, A. D., B. J. Wienhold, and A. L. Black. 2002. Tillage, nitrogen, and cropping system effects on soil carbon sequestration. *Soil Science Society of America Journal* 66:906–912.

Hättenschwiler, S., and P. Gasser. 2005. Soil animals alter plant litter diversity effects on decomposition. *Proceedings of the National Academy of Sciences of the United States of America* 102:1519–1524.

Herridge, D. F., M. B. Peoples, and R. M. Boddey. 2008. Global inputs of biological nitrogen fixation in agricultural systems. *Plant and Soil* 311:1–18.

Hoang, V. N., and M. Alauddin. 2010. Assessing the eco-environmental performance of agricultural production in OECD countries: the use of nitrogen flows and balance. *Nutrient Cycling in Agroecosystems* 87:353–368.

Hodge, A. 2006. Plastic plants and patchy soils. *Journal of Experimental Botany* 57:401–411.

Hodge, A., D. Robinson, and A. Fitter. 2000. Are microorganisms more effective than plants at competing for nitrogen? *Trends in Plant Science* 5:304–308.

Hodgen, P. J., R. B. Ferguson, J. F. Shanahan, and J. S. Schepers. 2009. Uptake of point source depleted ^{15}N fertilizer by neighboring corn plants. *Agronomy Journal* 101:99–105.

Hofmann, A., A. Heim, P. Gioacchini, A. Miltner, M. Gehre, and M. W. I. Schmidt. 2009. Mineral fertilization did not affect decay of old lignin and SOC in a ^{13}C-labeled arable soil over 36 years. *Biogeosciences* 6:1139–1148.

Howarth, R. W. 2008. Coastal nitrogen pollution: a review of sources and trends globally and regionally. *Harmful Algae* 8:14–20.

Howarth, R. W., A. Sharpley, and D. Walker. 2002. Sources of nutrient pollution to coastal waters in the United States: implications for achieving coastal water quality goals. *Estuaries* 25:656–676.

Intergovernmental Panel on Climate Change (IPCC). 2001. *Climate Change 2001; Synthesis Report.* Cambridge University Press, Cambridge, UK.

Kallenbach, C. M., and A. S. Grandy. 2011. Controls over soil microbial biomass responses to carbon amendments in agricultural systems: a meta-analysis. *Agriculture, Ecosystems, and Environment* 144:241–252.

Kaye, J. P., and S. C. Hart. 1997. Competition for nitrogen between plants and soil microorganisms. *Trends in Ecology and Evolution* 12:139–143.

Khan, S. A., R. L. Mulvaney, T. R. Ellsworth, and C. W. Boast. 2007. The myth of nitrogen fertilization for soil carbon sequestration. *The Journal of Environmental Quality* 36:1821–183.

Korsaeth, A., T. M. Henriksen, and L. R. Bakken. 2002. Temporal changes in mineralization and immobilization of N during degradation of plant material: implications for the plant N supply and nitrogen losses. *Soil Biology and Biochemistry* 34:789–799.

Korsaeth, A., L. Molstad, and L. R. Bakken. 2001. Modeling the competition for nitrogen between plants and microflora as a function of soil heterogeneity. *Soil Biology and Biochemistry* 33:215–226.

Kuzyakov, Y., J. K. Friedel, and K. Stahr. 2000. Review of mechanisms and quantification of priming effects. *Soil Biology and Biochemistry* 32:1485–1498.

Kwon, H. Y., and R. J. M. Hudson. 2010. Quantifying management-driven changes in organic matter turnover in an agricultural soil: an inverse modeling approach using historical data and a surrogate CENTURY-type model. *Soil Biology and Biochemistry* 42:2241–2253.

Lamersdorf, N. P., and W. Borken. 2004. Clean rain promotes fine root growth and soil respiration in a Norway spruce forest. *Global Change Biology* 10:1351–1362.

Liebman, M., L. R. Gibson, D. N. Sundberg, A. H. Heggenstaller, P. R. Westerman, C. A. Chase, R. G. Hartzler, F. D. Menalled, A. S. Davis, and P. M. Dixon. 2008. Agronomic and economic performance characteristics of conventional and low-external-input cropping systems in the Central Corn Belt. *Agronomy Journal* 100:600–610.

Loecke, T. D., and G. P. Robertson. 2009a. Soil resource heterogeneity in the form of aggregated litter alters maize productivity. *Plant and Soil* 325:231–241.

Loecke, T. D., and G. P. Robertson. 2009b. Soil resource heterogeneity in terms of litter aggregation promotes nitrous oxide production and slows decomposition. *Soil Biology and Biochemistry* 41:228–235.

Mazzoncini, M., T. B. Sapkota, P. Barberi, D. Antichi, and R. Risaliti. 2011. Long-term effect of tillage, nitrogen fertilization and cover crops on soil organic carbon and total nitrogen content. *Soil and Tillage Research* 114:165–174.

McSwiney, C. P., and G. P. Robertson. 2005. Nonlinear response of N_2O flux to incremental fertilizer addition in a continuous maize (*Zea mays* L.) cropping system. *Global Change Biology* 11:1712–1719.

Melillo, J. M., J. D. Aber, and J. F. Muratore. 1982. Nitrogen and lignin control of hardwood leaf litter dynamics in forest ecosystems. *Ecology* 63:621–626.

Milcu, A., S. Partsch, C. Scherber, W. W. Weisser, and S. Scheu. 2008. Earthworms and legumes control litter decomposition in a plant diversity gradient. *Ecology* 89:1872–1882.

Moll, R. H., E. J. Kamprath, and W. A. Jackson. 1982. Analysis and interpretation of factors which contribute to efficiency of nitrogen-utilization. *Agronomy Journal* 74:562–564.

Mulvaney R. L., S. A. Khan, and T. R. Ellsworth. 2009. Synthetic nitrogen fertilizers deplete soil nitrogen: a global dilemma for sustainable cereal production. *Journal of Environmental Quality* 38:2295–2314.

Neill, C. 2011. Impacts of crop residue management on soil organic matter stocks: a modeling study. *Ecological Modeling* 222:2751–2760.

Ni, B. L., M. Z. Liu, S. Y. Lu, L. H. Xie, and Y. F. Wang. 2011. Environmentally friendly slow-release nitrogen fertilizer. *Journal of Agricultural and Food Chemistry* 59:10169–10175.

O'Donnell, A. G., M. Seasman, A. Macrae, I. Waite, and J. T. Davies. 2001. Plants and fertilisers as drivers of change in microbial community structure and function in soils. *Plant and Soil* 232:135–145.

Osler, G. H. R., and M. Sommerkorn. 2007. Toward a complete soil C and N cycle: incorporating the soil fauna. *Ecology* 88:1611–1621.

Patzold, S., F. M. Mertens, L. Bornemann, B. Koleczek, J. Franke, H. Feilhauer, and G. Welp. 2008. Soil heterogeneity at the field scale: a challenge for precision crop protection. *Precision Agriculture* 9:367–390.

Poirier, V., D. A. Angers, P. Rochette, M. H. Chantigny, N. Ziadi, G. Tremblay, and J. Fortin. 2009. Interactive effects of tillage and mineral fertilization on soil carbon profiles. *Soil Science Society of America Journal* 73:255–261.

Ramirez, K. S., J. M. Craine, and N. Fierer. 2010. Nitrogen fertilization inhibits soil microbial respiration regardless of the form of nitrogen applied. *Soil Biology and Biochemistry* 42:2336–2338.

Reay, D. S., F. Dentener, P. Smith, J. Grace, and R. A. Feely. 2008. Global nitrogen deposition and carbon sinks. *Nature Geoscience* 1:430–437.

Robertson, G. P., and P. M. Groffman. 2007. Nitrogen transformations. In *Soil Microbiology, Biochemistry, and Ecology*, ed. E. A. Paul, 341–364. Springer, New York.

Robertson, G. P., E. A. Paul, and R. R. Harwood. 2000. Greenhouse gases in intensive agriculture: contributions of individual gases to the radiative forcing of the atmosphere. *Science* 289:1922–1925.

Robertson, G. P., and P. M. Vitousek. 2009. Nitrogen in agriculture: balancing the cost of an essential resource. *Annual Review of Environment and Resources* 34:97–125.

Ross, S. M., R. C. Izaurralde, H. H. Janzen, J. A. Robertson, and W. B. McGill. 2008. The nitrogen balance of three long-term agroecosystems on a boreal soil in western Canada. *Agriculture Ecosystems and Environment* 127:241–250.

Russell, A. E., C. A. Cambardella, D. A. Laird, D. B. Jaynes, and D. W. Meek. 2009. Nitrogen fertilizer effects on soil carbon balances in Midwestern U.S. agricultural systems. *Ecological Applications* 19:1102–1113.

Ryan, J., S. Masri, and M. Singh. 2009. Seasonal changes in soil organic matter and biomass and labile forms of carbon as influenced by crop rotations. *Communications in Soil Science and Plant Analysis* 40:188–199.

Sanginga, N. 2003. Role of biological nitrogen fixation in legume based cropping systems; a case study of West Africa farming systems. *Plant and Soil* 252:25–39.

Schimel, J. P., and J. Bennett. 2004. Nitrogen mineralization: challenges of a changing paradigm. *Ecology* 85:591–602.

Schmidt, M. W. I., M. S. Torn, S. Abiven, T. Dittmar, G. Guggenberger, I. A. Janssens, M. Kleber, I. Kogel-Knabner, J. Lehmann, D. A. C. Manning, P. Nannipieri, D. P. Rasse, S. Weiner, and S. E. Trumbore. 2011. Persistence of soil organic matter as an ecosystem property. *Nature* 478:49–56.

Senthilkumar, S., A. N. Kravchenko, and G. P. Robertson. 2009. Topography influences management system effect on total soil carbon and nitrogen. *Soil Science Society of America Journal* 73:2059–2067.

Singh, P., and N. Ghoshal. 2010. Variation in total biological productivity and soil microbial biomass in rainfed agroecosystems: impact of application of herbicide and soil amendments. *Agriculture Ecosystems and Environment* 137:241–250.

Six, J., S. D. Frey, R. K. Thiet, and K. M. Batten. 2006. Bacterial and fungal contributions to C-sequestration in agroecosystems. *Soil Science Society of America Journal* 70:555–569.

Smil, V. 1999. Nitrogen in crop production: an account of global flows. *Global Biogeochemical Cycles* 13:647–662.

Smith, R. G., K. L. Gross, and G. P. Robertson. 2008. Effects of crop diversity on agroecosystem function: crop yield response. *Ecosystems* 11:355–366.

Smith, R. G., D. A. Mortensen, and M. R. Ryan. 2010. A new hypothesis for the functional role of diversity in mediating resource pools and weed-crop competition in agroecosystems. *Weed Research* 50:37–48.

Snapp, S. S., M. J. Blackie, R. A. Gilbert, R. Bezner-Kerr, and G. Y. Kanyama-Phiri. 2010. Biodiversity can support a greener revolution in Africa. *Proceedings of the National Academy of Sciences of the United States of America* 107:20840–20845.

Snapp, S. S., S. M. Swinton, R. Labarta, D. Mutch, J. R. Black, R. Leep, J. Nyiraneza, and K. O'Neil. 2005. Evaluating cover crops for benefits, costs and performance within cropping system niches. *Agronomy Journal* 97:322–332.

Sogbedji, J. M., H. M. van Es, and K. L. Agbeko. 2006. Cover cropping and nutrient management strategies for maize production in western Africa. *Agronomy Journal* 98:883–889.

Strickland, M. S., C. Lauber, N. Fierer, and M. A. Bradford. 2009. Testing the functional significance of microbial community composition. *Ecology* 90:441–451.

Tarkalson, D. D., G. W. Hergert, and K. G. Cassman. 2006. Long-term effects of tillage on soil chemical properties and grain yields of a dryland winter wheat-sorghum/corn-fallow rotation in the great plains. *Agronomy Journal* 98:26–33.

Taylor, P. G., and A. R. Townsend. 2010. Stoichiometric control of organic carbon-nitrate relationships from soils to the sea. *Nature* 464:1178–1181.

Thiet, R., S. Frey, and J. Six. 2006. Do growth yield efficiencies differ between soil microbial communities differing in fungal:bacterial ratios? Reality check and methodological issues. *Soil Biology and Biochemistry* 38:837–844.

Turner, R. E., and N. N. Rabalais. 2003. Linking landscape and water quality in the Mississippi river basin for 200 years. *Bioscience* 53:563–572.

United States Environmental Protection Agency (EPA). 2009. *Inventory of U.S. Greenhouse Gas Emissions and Sinks: 1990–2007*. EPA 430-R-09-004. U.S. Environmental Protection Agency. Washington, DC.

Vanlauwe, B., K. Aihou, S. Aman, E. N. O. Iwuafor, B. K. Tossah, J. Diels, N. Sanginga, O. Lyasse, R. Merckx, and J. Deckers. 2001. Maize yield as affected by organic inputs and urea in the West African moist savanna. *Agronomy Journal* 93:1191–1199.

Van Noordwijk, M., P. C. Deruiter, K. B. Zwart, J. Bloem, J. C. Moore, H. G. Vanfaassen, and S. Burgers. 1993. Synlocation of biological-activity, roots, cracks and recent organic inputs in a sugar-beet field. *Geoderma* 56:265–276.

Waldrop, M. P., and M. K. Firestone. 2004. Altered utilization patterns of young and old soil C by microorganisms caused by temperature shifts and N additions. *Biogeochemistry* 67:235–248.

Wang, J. G., and L. R. Bakken. 1997. Competition for nitrogen during mineralization of plant residues in soil: microbial response to C and N availability. *Soil Biology and Biochemistry* 29:163–170.

Whitmore, A. P., and J. J. Schroder. 2007. Intercropping reduces nitrate leaching from under field crops without loss of yield: a modeling study. *European Journal of Agronomy* 27:81–88.

Wickings, K., and A. S. Grandy. 2011. The oribatid mite *Scheloribates moestus* (Acari: Oribatida) alters litter chemistry and nutrient cycling during decomposition. *Soil Biology and Biochemistry* 43:351–358.

Zingore, S., R. J. Delve, J. Nyamangara, and K. E. Giller. 2008. Multiple benefits of manure: the key to maintenance of soil fertility and restoration of depleted sandy soils on African smallholder farms. *Nutrient Cycling in Agroecosystems* 80:267–282.

chapter 7

The contribution of arbuscular mycorrhizal fungi to the success or failure of agricultural practices

David D. Douds Jr.
USDA-ARS Eastern Regional Research Center

Rita Seidel
The Rodale Institute

Contents

7.1 Introduction to the biology of arbuscular mycorrhizal fungi

Arbuscular mycorrhizal (AM) fungi are soilborne fungi that generally form a mutualistic symbiosis with the majority of crop plants. However, one agronomically important crop family that does not form arbuscular mycorrhizas is the Brassicaceae. Among the benefits to the plant host ascribed to the symbiosis are enhanced mineral nutrient uptake, water relations, and disease resistance. The fungus benefits from the receipt of fixed carbon through the roots of the host plant (Shachar-Hill et al., 1995). A brief discussion of the life cycle and function of AM fungi will help understand and predict both the impacts of farm management practices on AM fungi and the role of AM fungi in the success or failure of certain agronomic practices.

7.1.1 Life cycle of AM fungi

A generalized life cycle of AM fungi begins in the spring with the germination of a spore produced the previous autumn. Spores are induced to germinate when (1) any necessary dormancy period has been met (Tommerup, 1981), and (2) the soil reaches the proper temperature (e.g., 18–25°C; Daniels and Trappe, 1980) and is further stimulated (3) by increased CO_2 levels in the soil due to microbial or root respiration (Bécard and Piché, 1989a). The germ tube hyphae grow through the soil in "search" of a host root to colonize. Spores are capable of regerminating a number of times if initial attempts to find a root to colonize are unsuccessful. Growth during this "asymbiotic" or presymbiotic phase is supported by carbon reserves—mainly lipid, trehalose, and glycogen—in the spore (Bago et al., 1999). Other forms of inocula of AM fungi that can overwinter in agricultural soil are colonized root pieces containing AM fungus vesicles or spores and the extraradical mycelium (ERM) of the previous crop host.

Plant roots exude a number of chemical signals into the soil that form a type of communication with the germ tube hyphae (Nagahashi et al., 2010). More of these signals are released when the root is low in phosphorus (P) nutrition (Nagahashi and Douds, 2000). The AM fungus hyphae branch in response to these signals in a concentration-dependent manner: The hyphae produce more long branches at low concentrations of the signals and dense clusters of short branches in response to high concentrations of the signals (Nagahashi and Douds, 2000). These responses serve to increase the probability of contact with a host root.

Colonization begins with the formation of an appressorium and penetration through the root epidermis. Subsequent extracellular growth into the cortex is followed by growth into cortical cells via digestion through the cell wall and envagination of the plasma membrane (Genre and Bonfante, 2010). Once inside the cell, the hyphal tip branches profusely, forming a small, tree-like structure called the arbuscule. The arbuscule is the most likely candidate for the site of nutrient transfer between the symbionts due to the high density of cytoplasm and membrane surface area (Toth and Miller, 1984). Once the fungus begins to receive fixed carbon from the root, it then grows out into the soil, forming the ERM: the nutrient-absorbing organ of the mycorrhiza (Bécard and Piché, 1989b). Colonization of the root spreads, and new spores are produced when a certain colonized root length is achieved, or on senescence of the host plant, to begin the cycle again (Douds and Chaney, 1982; Gazey et al., 1992).

Colonization of roots is under the regulation of the host plant. There is now abundant evidence that much of the same genetic regulatory pathway that governs the formation of nodules in the legume-*Bradyrhizobium* symbiosis also operates to control the formation of arbuscular mycorrhizas (e.g., Oldroyd and Downie, 2006). Further, and more important for consideration in agriculture, the formation of the mycorrhiza is regulated by the availability of P in the soil and concomitant P status of the host plant. Colonization of roots is inhibited with increasing P status of the root (Jasper et al., 1979). This occurs through reduced exudation of hyphal branching signals and reduced carbohydrate availability within the root (Nagahashi and Douds, 2000; Treseder and Allen, 2002).

7.1.2 Functioning of the symbiosis

The most notable benefit plants receive from colonization by AM fungi is enhanced uptake of mineral nutrients that are immobile in the soil solution (i.e., P, Zn, and Cu). For example, Barber (1977) estimated that soluble forms of P can only diffuse through the soil 0.5 mm due to the formation of insoluble salts with cations. The typical zone of uptake for these nutrients for a nonmycorrhizal root is 1 or 2 mm, about the length of a root hair (Li et al., 1991). The extraradical hyphae of the mycorrhizal fungi extend upward of 14 cm away from the root and act, in effect, as extensions of the root system (Mozafar et al., 2001). The same forms of P that are normally available to roots are taken up by the ERM, where they are synthesized into polyphosphate granules and translocated into the root. There P is released to the interfacial apoplastic space between the membranes of both partners (Balestrini et al., 2007).

Other benefits plants receive from the symbiosis are enhanced water relations and disease resistance. Mycorrhizal plants have been found to continue to photosynthesize at lower soil moisture levels than paired nonmycorrhizal plants due to a number of reasons, including increases in overall plant health, better exploration of the soil volume; enhanced soil structure, which retains more water; increased stomatal conductance; and lower osmotic potential in leaf tissues due to increased solute concentrations (Augé, 2000). Enhanced disease resistance can occur not only due to nutritive effects but also due to selection of the ERM for soil bacteria that are antagonistic to fungal plant pathogens (Linderman, 2000).

The receipt of fixed carbon, in the form of hexose, from the host plant is essential to the fungus due to its obligate symbiotic nature. There is a metabolic division of labor among the structures of these coenocytic organisms (Shachar-Hill et al., 1995). Only the hyphae inside the root express the genes necessary for the uptake of hexose and synthesis of fatty acids. Extraradical hyphae are incapable of these metabolic events, and germinating spores have only limited carbon uptake and no de novo synthesis of lipids (Bago et al., 1999; Pfeffer et al., 1999).

Although AM fungi are generally considered to be mutualists, the cost of the symbiosis to the host plant, in the form of fixed carbon, can exceed benefits to net photosynthesis due to any impact of the symbiosis on nutrient uptake or water relations (Johnson et al., 1997). The metabolic demands of AM fungi can require 4–20% of host photosynthate (see Douds et al., 2000, for review). Experimentation with fungicide application to reduce AM fungus colonization has demonstrated mycorrhiza-mediated growth suppression of citrus in the field under P-sufficient conditions (Graham and Eissenstat, 1998). On the far negative end of the spectrum is tobacco stunt disease caused by the AM fungus *Glomus macrocarpum* (Hendrix et al., 1992).

Although a common viewpoint is that the over 150 described species of AM fungi more or less occupy the same niche (Dodd et al., 2000), there is mounting experimental evidence of functional diversity among these fungi. AM fungi have been shown to differ in competitiveness in colonization of roots (Bennett and Bever, 2009), in the distance away from the root in which the ERM forages for nutrients (Smith et al., 2000), and ramification of ERM in localized areas of nutrient availability (Gavito and Olsson, 2008). Functional complementarity achieved through colonization of a host root system by a community of AM fungi can result in enhanced plant growth over that occurring with colonization by only one AM fungus (Jansa et al., 2008).

7.2 Interaction of agricultural practices with communities of AM fungi

The goals of agriculture have progressed from subsistence to economies of scale to renewed emphasis on smaller farms and sustainability. With the move toward sustainability, many agricultural practices have been modified to better manage soil nutrients, organic matter, and water; reduce soil erosion; and control plant diseases and weed pressure. These practices have largely been adopted without full knowledge of their impact on the soil biological community or how the soil biological community may have an impact on the success or failure of these practices. Here, we examine three sustainable agricultural practices: reduced tillage, crop rotation, and overwintering cover crops; give a brief overview of their goals; and then show how AM fungus communities are impacted by these practices and contribute to the achievement of the goals of these practices.

7.2.1 Reduced tillage

7.2.1.1 Agronomic background

7.2.1.1.1 Brief history and definition. Soil tillage had its beginning thousands of years ago, although modern tools such as the moldboard plow were not introduced until the 18th and 19th centuries. The replacement of animal power by tractors allowed large areas of open grassland in the Great Plains of the United States to be turned into fields with row crops. Tillage of the former grassland combined with drought caused severe soil erosion and led to the Dust Bowl in the 1930s, making millions of acres of farmland unusable. Although the no-till movement first started in the 1950s and 1960s, made possible by the invention of herbicides such as 2,4-D® and Paraquat®, it was not a common practice at that time (Derpsch, 1997). The 1985 Farm Bill, which required farmers to prevent and reduce erosion, may have been a major incentive for conservation tillage to become more widely adopted in the United States (Brock et al., 2000).

In reduced or conservation tillage, crops are grown with minimal disturbance of the soil. Methods differ from each other mainly in the degree to which the soil is disturbed prior to planting. Even in "no-till" systems, the soil is opened to a certain degree (e.g., with coulters or disk openers) to plant the seed. By definition, conservation tillage leaves at least 30% of the soil covered by crop residues (Day et al., 1999).

7.2.1.1.2 Effects of reduced tillage. No-till practices were first introduced as a soil conservation tool and to decrease labor requirements and fuel use (Doren and Linn, 1994; Sims et al., 1994). Numerous studies have shown that soil is more protected from erosion and runoff in no-till systems (Gilley et al., 1997; Langdale et al., 1979; Shipitalo and Edwards, 1998; Williams et al., 2009), and that yields in no-till systems can be as good as or better than with conventional tillage (Day et al., 1999; Dick et al., 1991; Karlen et al., 1994; Teasdale et al., 2007). Soil carbon (Berner et al., 2008; Halpern et al., 2010; West and Post, 2002) and other soil quality parameters (aggregate stability, microbial activity, earthworm populations) can increase significantly after switching from conventional tillage to no-till (Karlen et al., 1994). Potential disadvantages of no-till are compaction, flooding or poor drainage, delays in planting because fields are too wet or too cold, and carryover of diseases or pests in crop residue.

When tillage is reduced, crop residues are not or only partially incorporated into the soil. The new crop is planted into the stubble left on the soil surface or into small strips of tilled soil; fertilizers and lime are generally placed on top of the soil. Weeds are usually controlled with herbicides rather than by tillage or cultivation. This increased dependence on herbicides for weed control (Day et al., 1999) is often considered unsustainable, possibly leading to herbicide resistance in certain weeds and increased leaching of pesticides into groundwater due to higher infiltration rates in no-till relative to tilled systems (Isensee and Sadeghi, 1995; Lyons et al., 1996; Shipitalo and Edwards, 1998). Herbicide resistance could cause farmers to revert to using tillage as weed control, thereby potentially undoing soil and water quality improvements made with no-till systems. In a study by Krutz et al. (2009), the authors evaluated the impact of cover crops and tillage on water, sediment, and herbicide loss in glyphosate-resistant cropping systems. Their results indicated that a rye cover crop (see Section 7.3) established in fields requiring tillage for weed control can potentially limit water, sediment, and preemergence herbicide losses to levels equivalent to or less than those of no-till.

"Standard no-till" with herbicides is not an option in organic production systems. To reduce frequency or intensity of tillage in organic systems, many farmers are exploring the option of terminating a cover crop mechanically by mowing, undercutting, or rolling instead of plowing. While plowing incorporates the cover crop into the soil, leaving it bare as a result, mowing, undercutting, and rolling all keep the cover crop on the surface to act as a weed-suppressing and moisture-conserving mulch. Flail mowing is usually the preferred method of cover crop mowing. It cuts low, right above ground level, and leaves an even layer of residue. Undercutting terminates a cover crop with sweeps or blades that travel just below the soil surface, cutting the plants below the crowns. Rolling is performed using a rolling drum with blunted blades that terminate the cover crop by rolling it into a mat without cutting the stems. Both undercutting and rolling keep the plants more or less intact and in place, thereby reducing decomposition rates and increasing the time the mulch stays on the soil surface and works to suppress weeds. Mowing chops the plant biomass into small pieces, increasing the surface area and the rate at which the cover crop breaks down. Cover crops are already a common feature of organic systems and fulfill numerous functions (see Section 7.3).

Recent field trials examining the effectiveness of the roller as a mechanical termination technique showed promising results. Cover crop rollers can successfully terminate annual crops, including cereal grains (rye, wheat, oats, and barley) and annual legumes (hairy vetch, winter pea, and crimson clover) without the use of any herbicides (Ashford and Reeves, 2003; Mirsky et al., 2009; Mischler et al., 2010). Although much of the interest for the roller comes from organic producers, it can also be used in conventional systems: Some studies have shown that the roller-crimper in combination with a burndown herbicide, such as glyphosate, can both increase the effectiveness of cover crop control and reduce the rate of herbicides needed to kill the cover crop (Ashford and Reeves, 2003; Curran et al., 2007).

Another advantage of the roller is the fairly small amount of energy required to operate it. Fuel need for the roller is similar to a cultipacker (a corrugated roller that is usually used as the last step in seedbed preparation to break up clods and firm the soil) and 10 times less than the energy required for mowing (Hunt, 1977, as cited in Ashford and Reeves, 2003). In an organic system, rolling a cover crop before soybeans can save farmers up to 5 gallons of fuel per acre by reducing tillage operations (Mutch, 2004), and when averaged over a 3-year corn-soybean-wheat rotation, no-till planting with a roller uses 25% less energy than traditional field operations for organic crops (Ryan et al., 2009).

Yield results and weed suppression for the roller-crimper system are also promising. In a field trial in Illinois, no-till soybeans grown after rye termination with a roller achieved similar yields to those in a chemically terminated cover crop while reducing residual weed biomass (Davis, 2010). Smith et al. (2011) conducted a field trial in North Carolina in which soybeans were no-till planted into a rolled or flail-mowed rye cover crop. Both treatments controlled weeds in the soybeans sufficiently (no herbicides were used), and yields were the same as in a weed-free treatment as long as dry rye biomass was high (>9,000 kg/ha).

7.2.1.2 Role of AM fungi

The presence or absence of tillage can have an impact on the AM fungus community structure and function in several ways. Soil disturbance affects the functioning of the ERM, the distribution of AM fungi throughout the soil profile, and the community of AM fungi colonizing roots of the crop plant.

The ERM of the arbuscular mycorrhiza serves two primary functions. First, it serves as the nutrient uptake organ of the symbiosis, exploring the soil for nutrients, which it then transports back to the intraradical phase for release to the host root. Second, it grows out into the soil to colonize other roots of the same or other potential host plants. Both of these functions are impacted by tillage-induced soil disturbance (Kabir, 2005). This has been demonstrated well in a series of greenhouse and growth chamber experiments conducted by Miller and coworkers (Evans and Miller, 1990; Fairchild and Miller, 1990). Corn (*Zea mays* L.) was grown in plastic cylinders containing field soil amended with various levels of P, ranging from 0 to 160 mg P g^{-1} soil. Three growth cycles were conducted in which shoots of successive plantings of maize were harvested after 3 weeks of growth. All soils were initially disturbed to fill the containers, but after the first harvest, half of the containers were replanted without disturbing the soil. In the other containers, the soil was removed, mixed, and repotted prior to sowing new seeds. Shoot growth was equivalent among paired disturbance treatments, within P addition level, for the first cycle (Figure 7.1). With successive cycles, however, the plants grown in undisturbed soil, at low P, exhibited progressively greater shoot weight than those grown in disturbed soil. The soil disturbance had an impact on the ability of the ERM of the previously grown seedling to colonize the subsequently grown plant. This was reflected in colonization parameters and shoot P concentrations (Fairchild and Miller, 1990). In addition, soil disturbance between planting had no effect on shoot growth in the high-P treatment (Figure 7.1). This was not surprising because AM fungus colonization is expected to provide little benefit for nutrient uptake in high-P soils. A high rate of AM fungus colonization is not necessary to receive the mycorrhiza-mediated contribution to the success of a reduced-tillage treatment. The important factor is the connection to the previously established, intact ERM and its nutrient-absorbing capacity (McGonigle et al., 1990; Miller, 2000). Maintenance of the attachment of the ERM to the roots of the previous crop is not essential to the survival and infectivity of the ERM (Jasper et al., 1989b). However, survival of the disturbed mycelium and the ability of disturbed ERM fragments to function as propagules decrease with time after disturbance and with increasing moisture level of the soil (Jasper et al., 1989a; Kabir et al., 1998).

Soil disturbance also has been shown to affect the interaction between AM fungi, *Bradyrhizobium japonicum*, and soybean plants in a greenhouse experiment (Goss and deVarennes, 2002). Soil disturbance had the previous effects on shoot growth, P uptake, and mycorrhiza development. In addition, the number of nodules and overall N fixation was greater for plants growing in the undisturbed versus disturbed soil.

Do these findings in the greenhouse or growth chamber translate to the field? Field experiments have demonstrated increased AM fungus colonization of maize, wheat, a *Vicia villosa* cover crop, and pasture grasses when sown into no-tilled versus tilled soils (Miller, 2000; Galvez et al., 1995; Galvez, Douds, Drinkwater, et al., 2001; Galvez, Wagoner, et al., 2001; Borie et al., 2006). These studies also showed increased early season P uptake by plants in no-tilled soils relative to tilled soils (Miller, 2000; Galvez, Douds, Drinkwater, et al., 2001; Galvez, Wagoner, et al., 2001), and this effect was lessened with increasing

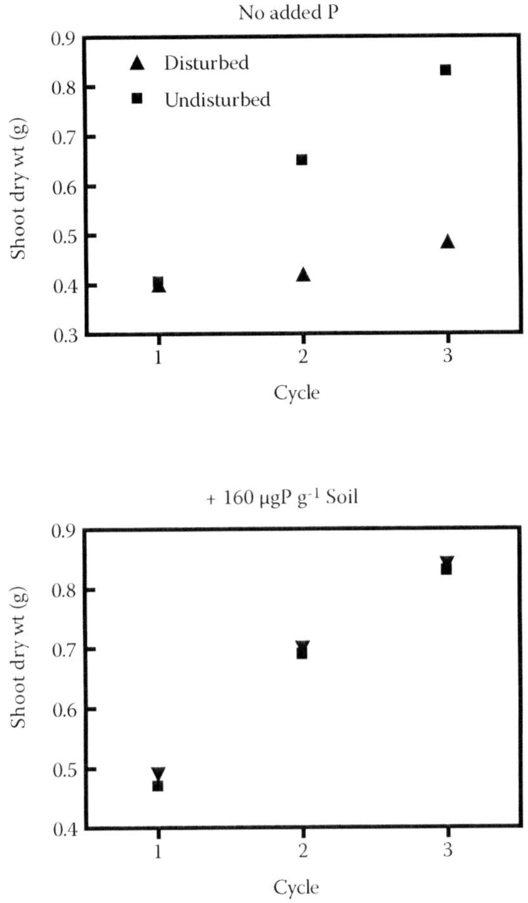

Figure 7.1 Shoot growth response of maize grown in greenhouse pots for three cycles of 3 weeks each. After shoot harvest, the soil in the pots was either removed and remixed or left undisturbed. All pots had the same population of native AM fungi at the start of the first cycle. (Adapted from Fairchild, G.L., and M.H. Miller. 1990. Vesicular-arbuscular mycorrhizas and the soil-disturbance-induced reduction of nutrient absorption in maize. III. Influence of phosphorous amendments to soils. *New Phytologist* 114:641–650.)

P application (McGonigle and Miller, 1996). Other factors present in the field environment, not fully explored in greenhouse studies, resulted in the enhancement of early season AM fungus colonization and P uptake not translating into increased plant growth and yield in no-tilled versus conventionally tilled soils. Despite the importance of early season P nutrition of crops such as maize (Grant et al., 2001), the cooler soil temperatures in no-tilled soils due to shading by crop residues contributes to reduced early season plant development.

As would be expected, disturbance of the soil such as by moldboard plowing, which inverts the soil column 135°, also affects the distribution of AM fungi within the soil profile. No-till soils under continuous wheat cultivation had more spores of AM fungi in the surface 80 mm of soil and fewer in the 80- to 150-mm layer than did soils from conventionally tilled plots due to mechanical inversion of the soil (Smith, 1978), even though the total number of spores was not affected. The density of propagules of AM fungi declines with soil depth (Jakobsen and Nielson, 1983; Oehl et al., 2005), and they are more numerous in

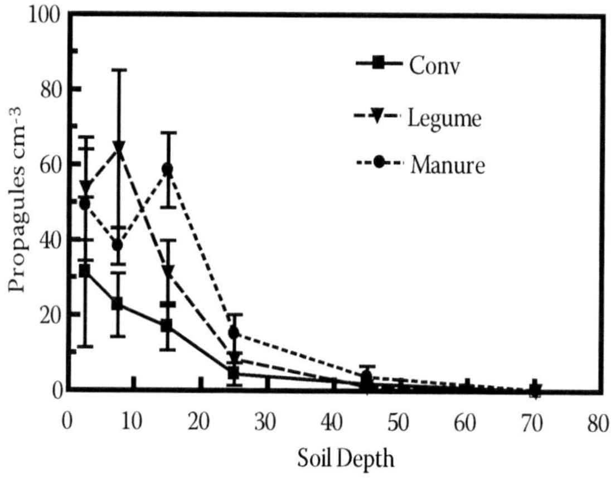

Figure 7.2 Levels of propagules of AM fungi at six soil depths: 0–5, 5–10, 10–20, 20–30, 30–60, and 60–80 cm. Soil samples were collected from three farming systems at the Rodale Institute: conventional, organic with legume cover crops, and organic with animal manure addition. Results of most probable number bioassays. Means of 3 years of data, each year having three replicates per farming system × depth combination. (D. D. Douds et al., unpublished results.)

the top 20 cm of soil than below (Figure 7.2). Inversion of soil with deep moldboard tillage would bury the propagule-dense soil and bring soil with fewer propagules up to the surface, contributing further to lower early season colonization of seedlings in tilled soils.

In addition to the quantitative effects tillage has on AM fungi, it has qualitative effects on the AM fungus community both in the soil and within roots of the crop plant. Generally, the diversity of the AM fungus community has been observed to decline with increasing disturbance, whether from native grassland/forest to cultivation (Helgason et al., 1998) or from no-till to conventional tillage (Boddington and Dodd, 2000). However, no change in diversity of AM fungus spore communities in response to tillage was observed in a wheat field in Argentina (Schalamuk et al., 2006). Cultivation can select for certain members of a genus versus others, which can be more prominent in fallow soils (Rosendahl and Matzen, 2008). Communities of AM fungi in tilled soils generally have been found to be dominated more by members of the genus *Glomus*, while representation by members of the genera *Gigaspora* and *Scutellospora* are greater in no-till or uncultivated soils (Jansa et al., 2002; Blaszkowski, 1993). This observation also was reflected in the AM fungi colonizing the roots of the crop. For example, tillage reduced the colonization of maize roots by *Scutellospora* spp. and increased colonization by *Glomus* spp. (Jansa et al., 2003). These differences may be due to differences in the tolerance of the ERM of various groups of AM fungi to disturbance or the impact of tillage on other factors that affect AM fungus species diversity (e.g., weed community structure) (Jansa et al., 2003). Members of the Gigasporaceae are sensitive to soil disturbance: Root pieces colonized by these fungi and ERM fragments generally do not function well as inoculum (Klironomos and Hart, 2002). These species rely more heavily on spores as inoculum than do *Glomus* spp. Members of *Glomus* and other genera of AM fungi produce lipid-rich vesicles as well as arbuscules within roots, which enable colonized root fragments from the previous crop or weed to function as propagules (Biermann and Linderman, 1983).

Qualitative impacts of tillage on the AM fungus community are potentially important from the standpoint of the functioning of the symbiosis. Functional diversity has been reported among AM fungus species (Smith et al., 2000; Stampe and Daehler, 2003). For example, some AM fungi produce more of the soil-aggregating compound glomalin than others (Wright and Upadhyaya, 1996) or differ in the distance from the root in which they explore for P (Smith et al., 2000).

7.2.2 Crop rotation

7.2.2.1 Agronomic background

7.2.2.1.1 Brief history and definition. Crop rotation is the planned succession of different crops planted in the same field. Farmers have been using crop rotations since Roman times, when a 3-year sequence of "food-feed-fallow" was common. Land was used to grow a food crop, planted for livestock feed, or left fallow to replenish its nutrients. This simple early rotation was improved over centuries by eliminating the fallow and adding legumes and animal manures into the cycle, which in turn protected the soil from erosion and increased yields ("Crop Rotation History," 2011). However, the advent of chemical fertilizers and pesticides after World War II caused cropping systems in industrialized countries to become increasingly simplified, with large tracts of land being taken up by monocultures. In more recent years, farmers have been reverting from monocultures to systems that include several different crops. Reasons to do so might originate from an economic need to diversify in order not to depend on one commodity alone or could be driven by efforts to combat problems that are often attributed to monocultures, such as declining yields and increased pest pressures. Rising interest in sustainable agricultural practices and organic systems is probably also a factor. The U.S. Department of Agriculture National Organic Program (NOP) Standards require, for example, that "the producer must implement a crop rotation, including but not limited to sod, cover crops, green manure crops and catch crops." The NOP further instructs that the rotation must provide ecosystem functions, including maintaining or improving soil organic matter content, providing for pest management in annual and perennial crops, managing deficient or excess plant nutrients, and providing erosion control (http://www.ams.usda.gov/nop/).

Rotations can be very simple, alternating between two different crops, for example, or last more than 5 or 6 years. In general, farmers will alternate between heavy feeders and light feeders (nonlegumes and legumes), cool and warm season crops, annuals and perennials, or row crops, drilled crops, and sod-forming crops. In vegetable production systems, growers often rotate according to plant families or root crops with nonroot crops. The exact sequence and type of crops will depend on an area's climate, culture, or market trends but will also be influenced by the grower's preference or need for certain crops.

7.2.2.1.2 Effects of rotations. Numerous studies have shown that longer rotations can have many benefits. Smith et al. (2008) conducted a field experiment in Michigan comparing continuous monocultures of three row crops (maize, soybean, and winter wheat) and 2- and 3-year rotations with and without cover crops (no cover crop, one or two legume/small-grain species), encompassing a range of crop diversity from one to six species. All rotations were grown without chemical inputs. Maize grain yields increased linearly with the number of crops in the rotation and were highest in the rotation with six crop species (more than twice as high as in monocultured maize). Soybean and wheat yields were also significantly different between treatments but did not follow a linear trend like maize.

A long-term crop rotation experiment in Hungary tested the effect of seven crop sequences and five fertilization treatments on maize and wheat yields. Rotations included maize and wheat monocultures as well as two-crop, three-crop, and four-crop rotations (maize, wheat, in rotation with spring barley, alfalfa, or peas). Grain yields for both maize and wheat were always lower in a monoculture than in a crop rotation, and the four-crop rotation had the highest yields. Yield reductions in monocultured corn seemed to be mostly caused by lack of moisture, whereas the reduction in wheat yield in a monoculture could be mainly attributed to pathogenic factors stimulated by the weather (Berzsenyi et al., 2000). Many other studies have tested the effect of rotations on diseases (e.g., soilborne pathogens) (Abawi, 1989; Glynne, 1965; Trivedi and Barker, 1986).

Two or more crops in a rotation can reduce weed populations compared to monocropping, mainly because crops with different growth habits disrupt the life cycle of different weed species (Benoit et al., 2003; Kegode et al., 1999; Liebman and Dyck, 1993; Schreiber, 1992; Teasdale et al., 2004). This is especially critical in organic cropping systems, in which weeds cannot be controlled by herbicides. Porter et al. (2003) showed that weed control, yields, and soil quality in an organic 4-year rotation with hay were better than in an organic 2-year maize-soybean rotation. Diverse crop rotations can also increase soil carbon (West and Post, 2002) and decrease nutrient leaching (Sanchez et al., 2004).

7.2.2.2 Role of AM fungi

The interaction of crop rotation and the indigenous AM fungus community is important from an agronomic viewpoint in several ways: the quantitative impact that the mycotrophic status of one crop has on the AM fungus population available for the next crop, the change each crop imparts on the relative abundance of members of the AM fungus community, and the impact of that new community on the growth of the proceeding crop.

From a mycorrhizocentric viewpoint, crop plants can be categorized as mycotrophic (hosts) or nonmycotrophic (nonhosts). Among angiosperms, the primary nonmycotrophic species are members of the Brassicaceae and Chenopodiaceae. Therefore, many common vegetable crops and weedy plants are not hosts for AM fungi and therefore will not support their growth and sporulation. As expected, when part of a crop rotation, these plants depress populations of AM fungi and have been shown to have negative impacts on the growth of proceeding mycotrophic crops (Arihara and Karasawa, 2000; Karasawa et al., 2001; Panja and Chaudhuri, 2004). As with the impacts of tillage on mycorrhiza-mediated crop response, the degree of this response is dependent on the available P level of the soil (Ryan et al., 2002).

Among host plants, there are differences in the level of colonization by and sporulation of specific species of AM fungi that they will support. Work on this topic began with experiments to find good general hosts for inoculum production in the greenhouse (Hetrick and Bloom, 1986; Struble and Skipper, 1985). These experiments showed that certain AM fungi sporulated better in the presence of some plants than others, and this phenomenon has been seen in the field (Johnson et al., 1991). Spores of the AM fungus *Gigaspora gigantea* were much more numerous after maize than small grains, soybean, or bell peppers (Douds et al., 1993, 1997; Figure 7.3). This represents a qualitative effect on the relative abundance of AM fungus species within the community of AM fungi to which the next crop is exposed. Coupled with the functional diversity of AM fungi, these fluctuations in abundance within the community can have impacts, subtle or otherwise (Hendrix et al., 1992), on growth of the proceeding crop. These impacts are difficult to predict when an AM fungus community may have as many as 26 members (Ellis et al., 1992), and little if anything is known about their individual functional diversities.

One agricultural phenomenon in which the influence of a crop on its attendant AM fungus community has a significant role is the yield decline noted for crops when grown

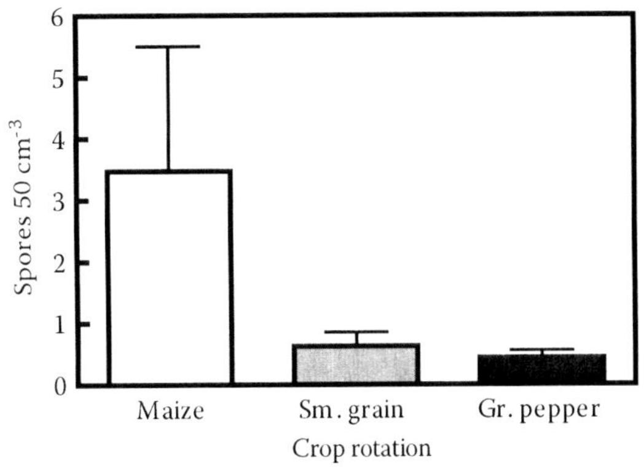

Figure 7.3 Effect of crop rotation on populations of the AM fungus *Gigaspora gigantea*. Spores were isolated from the soil in the autumn after final crop harvest. Each bar represents the mean of 3 years, 48 soil samples per year. (Adapted from Douds, D.D., L. Galvez, M. Franke-Snyder, C. Reider, and L.E. Drinkwater. 1997. Effect of compost addition and crop rotation point upon VAM fungi. *Agriculture Ecosystems and Environment* 65:257–266.)

in continuous monoculture (i.e., in the absence of crop rotation). For example, the average increase in yield of wheat when grown in rotation, over that in continuous monoculture, ranges from 14% in North America to 33% in Australia (Kirkegaard et al., 2008). Crop monocultures can produce AM fungus communities that do not enhance the growth of that crop (Schenck et al., 1989; Johnson et al., 1992). Johnson et al. (1992) proposed a model that explains how this can occur. Basically, the AM fungi that proliferate on a given host are not necessarily the most effective symbionts from the phytocentric perspective. They are very effective at procuring carbon from that particular host for their own reproduction. The disproportionate representation of these species within the AM fungus community increases with successive cropping cycles of that plant, and crop vigor declines. Crop rotation reduces the relative abundance of these fungi and increases that of other members of the community. Interestingly, the community of AM fungi that was detrimental to maize, for example, was shown to be beneficial to soybean and vice versa (Johnson et al., 1992). Crop rotation therefore presents a win-win situation. There also is evidence that weeds can be beneficial in maintaining levels of potentially underrepresented members of the AM fungus community, and that this can enhance the efficacy of the community in enhancing crop growth (Feldmann and Boyle, 1999). This fits with a generalization that the diversity of the AM fungus community is proportional to that of the associated plant community (van der Heijden et al., 1998).

7.2.3 *Overwintering cover crops*

7.2.3.1 *Agronomic background*

7.2.3.1.1 History and definition. Cover crops are usually not harvested but grown between permanent crops such as trees and vines or in fields before or after main crops. They have been used for centuries but are often missing in "modern" farming systems. Without cover crops, farmers might encounter erosion problems (especially if they are tilling their soil) and nutrient management issues (lack of nutrients, especially nitrogen, as well as leaching of nutrients).

Table 7.1 Aboveground Biomass and Nitrogen Input from Overwintering Cover Crops Grown at the Rodale Institute, Kutztown, Pennsylvania

Cover crop	Seeding rate (kg ha^{-1})	Planting date	Sampling date	Biomass kg ha^{-1}	Biomass %N	Biomass N input (kg N ha^{-1})	Replicate years
Winter rye	157	Early November	Mid-April	114–1,147	2.4–4.5	5–27	4
Winter rye	157	Early November	Early May	240–2,283	1.6–3.6	7–57	7
Red clover	11	Late March	Early May	1,185–3,710	2.7–4.1	43–135	8
Hairy vetch	25–35	Late August	Early May	2,581–5,595	3.5–6.2	116–281	16

Cover crops can be annual or perennial species and primarily include legumes, grasses, and brassicas. A winter cover crop is planted in late summer or fall in temperate climates to provide soil cover during the nongrowing season. A range of species is available to use as winter cover crops (Table 7.1). Small grains (barley, oats, rye, and wheat) grow rapidly, they have good winter hardiness, and seed is readily available. With their fast growth, they are strong competitors against weeds, and some (such as rye) can be allelopathic, emitting chemicals that inhibit weed seed germination. Legumes, such as clovers, vetches, and peas, also may be used as winter cover crops. They are less winter hardy than grasses, grow less rapidly, and are not as effective in preventing erosion or reducing leaching loss of leftover nitrogen. However, they add significant amounts of nitrogen to the soil (Table 7.1), which is made available gradually to the following crop, a nitrogen availability pattern that is more adapted to plant growth and needs than most mineral fertilizers (Sarrantonio, 2003).

Adding cover crops to a farm plan may be a challenge; therefore, a farmer needs first to identify the main reason for planting a cover crop (e.g., to prevent erosion or to provide nitrogen) and then determine which cover crop best fulfills that criterion and where it can fit into the rotation. Typical planting and termination dates of the chosen cover crop have to be coordinated with the planting and harvesting dates of the cash crop to ensure a wide enough growing window for the cover crop. In northern regions, the cover crop needs to be cold tolerant to survive hard winters. Hairy vetch and most small grains are able to withstand low temperatures, and winter rye can be planted later in the fall than most other grasses. Legumes and grasses can also be planted in a mix and can either be established by aerial seeding into maturing cash crops in the fall or by drilling or broadcasting seed immediately following harvest.

7.2.3.1.2 Effects of cover crops. Similar to conservation tillage and crop rotation, the purposes of cover crops in any farming system are multifold: They prevent soil erosion by wind or water (Hargrove, 1991); increase yields, especially if legumes are used (Kramberger et al., 2009); reduce nitrate leaching and conserve water resources (Hargrove, 1991; Salmeron et al., 2010); enhance soil organic matter and nitrogen storage (Harper et al., 1995; Stivers and Shennan, 1991); and reduce insect and pathogen damage (Fritz et al., 1995; Laub and Luna, 1992). Although conservation tillage and cover crops overlap in some of their benefits (for example, erosion reduction, soil carbon enhancement), a trial by Veenstra et al. (2006) showed that of the two practices cover crops had the greater impact. In a cotton-tomato rotation in California, cover cropping increased soil organic matter, regardless of the tillage practice, while conservation tillage had effects that were potentially detrimental to plant growth: potassium redistribution to the subsurface of the soil, reduced organic matter, and increased salt concentration.

7.2.3.2 *Role of AM fungi*

One can consider the interval between cash crops as a part of the crop rotation. Whether a farmer leaves the soil bare or utilizes a cover crop can have an impact on the functioning of the AM symbiosis and, potentially, the growth of the proceeding cash crop.

Bare fallow in temperate climates is sometimes a practical choice (e.g., when the harvest of a cash crop such as maize occurs too late to plant a cover crop). In Mediterranean or other dry climates, a long, bare fallow of 10 or more months between crops has been utilized to recharge soil moisture and allow for mineralization of N (Thompson, 1991). Fallow soil syndrome, or long-fallow disorder, can occur if the proceeding crop is dependent on mycorrhizal fungi. Such crops may exhibit both P and Zn deficiency and reduced growth and yield despite high or adequate soil test P values (Thompson, 1996; Wetterauer and Killorn, 1996). Among the crops affected by this phenomenon are linseed (*Linum usitatissimum*) (Thompson 1996), sunflower (*Helianthus annuum*) (Thompson, 1987), cotton (*Gossipium hirsutum*) (Hulugalle et al., 1998), and maize (Wetterauer and Killorn, 1996). AM fungus colonization of roots of crops such as these is reduced due to the long fallow, and a remedy may require application of both P and Zn fertilizer (Thompson, 1996). Twice the normal rate of starter P fertilizer is recommended for maize as a remedy for fallow soil syndrome (Wetterauer and Killorn, 1996).

The negative impact of bare fallow on AM fungi can be predicted with knowledge of their unique biology. First, germination of resting spores of AM fungi is not dependent on the presence of a host plant (Douds and Schenck, 1991). As stated, germination can occur after any dormancy period is met and when soil moisture, temperature, and CO_2 levels are sufficient. Therefore, equal periods of bare fallow would be expected to be more detrimental to AM fungus populations in temperate than in dry climates. Next, given the obligate symbiotic nature of the germinating spores of AM fungi, the spore is dependent on its stored carbon reserves, mainly in the form of lipid, for growth. Although an AM fungus spore is capable of repeated germination if it is unable to find a host root to colonize, successive germinations progressively deplete its energy reserves, resulting in diminished vigor of the AM fungus population as a whole prior to establishment of the cash crop. Long-fallow disorder is exacerbated by periodic rainfall because it may stimulate growth of otherwise-dormant AM fungus spores and ERM (Pattinson and McGee, 1997). Even the relatively short fallow in temperate climates can be detrimental to the indigenous population of AM fungi when uncharacteristically mild fall or spring weather, outside the cropping season, stimulates metabolic activity of fungal structures.

In addition to the benefits listed, overwintering cover crops are a remedy to fallow soil syndrome. Planting a mycotrophic cover crop during an otherwise-fallow period gives the indigenous population of AM fungi a host plant to colonize and replenish carbon reserves and expand or reestablish the ERM network. An overwintering cover crop of hairy vetch (*Vicia villosa*), planted after wheat harvest, increased the level of AM fungus inoculum in the soil versus that of uncovered soil prior to sowing maize in the spring (Galvez et al., 1995). Use of cover crops such as winter wheat, oats, rye, and even the weed dandelion (*Taraxacum officinale*) resulted in enhanced colonization of succeeding crops of maize and increased yield over plots without the cover crop (Boswell et al., 1998; Kabir and Koide, 2000, 2002).

7.3 Summary

Good farming practices are essential for long-term agricultural sustainability. Farmers now include management practices such as reduced tillage, crop rotation, and overwintering cover crops with the goals of reducing soil erosion, managing nutrient availability, building soil organic matter, controlling weeds, and maintaining yields. What was not considered when these practices were developed was the role of soil biology in the success of these practices or the effect of these practices on soil biology. An important component of the soil biological community is AM fungi. AM fungi form a mutualistic symbiosis with the majority of crop plants. Among the benefits to the plant ascribed to the symbiosis are enhanced nutrient uptake, water relations, and disease resistance. Studies of the biology and function of these organisms help to answer the question of how AM fungi are impacted by, and what their roles are in the success of, these farming practices.

Reduced tillage is increasingly practiced to enhance soil quality through reducing soil erosion and building levels of soil organic matter. It also enhances the dual functioning of the extraradical phase of AM fungi as both the nutrient-absorbing organ of the symbiosis and an important source of inoculum for the new crop. Crop rotations not only combat yield decline in continuous monocultures but also help guard against the characteristic of AM fungi that those that proliferate on a given host plant species are not necessarily those that enhance its growth. Overwintering cover crops are planted to retain or replenish soil N, retard soil erosion, and compete with weeds. An added benefit is that they can act as host plants for AM fungi and in so doing boost their populations. The cover crop provides host roots to colonize and from which to procure fixed carbon during what would otherwise be a bare fallow period for these obligate symbionts. Therefore, understanding the soil biological underpinnings of farming practices can aide in their success and should be considered in developing the sustainable agricultural practices of the future.

References

Abawi, G.S. 1989. Root rot. In *Bean Production Problems: Disease, Insect, Soil and Climatic Constraints of Phaseolus vulgaris*, eds. H.F. Schwartz and M.A. Pastor-Corrales, 105–157. CIAT, Cali, Colombia.

Arihara, J., and T. Karasawa. 2000. Effect of previous crops on arbuscular mycorrhizal formation and growth of succeeding maize. *Soil Science and Plant Nutrition* 46:43–51.

Ashford, D.L., and D.W. Reeves. 2003. Use of a mechanical roller-crimper as an alternative kill method for cover crops. *American Journal of Alternative Agriculture* 18:37–45.

Augé, R.M. 2000. Stomatal behavior of arbuscular mycorrhizal plants. In *Arbuscular Mycorrhizas: Physiology and Function*, eds. Y. Kapulnik and D.D. Douds, Jr., 201–237. Kluwer, Dordrecht, the Netherlands.

Bago, B., P.E. Pfeffer, D.D. Douds, J. Brouillette, G. Bécard, and Y. Shachar-Hill. 1999. Carbon metabolism in spores of the arbuscular mycorrhizal fungus *Glomus intraradices* as revealed by NMR spectroscopy. *Plant Physiology* 121:263–271.

Balestrini, R., J. Gomez-Ariza, L. Lanfranco, and P. Bonfante. 2007. Laser micro dissection reveals that transcripts for fire plant and one fungal phosphate transporter genes are contemporaneously present in arbusculated cells. *Molecular Plant-Microbe Interactions* 20:1055–1062.

Barber, S.A. 1977. Application of phosphate fertilizers: methods, rates and time of application in relation to the phosphorus status of soils. *Phosphorus Agriculture* 70:109–115.

Bécard, G., and Y. Piché. 1989a. Fungal growth stimulation by CO_2 and root exudates in vesicular-arbuscular mycorrhizal symbiosis. *Applied and Environmental Microbiology* 55:2320–2325.

Bécard, G., and Y. Piché. 1989b. New aspects on the acquisition of biotrophic status by a vesicular-arbuscular mycorrhizal fungus, *Gigaspora margarita*. *New Phytologist* 112:77–83.

Bennett, A.E., and J.D. Bever. 2009. Trade-offs between arbuscular mycorrhizal fungal competitive ability and host growth promotion in *Plantago lancelota*. *Oecologia* 160:807–816.

Benoit, D.L., G. Leroux, and S. Banville. 2003. Influence of carrot/onion/barley cropping sequence on the weed seed bank and field flora in an organic soil in Quebec, Canada. *Aspects of Applied Biology* 69:69–75.

Berner, A., I. Hildermann, A. Fliessbach, L. Pfiffner, U. Niggli, and P. Mäder. 2008. Crop yield and soil fertility response to reduced tillage under organic management. *Soil and Tillage Research* 101:89–96.

Berzsenyi, Z., B. Gyrffy, and D. Lap. 2000. Effect of crop rotation and fertilization on maize and wheat yields and yield stability in a long-term experiment. *European Journal of Agronomy* 13:225–244.

Biermann, B., and R.G. Linderman. 1983. Use of vesicular-arbuscular mycorrhizal roots, intraradical vesicles and extraradical vesicles as inoculum. *New Phytologist* 95:97–105.

Blaszkowski, J. 1993. Comparative studies of the occurrence of arbuscular fungi and mycorrhiza (Glomales) in cultivated and uncultivated soils of Poland. *Acta Mycologia* 28:93–140.

Boddington, C.L., and J.C. Dodd. 2000. The effect of agricultural practices on the development of indigenous arbuscular mycorrhizal fungi. I. Field studies in an Indonesian ultisol. *Plant and Soil* 218:137–144.

Borie, F., R. Rubio, J.L. Rouanet, A. Morales, G. Borie, and C. Rojas. 2006. Effects of tillage systems on soil characteristics, glomalin and mycorrhizal propagules in a Chilean Ultisol. *Soil and Tillage Research* 88:253–261.

Boswell, E.P., R.T. Koide, D.L. Shumway, and H.D. Addy. 1998. Winter wheat cover cropping, vesicular arbuscular mycorrhizal fungi and maize growth and yield. *Agricultural Ecosystems and Environment* 67:55–65.

Brock, B.G., J.H. Canterberry, and G.C. Naderma. 2000. Ten milestones in conservation tillage: history and role in the North Carolina conservation program. In *Proceedings of the 43rd Annual Meeting of the Soil Science Society of North Carolina*, ed. J.L. Sutherland, 13–18. 18–19 January, Raleigh, NC. SSSNC, Raleigh, NC.

Crop rotation history. 2011. http://science.jrank.org/pages/1870/Crop-Rotation-History.html.

Curran, W., S. Mirsky, and M. Ryan. 2007. Effectiveness of a roller-crimper for control of winter annual cover crops. *Proceedings of the Northeast Weed Science Society* 61:29.

Daniels, B.A., and J.M. Trappe. 1980. Factors affecting spore germination of the vesicular-arbuscular mycorrhizal fungus, *Glomus epigaeus*. *Mycologia* 77:457–471.

Davis, A.S. 2010. Cover-crop roller-crimper contributes to weed management in no-till soybean. *Weed Science* 58:300–309.

Day, J.C., C.B. Hallahan, C.L. Sandretto, and W.A. Lindamood. 1999. Pesticide use in U.S. corn production: does conservation tillage make a difference? *Journal of Soil and Water Conservation* 54:477–484.

Derpsch, R. 1997. Historical review of no-tillage cultivation of crops. MAG-GTZ Soil Conservation Project, Casilla de Correo 1859, Asunción, Paraguay. http://www.rolf-derpsch.com/notill.htm.

Dick, W.A., E.L. McCoy, W.M. Edwards, and R. Lal. 1991. Continuous application of no-tillage in Ohio soils. *Agronomy Journal* 83:65–73.

Dodd, J.C., C.L. Boddington, A. Rodriguez, C. Gonzalez-Gomez, and I. Mansur. 2000. Mycelium of arbuscular mycorrhizal fungi (AMF) from different genera: form, function, and detection. *Plant and Soil* 226:131–151.

Doren, J.W., and D.M. Linn. 1994. Microbial ecology of conservation management systems. In *Soil Biology: Effects on Soil Quality. Advances in Soil Science*, ed. J.I. Hatfield and B.A. Stewart, 1–28. Lewis, Boca Raton, FL.

Douds, D.D., and W.R. Chaney. 1982. Correlation of fungal morphology and development to host growth in a green ash mycorrhiza. *New Phytologist* 92:519–526.

Douds, D.D., L. Galvez, M. Franke-Snyder, C. Reider, and L.E. Drinkwater. 1997. Effect of compost addition and crop rotation point upon VAM fungi. *Agriculture Ecosystems and Environment* 65:257–266.

Douds, D.D., R.R. Janke, and S.E. Peters. 1993. VAM fungus spore populations and colonization of roots of maize and soybean under conventional and low-input sustainable agriculture. *Agriculture Ecosystems and Environment* 43:325–335.

Douds, D.D., P.E. Pfeffer, and Y. Shachar-Hill. 2000. Carbon partitioning, cost, and metabolism of arbuscular mycorrhizas. In *Arbuscular Mycorrhizas: Physiology and Function*, eds. Y. Kapulnik and D.D. Douds, Jr., 107–129. Kluwer, Dordrecht, the Netherlands.
Douds, D.D., and N.C. Schenck. 1991. Germination and hyphal growth of VAM fungi during and after storage in soil at five matric potentials. *Soil Biology and Biochemistry* 23:177–183.
Ellis, J.R., W. Roder, and S.C. Mason. 1992. Grain sorghum-soybean rotation and fertilization influence on vesicular-arbuscular mycorrhizal fungi. *Soil Science Society of America Journal* 56:783–794.
Evans, D.G., and M.H. Miller. 1990. The role of the external mycelial network in the effect of soil disturbance upon vesicular-arbuscular mycorrhizal colonization of maize. *New Phytologist* 114:65–72.
Fairchild, G.L., and M.H. Miller. 1990. Vesicular-arbuscular mycorrhizas and the soil-disturbance-induced reduction of nutrient absorption in maize. III. Influence of phosphorous amendments to soils. *New Phytologist* 114:641–650.
Feldmann, F., and C. Boyle. 1999. Weed-mediated stability of arbuscular mycorrhizal effectiveness in maize monocultures. *Journal of Applied Botany* 73:1–5.
Fritz, V.A., R.R. Allmaras, F.L. Pfleger, and D.W. Davis. 1995. Oat residue and soil compaction influences on common root rot (*Aphanomyces euteiches*) of peas in a fine-textured soil. *Plant and Soil* 171:235–244.
Galvez, L., D.D. Douds, L.E. Drinkwater, and P. Wagoner. 2001. Effect of tillage and farming system upon VAM fungus populations and mycorrhizas and nutrient uptake of maize. *Plant and Soil* 228:299–308.
Galvez, L., D.D. Douds, and P. Wagoner. 2001. Tillage and farming system affect AM fungus populations, mycorrhiza formation, and nutrient uptake by winter wheat in a high P soil. *American Journal of Alternative Agriculture* 16:152–160.
Galvez, L., D.D. Douds, P. Wagoner, L.R. Longnecker, L.E. Drinkwater, and R.R. Janke. 1995. An overwintering cover crop increases inoculum of VAM fungi in agricultural soil. *American Journal of Alternative Agriculture* 10:152–156.
Gavito, M.E., and P.A. Olsson. 2008. Foraging strategies of the external mycelium of the arbuscular mycorrhizal fungi *Glomus intraradices* and *Scutellospora calospora*. *Applied Soil Ecology* 39:282–290.
Gazey, C., L.K. Abbott, and A.D. Robson. 1992. The rate of development of mycorrhizas affects the onset of sporulation and production of external hyphae by two species of *Acaulospora*. *Mycological Research* 96:643–650.
Genre, A., and P. Bonfante. 2010. The making of symbiotic cells in arbuscular mycorrhizal roots. In *Arbuscular Mycorrhizas: Physiology and Function*, 2nd ed., eds. H. Koltai and Y. Kapulnik, 57–72. Springer, Dordrecht, the Netherlands.
Gilley, J.E., J.W. Doran, and D.L. Karlen. 1997. Runoff, erosion, and soil quality characteristics of a former Conservation Reserve Program site. *Journal of Soil and Water Conservation* 52:189–193.
Glynne, M.D. 1965. Crop sequence in relation to soil-borne pathogens. In *Ecology of Soil-borne Plant Pathogens, Prelude to Biological Control*, eds. K.F. Baker and W.C. Snyder, 423–433. University of California Press, Berkeley.
Goss, M.J., and A. deVarennes. 2002. Soil disturbance reduces the efficacy of mycorrhizal associations for early soybean growth and N_2 fixation. *Soil Biology and Biochemistry* 34:1167–1173.
Graham, J.H., and D.M. Eissenstat. 1998. Field evidence for the carbon cost of citrus mycorrhizas. *New Phytologist* 140:103–110.
Grant, C.A., D.N. Flaten, D.J. Tomasiewicz, and S.C. Sheppard. 2001. The importance of early season phosphorous nutrition. *Canadian Journal of Plant Science* 81:211–224.
Halpern, M.T., J.K. Whalen, and C.A. Madramootoo. 2010. Long-term tillage and residue management influences soil carbon and nitrogen dynamics. *Soil Science Society of America Journal* 74:1211–1217.
Hargrove, W.L. 1991. *Cover Crops for Clean Water.* Soil Water Conservation Society, Ankeny, IA.
Harper, L.A., P.F. Hendrix, G.W. Langdale, and D.C. Coleman. 1995. Clover management to provide optimum nitrogen and soil water conservation. *Crop Science* 35:176–182.
Helgason, T., J. Merryweather, A. Fritter, and P. Young. 1998. Host preference and community structure of arbuscular mycorrhizal (AM) fungi in a semi-natural woodland. In *Proceedings of the Second International Conference on Mycorrhiza*, 81. 5–10 July, Uppsala, Sweden.

Hendrix, J.W., K.J. Jones, and W.C. Nesmith. 1992. Control of pathogenic mycorrhizal fungi in maintenance of soil productivity by crop rotation. *Journal of Production Agriculture* 5:383–386.

Hetrick, B., A. Daniels, and J. Bloom. 1986. The influence of host plant on production and colonization ability of vesicular-arbuscular mycorrhizal spores. *Mycologia* 78:32–36.

Hulugalle, N.R., P.C. Entwistle, J.L. Cooper, S.J. Allen, and D.B. Nehl. 1998. Effect of long-fallow on soil quality and cotton lint yield in an irrigated self-mulching, grey vertosol in the central-west of New South Wales. *Australian Journal of Soil Research* 36:621–639.

Hunt, D. 1977. *Farm Power and Machinery Management—Laboratory Manual and Workbook.* Iowa State University, Ames, IA, pp. 46–47.

Isensee, A.R., and A.M. Sadeghi. 1995. Long-term effects of tillage and rainfall on herbicide leaching to shallow ground water. *Chemosphere* 30:671–685.

Jakobsen, I., and N.E. Nielsen. 1983. Vesicular-arbuscular mycorrhiza in field-grown crops. I. Mycorrhizal infection in cereals and peas at various times and soil depths. *New Phytologist* 93:401–413.

Jansa, J., A. Mozofar, T. Anken, R. Ruh, I.R. Sanders, and E. Frossard. 2002. Diversity and structure of AMF communities as affected by tillage in a temperate soil. *Mycorrhiza* 12:225–234.

Jansa, J., A. Mozofar, G. Kuhn, T. Anken, R. Ruh, I.R. Sanders, and E. Frossard. 2003. Soil tillage affects the community structure of mycorrhizal fungi in maize roots. *Ecological Applications* 13:1164–1176.

Jansa, J., F.A. Smith, and S.E. Smith. 2008. Are there benefits of simultaneous root colonization by different arbuscular mycorrhizal fungi? *New Phytologist* 177:779–789.

Jasper, D.A., L.K. Abbott, and A.D. Robson. 1989a. Soil disturbance reduces the infectivity of external hyphae of vesicular-arbuscular mycorrhizal fungi. *New Phytologist* 112:93–99.

Jasper, D.A., L.K. Abbott, and A.D. Robson. 1989b. Hyphae of a vesicular-arbuscular mycorrhizal fungus maintain infectivity in dry soil, except when the soil is disturbed. *New Phytologist* 112:101–107.

Jasper, D.A., A.D. Robson, and L.K. Abbott. 1979. Phosphorous and the formation of vesicular-arbuscular mycorrhizas. *Soil Biology and Biochemistry* 11:501–505.

Johnson, N.C., P.J. Copeland, R.K. Crookston, and F.L. Pfleger. 1992. Mycorrhizae: possible explanation for yield decline with continuous corn and soybean. *Agronomy Journal* 84:387–390.

Johnson, N.C., J.H. Graham, and F.A. Smith. 1997. Functioning of mycorrhizal associations along the mutualism-parasitism continuum. *New Phytologist* 135:575–585.

Johnson, N.C., F.L. Pfleger, R.K. Crookston, and S.R. Simmons. 1991. Vesicular-arbuscular mycorrhizas respond to corn and soybean cropping history. *New Phytologist* 117:657–664.

Kabir, Z. 2005. Tillage or no tillage: impact on mycorrhizae. *Canadian Journal of Plant Science* 85:23–29.

Kabir, Z., and R.T. Koide. 2000. The effect of dandelion or a cover crop on mycorrhiza inoculum potential, soil aggregation and yield of maize. *Agriculture Ecosystems and Environment* 78:167–174.

Kabir, Z., and R.T. Koide. 2002. Effect of autumn and winter mycorrhizal cover crops on soil properties, nutrient uptake and yield of sweet corn in Pennsylvania, USA. *Plant and Soil* 238:205–215.

Kabir, Z., I.P. O' Halloran, J.W. Fyles, and C. Hamel. 1998. Dynamics of mycorrhizal symbiosis of corn (*Zea mays L.*): effects of host physiology, tillage practice and fertilization on spatial distribution of extra-radical mycorrhizal hyphae in the field. *Agriculture Ecosystems and Environment* 68:151–163.

Karasawa, T., Y. Kasahara, and M. Takebe. 2001. Variable response of growth and arbuscular mycorrhizal colonization of maize plants to preceding crop in various types of soil. *Biology Fertility of Soils* 33:286–293.

Karlen, D.L., N.C. Wollenhaupt, D.C. Erbach, E.C. Berry, J.B. Swan, N.S. Eash, and J.L. Jordahl. 1994. Long-term tillage effects on soil quality. *Soil and Tillage Research* 32:313–327.

Kegode, G.O., F. Forcella, and S. Clay. 1999. Influence of crop rotation, tillage and management inputs on weed seed production. *Weed Science* 47:175–183.

Kirkegaard, J., O. Christen, J. Krupinsky, and D. Layzell. 2008. Break crop benefits in temperate wheat production. *Field Crops Research* 107:185–195.

Klironomos, J.N., and M.M. Hart. 2002. Colonization of roots by arbuscular mycorrhizal fungi using different sources of inoculum. *Mycorrhiza* 12:181–184.

Kramberger, B., A. Gselman, M. Janzekovic, M. Kaligaric, and B. Bracko. 2009. Effects of cover crops on soil mineral nitrogen and on the yield and nitrogen content of maize. *European Journal of Agronomy* 31:103–109.

Krutz, L.J., M.A. Locke, and R.W. Steinriede, Jr. 2009. Interactions of tillage and cover crop on water, sediment, and pre-emergence herbicide loss in glyphosate-resistant cotton: implications for the control of glyphosate-resistant weed biotypes. *Journal of Environmental Quality* 38:1240–1247.

Langdale, G.W., A.P. Barnett, R.A. Leonard, and W.G. Fleming. 1979. Reduction of soil erosion by the no-till system in the southern piedmont. *Transactions of the American Society of Agricultural Engineers* 22:82–86, 92.

Laub, C.A., and J.M. Luna. 1992. Winter cover crop suppression practices and natural enemies of armyworm (*Lepidoptera Noctuidae*) in no-till corn. *Environmental Entomology* 21:41–49.

Li, X.-L., E. George, and H. Marschner. 1991. Extension of the phosphorous depletion zone in VA-mycorrhizal white clover in a calcareous soil. *Plant and Soil* 136:41–48.

Liebman, M., and E. Dyck. 1993. Crop rotation and intercropping strategies for weed management. *Ecological Applications* 3:92–122.

Linderman, R.G. 2000. Effects of mycorrhizas on plant tolerance to disease. In *Arbuscular Mycorrhizas: Physiology and Function*, eds. Y. Kapulnik and D.D. Douds, Jr., 345–366. Kluwer, Dordrecht, the Netherlands.

Lyons, D.J., S.D. Miller, and G.A. Wicks. 1996. The future of herbicides in weed control systems of the great plains. *Journal of Production Agriculture* 9:209–212.

McGonigle, T.P., D.G. Evans, and M.H. Miller. 1990. Effect of degree of soil disturbance on mycorrhizal colonization and phosphorous absorption by maize in growth chamber and field experiments. *New Phytologist* 116:629–636.

McGonigle, T.P., and M.H. Miller. 1996. Mycorrhizae, phosphorous absorption and yield of maize in response to tillage. *Soil Science Society of American Journal* 60:1856–1861.

Miller, M.H. 2000. Arbuscular mycorrhizae and the phosphorous nutrition of maize: a review of Guelph studies. *Canadian Journal of Plant Science* 80:47–52.

Mirsky, S.B., W.S. Curran, D.A. Mortensen, M.R. Ryan, and D.L. Shumway. 2009. Control of cereal rye with a roller/crimper as influenced by cover crop phenology. *Agronomy Journal* 101:1589–1596.

Mischler, R.A., S.W. Duiker, W.S. Curran, and D. Wilson. 2010. Hairy vetch management for no-till organic corn production. *Agronomy Journal* 102:355–362.

Mozafar, A., J. Jansa, R. Ruh, T. Anken, I. Sanders, and E. Frossard. 2001. Functional diversity of AMF co-existing in agricultural soils subjected to different tillage. In *Proceeding of the Third International Conference on Mycorrhizas*, 132. 8–13 July, Adelaide, South Australia.

Mutch, D. 2004. New weed control strategies for low input and organic soybean. http://fieldcrop.msu.edu/documents/GR04-069.pdf.

Nagahashi, G., and D.D. Douds. 2000. Partial separation of root exudates components and their effects upon the growth of germinated spores of AM fungi. *Mycologia Research* 104:1453–1464.

Nagahashi, G., D.D. Douds, and Y. Ferhatoglu. 2010. Functional categories of root exudate compounds and their relevance to AM fungal growth. In *Arbuscular Mycorrhizas: Physiology and Function*, 2nd ed., eds. H. Koltai and Y. Kapulnik, 33–56. Springer, Dordrecht, the Netherlands.

Oehl, F., E. Sieverding, K. Ineichen, E.A. Ris, and T. Boller. 2005. Community structure of arbuscular mycorrhizal fungi at different soil depths in extensively and intensively managed agro eco systems. *New Phytologist* 165:273–283.

Oldroyd, E.D., and J. Downie. 2006. Nuclear calcium changes at the core of symbiosis signaling. *Current Opinion in Plant Biology* 9:351–357.

Panja, B.N., and S. Chaudhuri. 2004. Exploitation of soil arbuscular mycorrhizal potential for AM-dependent mandarin orange plants by pre-cropping with mycotrophic crops. *Applied Soil Ecology* 26:249–255.

Pattinson, G.S., and P.A. McGee. 1997. High densities of arbuscular mycorrhizal fungi maintained during long fallow in soils used to grow cotton except when soil is wetted periodically. *New Phytologist* 136:571–580.

Pfeffer, P.E., D.D. Douds, G. Bécard, and Y. Shachar-Hill. 1999. Carbon uptake and the metabolism and transport of lipids in an arbuscular mycorrhiza. *Plant Physiology* 120:587–598.

Porter, P.M., D.R. Huggins, C.A. Perillo, S.R. Quiring, and R.K. Crookston. 2003. Organic and other management strategies with two- and four-year crop rotations in Minnesota. *Agronomy Journal* 95:233–244.

Rosendahl, S., and B. Matzen. 2008. Genetic structure of arbuscular mycorrhizal populations in fallow and cultivated soils. *New Phytologist* 179:1154–1161.

Ryan, M.R., D.A. Mortensen, G.G.T. Camargo, and T.L. Richard. 2009. Rotational no-tillage management for reducing the carbon footprint of organic corn-soybean based cropping systems. In *Proceedings of the Annual Meeting of the Agronomy Society of America*, 1–5 November, Pittsburgh, PA.

Ryan, M.H., R.M. Norton, J.A. Kirkegaard, K.M. McCormick, S.E. Knights, and J.F. Angus. 2002. Decreasing mycorrhizal colonization does not improve growth and nutrition of wheat on vertisols in south-eastern Australia. *Australian Journal of Agricultural Research* 53:1173–1181.

Salmeron, M., J. Cavero, D. Quilez, and R. Isla. 2010. Winter cover crops affect monoculture maize yield and nitrogen leaching under irrigated Mediterranean conditions. *Agronomy Journal* 102:1700–1709.

Sanchez, J.E., R.R. Harwood, T.C. Willson, K. Kizilkaya, J. Schmeenk, E. Parker, E.A. Paul, B.D. Knezek, and G.P. Robertson. 2004. Managing soil carbon and nitrogen for productivity and environmental quality. *Agronomy Journal* 96:769–775.

Sarrantonio, M. 2003. Soil responses to surface-applied residues of varying carbon-nitrogen ratios. *Biology and Fertility of Soils* 37:175–183.

Schalamuk, S., S. Velazquez, H. Chidichimo, and M. Cabello. 2006. Fungal spore diversity of arbuscular mycorrhizal fungi associated with spring wheat: effects of tillage. *Mycologia* 98:16–22.

Schenck, N.C., J.O. Sigueira, and E. Oliveira. 1989. Changes in the incidence of VA mycorrhizal fungi with changes in ecosystems. In *Interrelationships between Microorganisms and Plants in Soil*, eds. V. Vancura and F. Kunc, 125–129. Elsevier, New York.

Schreiber, M.M. 1992. Influence of tillage, crop rotation and weed management on giant foxtail population dynamics and corn yield. *Weed Science* 40:645–653.

Shachar-Hill, Y., P.E. Pfeffer, D. Douds, S.F. Osman, L.W. Doner, and R.G. Ratcliffe. 1995. Partitioning of intermediary carbon metabolism in vesicular-arbuscular mycorrhizal leek. *Plant Physiology* 108:7–15.

Shipitalo, M.J., and W.M. Edwards. 1998. Runoff and erosion control with conservation tillage and reduced-input practices on cropped watersheds. *Soil and Tillage Research* 46:1–12.

Sims, G.K., D.D. Buhler, and R.F. Turco. 1994. Residue management impact on the environment. In *Managing Agricultural Residues*, ed. P.W. Under, 77–98. Lewis, Boca Raton, FL.

Smith, A., C. Reberg-Horton, G. Place, A. Meijer, C. Arellano, and J. Mueller. 2011. Rolled rye mulch for weed suppression in organic no-tillage soybeans. *Weed Science* 59:224–231.

Smith, F.A., I. Jakobsen, and S.E. Smith. 2000. Spatial differences in acquisition of soil phosphate between two arbuscular mycorrhizal fungi in symbiosis with *Medicago truncatula*. *New Phytologist* 147:357–366.

Smith, R.G., K.L. Gross, and G.P. Robertson. 2008. Effects of crop diversity on agroecosystem function: crop yield response. *Ecosystems* 11:355–366.

Smith, T.F. 1978. A note on the effect of soil tillage on the frequency and vertical distribution of spores of vesicular-arbuscular endophytes. *Australian Journal of Soil Research* 16:359–361.

Stampe, E.D., and C.C. Daehler. 2003. Mycorrhizal species identity affects plant community structure and invasion: a microcosm study. *Oikos* 100:362–372.

Stivers, L.J., and C. Shennan. 1991. Meeting the nitrogen needs of processing tomatoes through winter cover cropping. *Journal of Production Agriculture* 4:330–335.

Struble, J.E., and H.D. Skipper. 1985. Spore production by five VAM fungi as influenced by plant species. In *Proceedings of the 6th North American Conference on Mycorrhizas*, 251. 25–29 June, Bend, OR.

Teasdale, J.R., C.B. Coffman, and R.W. Mangum. 2007. Potential long-term benefits of no-tillage and organic cropping systems for grain production and soil improvement. *Agronomy Journal* 99:1297–1305.

Teasdale, J.R., R.W. Mangum, J. Radhakrishnan, and M.A. Cavigelli. 2004. Weed seedbank dynamics in three organic farming crop rotations. *Agronomy Journal* 96:1429–1435.

Thompson, J.P. 1987. Decline of vesicular-arbuscular mycorrhizae in long-fallow disorder of field crops and its expression in phosphorus deficiency of sunflower. *Australian Journal of Agricultural Research* 38:847–867.

Thompson, J.P. 1991. Improving the mycorrhizal condition of the soil through cultural practices and effects on growth and phosphorous uptake by plants. In *Phosphorous Nutrition of Grain Legumes in the Semi-arid Tropics*, eds. C. Johansen, K.K. Lee, and K.L. Sahrawat, 117–138. International Crops Research Institute for the Semi-Arid Tropics. Patancheru, Andhra Pradesh, India.

Thompson, J.P. 1996. Correction of dual phosphorous and zinc deficiencies of linseed (*Linum usitatissimum L.*) with cultures of vesicular-arbuscular mycorrhizal fungi. *Soil and Biology Biochemistry* 28:941–951.

Tommerup, I.C. 1981. Survival mechanisms of VA mycorrhizal fungi. In *Proceedings of the Fifth North American Conference on Mycorrhizae Program and Abstracts*, 16. 16–21 August, Montreal, Quebec, Canada.

Toth, R., and R.M. Miller. 1984. Dynamics of arbuscule development and degeneration in a *Zea mays* mycorrhiza. *American Journal of Botany* 71:449–460.

Treseder, K.K., and M.F. Allen. 2002. Direct nitrogen and phosphorous limitation of arbuscular mycorrhizal fungi: a model and field test. *New Phytologist* 155:507–515.

Trivedi, P.C., and K.R. Barker. 1986. Management of nematodes by cultural practices. *Nematropica* 16:213–236.

Van der Heijden, M.G.A., J.N. Klironomos, M. Ursic, P. Moutoglis, R. Streitwolf-Engel, T. Boller, A. Wiemken, and I.R. Sanders. 1998. Mycorrhizal fungus diversity determines plant biodiversity, ecosystem variability and productivity. *Nature* 396:69–72.

Veenstra, J.J., W.R. Horwath, J.P. Mitchell, and D.S. Munk. 2006. Conservation tillage and cover cropping influence soil properties in San Joaquin Valley cotton-tomato crop. *California Agriculture* 60:146–153.

West, T.O., and W.M. Post. 2002. Soil organic carbon sequestration rates by tillage and crop rotation: a global data analysis. *Soil Science Society of America Journal* 66:1930–1946.

Wetterauer, D.G., and R.J. Killorn. 1996. Fallow- and flooded-soil syndromes: effects on crop production. *Journal of Production Agriculture* 9:39–41.

Williams, J.D., H.T. Gollany, M.C. Siemens, S.B. Wuest, and D.S. Long. 2009. Comparison of runoff, soil erosion, and winter wheat yields from no-till and inversion tillage production systems in northeastern Oregon. *Journal of Soil and Water Conservation* 64:43–52.

Wright, S.F., and A. Upadhyaya. 1996. Extraction of an abundant and unusual protein from soil and comparison with hyphal protein of arbuscular mycorrhizal fungi. *Soil Science* 161:575–586.

chapter 8

Effects of the cultivation of genetically modified Bt crops on nontarget soil organisms

Tanya E. Cheeke
Portland State University

Contents

8.1 Introduction

Genetically modified (GM) crops were first commercially introduced in 1996 and are now cultivated in at least 25 countries (Table 8.1; James, 2010). Historically, the primary types of GM crops cultivated were herbicide tolerant (i.e., Roundup Ready® soybean) and insect resistant (genetically engineered to express a gene from *Bacillus thuringiensis,* i.e., *Bt* maize), but more recently new GM cultivars have been developed that offer stacked traits (herbicide tolerance plus resistance to multiple insect pests); increased stress tolerance (e.g., salt-stress- or drought-tolerant varieties); improved nutrient status; physiological enhancements (longer storage, delayed ripening); and even pharmaceutical crops engineered to produce drugs, human growth hormones, and other products of medical interest (Stotzky

Table 8.1 Global Area of Genetically Modified Crops in 2010: By Country (million hectares)

Rank	Country	Area (million hectares)	Biotech crops
1	United States	66.8	Maize, soybean, cotton, canola, sugar beet, alfalfa, papaya, squash
2	Brazil	25.4	Soybean, maize, cotton
3	Argentina	22.9	Soybean, maize, cotton
4	India	9.4	Cotton
5	Canada	8.8	Canola, maize, soybean, sugar beet
6	China	3.5	Cotton, tomato, poplar, papaya, sweet pepper
7	Paraguay	2.6	Soybean
8	Pakistan	2.4	Cotton
9	South Africa	2.2	Maize, soybean, cotton
10	Uruguay	1.1	Soybean, maize
11	Bolivia	0.9	Soybean
12	Australia	0.7	Cotton, canola
13	Philippines	0.5	Maize
14	Myanmar	0.3	Cotton
15	Burkina Faso	0.3	Cotton
16	Spain	0.1	Maize
17	Mexico	0.1	Cotton, soybean
18	Columbia	<0.1	Cotton
19	Chile	<0.1	Maize, soybean, canola
20	Honduras	<0.1	Maize
21	Portugal	<0.1	Maize
22	Czech Republic	<0.1	Maize, potato
23	Poland	<0.1	Maize
24	Egypt	<0.1	Maize
25	Slovakia	<0.1	Maize
26	Costa Rica	<0.1	Cotton, soybean
27	Romania	<0.1	Maize
28	Sweden	<0.1	Potato
29	Germany	<0.1	Potato

Source: James, C. 2010. Global status of commercialized biotech/GM crops: 2010. ISAAA brief 42-2010: executive summary. http://isaaa.org/resources/publications/briefs/42/executivesummary/default.asp. With permission.

and Saxena, 2009). Since the commercial introduction of GM plants, the acreage dedicated to GM crop cultivation has increased each year, such that the majority of all major crop plants grown in the United States—soybean, cotton, and maize—are genetically engineered (U.S. Department of Agriculture [USDA], 2010). Developing countries also continue to increase their share of global GM crop production and now account for almost half (46%) of the global hectarage of GM crops (James, 2010). This rapid and widespread adoption of GM crops has led to a dramatic shift in the agricultural landscape since the mid-1990s and has raised questions about the impact of agricultural biotechnology on nontarget organisms in the soil environment.

Although some GM crops can provide a variety of agricultural benefits, there may also be potential risks to nontarget organisms. Some of the key scientific concerns regarding the widespread cultivation of GM crops include (1) the potential for gene flow from transgenic plants to related species (e.g., Mercer and Wainwright, 2008; Pineyro-Nelson et al., 2009); (2) persistence of GM plant material in the environment (e.g., Saxena and Stotzky, 2001b; Zwahlen, Hilbeck, Gugerli, et al., 2003; Stotzky, 2004; Flores et al., 2005; Tarkalson et al., 2008); (3) the evolution of pest resistance (e.g., Gould et al., 2002; Abel and Adamczyk, 2004; Huang et al., 2007; Gao et al., 2010); (4) risks to the environment associated with changes in the agricultural landscape or farming practices associated with the adoption of GM crops (e.g., Krogh et al., 2007; Lupwayi et al., 2007; Watrud et al., 2011); and (5) the risk to nontarget organisms, including mammals, birds, fish, insects, and soil organisms (e.g., Stotzky, 2000; Adamczyk and Hardee, 2002; Kowalchuk et al., 2002; Clark et al., 2005; Rosi-Marshall et al., 2007; Thies and Devare, 2007; Icoz and Stotzky, 2008b; Lang and Otto, 2010; Then, 2010; Gatehouse et al., 2011).

This chapter summarizes the environmental risk assessment research of *Bt* crops to date in soil and offers suggestions on how to examine and understand better the effects of these types of GM crops on soil organisms. The chapter includes discussion of the following: an introduction to *Bt* crops (what they are, where they are grown, how they are engineered, etc.); how *Bt* and other transgenic crops are regulated in the United States; the fate of *Bt* plant material in soil (how *Bt* toxin enters soil, binding properties, potential differences in degradation rates of transgenic plant material, etc.); effects of *Bt* crops on soil organisms, including bacteria, fungi, protozoa, nematodes, and soil invertebrates (earthworms, microarthropods, insects); and a brief discussion of the escape and introgression of transgenes as this could also have nontarget effects on soil organisms. The question of whether *Bt* crops can contribute to the sustainability of agroecosystems is also discussed. The chapter concludes with a summary and recommendations for future research directions. The information here follows many excellent reviews evaluating effects of transgenic crop cultivation on nontarget organisms in the soil environment (e.g., Stotzky, 2000, 2002, 2004; Giovannetti, 2003; Kowalchuk et al., 2003; Saxena and Stotzky, 2003; Motavalli et al., 2004; Giovannetti et al., 2005; Liu et al., 2005; O'Callaghan et al., 2005; Thies and Devare, 2007; Icoz and Stotzky, 2008b; Liu, 2010; Saxena et al., 2010). Although herbicide tolerance is often incorporated into *Bt* cultivars as a stacked trait, nontarget effects of GM herbicide-tolerant (HT) crops on soil organisms are not included in this review because the genetic insertion has been shown to have no direct effect on soil organisms (although the use of this technology may have indirect effects on soil organisms as a result of changes in agricultural practices) (e.g., Siciliano and Germida, 1999; Dunfield and Germida, 2003, 2004; Kowalchuk et al., 2003; Krogh et al., 2007; Griffiths et al., 2008; reviewed by Lundgren et al., 2009; Watrud et al., 2011).

8.2 Bt crops

Insect-resistant crops are genetically engineered to express insecticidal toxins derived from the spore-forming soil bacterium *Bacillus thuringiensis* (*Bt*). Naturally occurring *Bt* soil organisms produce insecticidal crystalline proteins (called Cry proteins) during sporulation that are toxic to the larvae of certain insects (Hofte and Whiteley, 1989; reviewed in Schnepf et al., 1998; USDA, 2010). To date, more than 60 different Cry proteins have been identified that exhibit a high degree of specificity toward Lepidoptera (e.g., moths and butterflies); Coleoptera (e.g., beetles); Diptera (e.g., flies and mosquitoes); Homoptera (e.g., cicadas, leafhoppers, aphids, scales); Hymenoptera (e.g., wasps, bees, ants, sawflies); Orthoptera (e.g., grasshoppers, crickets, and locusts); Mallophaga (e.g., lice); and nematodes (reviewed in Schnepf et al., 1998; Federici, 2002; Stotzky, 2002; Lee et al., 2003; Icoz and Stotzky, 2008b; Sanchis, 2011). Some *Bt* proteins have even been used for targeted treatment of some types of cancer cells in humans (e.g., Ito et al., 2004; Yamashita et al., 2005; Ohba et al., 2009; Tharakan et al., 2009; Nagamatsu et al., 2010; Poornima et al., 2010; Wong et al., 2010).

Formulations of *Bt* carrying the parasporal crystals have been used as a natural insecticide in agricultural systems since the 1930s (Hofte and Whiteley, 1989; reviewed in Beegle and Yamamoto, 1992; Sanchis, 2011), but success is often compromised by the poor survival of the natural form of *Bt* in the environment (Griego and Spence, 1978; West, 1984; West et al., 1985; Clark et al., 2005). Moreover, the *Bt* toxin present in the soil bacteria is not activated until cleaved by alkaline hydrolysis in the gut of a susceptible insect larva (Hofte and Whiteley, 1989), and activation may also require the presence of indigenous bacteria in the midgut of susceptible insects (Broderick et al., 2006, 2009; reviewed by Then, 2010). The *Bt* gene that is genetically engineered into plants, however, is truncated and constitutively produces only the preactivated Cry protein in the cells of the GM plant (e.g., Shu et al., 2002; Xu et al., 2006).

At present, the two major crops that contain genes coding for insecticidal *Bt* toxin are *Bt* maize and *Bt* cotton. Other *Bt* crops that have been developed include *Bt* potato, *Bt* tobacco, *Bt* spruce, *Bt* tomato, *Bt* rice, *Bt* eggplant, *Bt* sunflower, and *Bt* canola, although not all of these are presently commercially available. In 2010, 86% of the maize and 93% of the cotton cultivated in the United States was genetically modified to express herbicide tolerance, insect resistance, or some combination of inserted traits (USDA, 2010; Table 8.2), making up 26% and 49% of the global GM crop acreage, respectively (James, 2010). The dramatic rise in the adoption rate of GM crops resulted primarily from the development of GM varieties containing "stacked traits" or "pyramided traits" (as opposed to single traits in one variety or hybrid). The term *stacked trait* refers to a plant that has been engineered to express multiple toxins against different pests (e.g., protection against European corn borer and corn root worm) or contains multiple plant protection properties (e.g., herbicide tolerance plus insect resistance), whereas a *pyramided trait* is one in which multiple toxins are expressed to target the same pest (U.S. Environmental Protection Agency [EPA], 2011). In 2009, 75% of the GM maize hybrids in the United States were engineered with double or triple stacked traits (James, 2010). One of the newest GM maize hybrids, SmartStax®, was engineered to express eight different genes coding for pest resistance and herbicide tolerance and produces six different types of Cry proteins—Cry1A.105, Cry2Ab2, Cry1F, Cry3Bb1, Cry34Ab1, and Cry35Ab1—to protect plants against 13 different insect pests (European corn borer, southwestern corn borer, southern cornstalk borer, corn earworm, fall armyworm, stalk borer, lesser corn stalk borer, sugarcane borer, western bean cutworm, black cutworm, western corn rootworm, northern corn rootworm, Mexican corn

Table 8.2 Adoption of Genetically Modified (GM) Plant Varieties by Major Crop (Maize, Upland Cotton, and Soybean) in the United States, 2000–2011 (percentage of all corn, cotton, and soybean planted by year)

Crop	2000	2001	2002	2003	2004	2005	2006	2007	2008	2009	2010	2011
Insect-resistant (*Bt*) only												
Maize[a]	18	18	22	25	27	26	25	21	17	17	16	16
Cotton[b]	15	13	13	14	16	18	18	17	18	17	15	17
Soybean[c]	N/A	N/A	N/A	N/A	N/A	N/A	N/A	N/A	N/A	N/A	N/A	N/A
Herbicide-tolerant only												
Maize[a]	6	7	9	11	14	17	21	24	23	22	23	23
Cotton[b]	26	32	36	32	30	27	26	28	23	23	20	15
Soybean[c]	54	68	75	81	85	87	89	91	92	91	93	94
Stacked gene varieties (multiple insect-resistance traits or insect resistance + herbicide tolerance)												
Maize[a]	1	1	2	4	6	9	15	28	40	46	47	49
Cotton[b]	20	24	22	27	30	34	39	42	45	48	58	58
Soybean[c]	N/A	N/A	N/A	N/A	N/A	N/A	N/A	N/A	N/A	N/A	N/A	N/A
All GM varieties												
Maize[a]	25	26	34	40	47	52	61	73	80	85	86	88
Cotton[b]	61	69	71	73	76	79	83	87	86	88	93	90
Soybean[c]	54	68	75	81	85	87	89	91	92	91	93	94

Source: United States Department of Agriculture, National Agricultural Statistics Service, Acreage, July 1, 2011. http://www.ers.usda.gov/Data/BiotechCrops/ExtentofAdoptionTable1.htm.

[a] Primarily grown in Illinois, Indiana, Iowa, Kansas, Michigan, Minnesota, Missouri, Nebraska, North Dakota, Ohio, South Dakota, Texas, and Wisconsin.

[b] Primarily grown in Alabama, Arkansas, California, Georgia, Louisiana, Mississippi, Missouri, North Carolina, Tennessee, and Texas.

[c] Primarily grown in Arkansas, Illinois, Indiana, Iowa, Kansas, Michigan, Missouri, Nebraska, North Dakota, Ohio, South Dakota, and Wisconsin.

rootworm) (EPA, 2009). Future GM crop varieties are expected to include multiple traits for pest resistance and tolerance to herbicides and drought, as well as nutritional traits such as high omega-3 oil in soybean or enhanced nutrient content of other staple crops (James, 2010; Monsanto, 2011). Stacked and pyramided traits have become important features of GM crops and will continue to be adopted by farmers worldwide.

8.2.1 *How Bt crops are genetically engineered*

Genetically modified plants can be engineered to express a variety of novel traits (usually with DNA obtained from another type of organism) that confer protection against insect pests, tolerance to herbicides, increase vitamin content or nutrient status, improve drought tolerance, produce pharmaceuticals, or impart almost any other characteristic that is deemed to be agriculturally or commercially important. The process of genetic engineering involves the identification and isolation of desired genes (e.g., the genes in strains of the soil bacterium *B. thuringiensis* that produce *Bt* proteins); the selection of a host plant (often called the parental line or the parental isoline); insertion of the foreign gene into the host plant cells (called transformation, often using a bacterial vector, such as *Agrobacterium tumefaciens*, electroporation, or microprojectile bombardment to insert the foreign DNA physically into the host cell); screening for successful transformants; and then regeneration of the whole transformed plants. The individual transformation events are indicated in the nomenclature of the GM plant, for example, *Bt* corn Event 11 (often abbreviated to *Bt* 11 or Event *Bt* 11). The nontransgenic parental cultivar from which the transgenic line was engineered is called the nontransgenic isoline. The isoline is a plant line that is nearly genetically identical to its parental base hybrid except for the genetic insertion(s) (e.g., *Bt* 11 and its nontransgenic parental isoline Providence).

Although it is usually clear through selective screening which plant cells have taken up the foreign genes, it is seldom known exactly where in the host plant genome new DNA has been incorporated. Consequently, unintended pleiotropic effects as a result of the genetic insertion can occur (e.g., Sheveleva et al., 1998; reviewed in Wang et al., 2003, and in Giovannetti et al., 2005), and whole transgenic plants must be carefully evaluated for plant performance and undesirable phenotypic characteristics before advancing to the GM crop registration stage (see the next section on how *Bt* and other GM plants are regulated in the United States). While obviously abnormal phenotypes are eliminated during this screening process, it is possible that certain types of pleiotropic effects, such as those that influence the physiology (e.g., sugar allocation, enzyme activity in roots, lignin content) of individual insertion events, may not be detected in some plant lines.

8.2.2 *How Bt and other GM plants are regulated in the United States*

The United States of America is the world's largest producer of GM crops, with 66.8 million hectares cultivated with GM crops in 2010, more than twice the biotech cultivation area of any other country (Table 8.1). The U.S. government oversees the development, deployment, and safety of transgenic organisms through three separate agencies: the USDA Animal and Plant Health Inspection Service (USDA-APHIS), the EPA, and the Food and Drug Administration (FDA). Together, these government agencies review potential areas of concern regarding the introduction of new GM crops, including the potential for gene flow, resistance management, effects on nontarget organisms and the environment, and the introduction or elevation of potential allergens. The specific regulatory duties of each agency are as follows:

1. The USDA is responsible for regulating the introduction (importation, interstate transport, and field release) of GM organisms (plants, insects, microbes, etc.) or any other organism that is, or could be, a plant pest. The Biotechnology Regulatory Service (BRS) of APHIS is the regulatory body within the USDA responsible for deciding whether a genetically engineered organism is as safe for the environment as its traditionally bred counterpart.
2. The EPA regulates and assesses the environmental impact of certain types of GM organisms with pesticidal properties (e.g., insect resistance, disease resistance, and certain plant growth regulator-expressing products) through the Federal Insecticide, Fungicide, Rodenticide Act (FIFRA). The EPA regulates the gene and its product, not the plant (as does APHIS and FDA), as plant-incorporated protectants (PIPs). Biotechnology companies submit their data to the EPA for permitting and approval before these crops can be grown commercially. The data requirements for EPA approval include product characterization, mammalian toxicity, allergenicity potential, effects on nontarget organisms, and environmental fate, and for *Bt* products include insect resistance management strategies (EPA, 2010). EPA also sets tolerances

under the Federal Food Drug and Cosmetic Act (FFDCA) for residues of pesticidal substances in or on food and feed crops like *Bt* maize or cotton (C. Wozniak, U.S. Environmental Protection Agency Biopesticides and Pollution Prevention Division, Washington, DC, personal communication).

3. The FDA examines the toxicity and allergenicity of GM plants under the FFDCA. This includes oversight of food plants that contain transgenic proteins that are not normally found in that food source (e.g., an animal gene expressed in plants). When the FDA conducts safety evaluations of GM crops, it evaluates both the final product and the techniques used to develop it. The FDA's process, however, is voluntary; the FDA's approach is a comparative one based on a compositional analysis of the GM plant and its non-GM counterpart (C. Wozniak, U.S. Environmental Protection Agency Biopesticides and Pollution Prevention Division, Washington, DC, personal communication). The FDA also enforces the tolerances set by the EPA in the event tolerance limits are exceeded.

It should be noted and emphasized here that the U.S. regulatory agencies do not generally conduct risk assessment studies themselves; the regulatory agencies review the data that biotechnology companies provide (which is often from work conducted by independent third-party laboratories), and to get a new GM plant approved for commercial use, the product must show minimal risk to nontarget organisms, human health, and the environment. It takes many years (usually > 6) for seed companies to go from the discovery phase to commercial sales and distribution (Monsanto, 2011). Usually, early contact is made with the regulatory agencies, and a reiterative dialogue between the biotechnology company and the regulatory agencies takes place early in the registration process (C. Wozniak, U.S. Environmental Protection Agency Biopesticides and Pollution Prevention Division, Washington, DC, personal communication, October 3, 2011). Data are then submitted to the EPA PIP products division for an experimental use permit a few years before registration occurs (C. Wozniak, U.S. Environmental Protection Agency Biopesticides and Pollution Prevention Division, Washington, DC, personal communication, October 3, 2011). Once a new crop line advances to the prelaunch phase, there is a 90% chance that it will be introduced into the commercial marketplace (Monsanto, 2011). In this final phase of development, the regulatory data are submitted to the EPA, large quantities of seeds are generated, and the premarket advertising campaign begins. After health and safety data have passed EPA guidelines, the product is approved for commercial sales, and the product is launched.

Some types of GM crops that receive EPA approval have restrictions that limit how and where a particular crop can be grown. For example, to plant *Bt* crops, farmers are required to keep 5–20% of their land (depending on the *Bt* product in use) in a non-*Bt* refuge to help minimize the rate of resistance of targeted pests (EPA, 2010). The EPA also monitors the potential for gene flow between GM plants and wild-type relatives and generally does not approve GM crops that have wild-type relatives that grow in close proximity to GM crop regions. For the major GM crops that have been developed thus far (corn, soy, canola, cotton), there is little risk for cross-pollination in the United States as the wild relatives of most of these crops are found primarily in tropical areas. In the case of *Bt* cotton, however, where the potential for gene flow to wild cotton relatives does exist in the United States (i.e., Hawaii, Florida south of State Route 60, Puerto Rico, and the U.S. Virgin Islands), the EPA has restricted sales and distribution of *Bt* cotton within these areas (EPA, 2010). However, even when there is little chance for gene flow between transgenic and feral or indigenous sexually compatible wild relatives (SCWRs) in the United States, there is a real possibility of gene flow between GM and wild-type varieties in other regions of the world. For example, gene flow has been detected between *Bt* maize and native maize landraces in Mexico (Quist and Chapela, 2001; Pineyro-Nelson et al., 2009), and despite limits on where *Bt* cotton can be cultivated in the United States, *Bt* cotton is also grown in at least 12 other countries, several of which are in tropical regions associated with areas that are within the center of origin for New World cotton species (Table 8.1).

As more and different types of GM crops are developed, the likelihood of gene transfer between SCWR and GM crops is expected to increase, especially if grown in regions where the level of government oversight is not as extensive as it is in the United States. Thus, while the safety testing and approval process can minimize much of the environmental and health/safety risk of GM crops in the United States, it is still possible for some GM crops to have nontarget effects in the environment or on organisms that have either not been tested or may have unexpected effects under certain environmental conditions. It is therefore important that GM crops continue to be evaluated for nontarget effects under a variety of environmental and experimental scenarios, even after they have been approved for commercial use in the United States.

8.3 *Fate and persistence of Bt protein in soil*

Numerous studies have shown that insecticidal *Bt* proteins are released from transgenic plants into soil through root exudates (e.g., Saxena et al., 1999; Saxena and Stotzky, 2000, 2001a; Saxena et al., 2002a, 2002b, 2004; Stotzky, 2004; Icoz and Stotzky, 2008a; Li et al., 2009), pollen (Losey et al., 1999; Zangerl et al., 2001), and plant residue decomposition (Zwahlen, Hilbeck, Gugerli, et al., 2003). Once in soil, the *Bt* toxins bind to clay particles (Tapp et al., 1994; Tapp and Stotzky, 1995) and humic acids (Crecchio and Stotzky, 1998) and can retain their insecticidal properties for at least up to 234 days (Tapp and Stotzky, 1998). In laboratory studies, Cry1Ab protein from *Bt* maize root exudates persisted in soil for at least 180 days and for 350 days in soil amended with *Bt* maize plant material (Saxena and Stotzky, 2002). In a greenhouse pot study in which one *Bt* cotton line, two stacked *Bt* and cowpea trypsin inhibitor (*Bt* + *CpTI*) cotton lines, and their non-GM isolines were consecutively cultivated for 4 years, Cry1Ac and *CpTI* proteins persisted in soil throughout the experiment (Chen et al., 2011), supporting a previous study in which 41% and 60% of the introduced amounts of *Bt* protein from stems and leaves of two *Bt* cottons (Events *Bt*-Zk and *Bt*-GK, respectively) incorporated into soil under laboratory conditions remained after 56 days (Sun et al., 2007). In field studies, Cry1Ab protein from transgenic maize litter has been shown to persist for at least 8 months (Zwahlen, Hilbeck, Gugerli, et al., 2003), although *Bt* protein in soil does not appear to accumulate over time (e.g., Hopkins and Gregorich, 2003; Baumgarte and Tebbe, 2005; Icoz and Stotzky, 2008a).

A higher lignin content has been reported in some *Bt* crops, including several different lines of *Bt* maize (Saxena and Stotzky, 2001c; Flores et al., 2005; Poerschmann et al., 2005; Fang et al., 2007). Higher lignin content has also been reported for *Bt* tobacco, *Bt* cotton, *Bt* canola, *Bt* potato, and *Bt* rice, although these differences were not statistically significant when compared with the non-*Bt* isolines (Flores et al., 2005). The slower decomposition of *Bt* organic material in soil, in some cases, has been attributed to higher lignin in transgenic plant residues (Saxena and Stotzky, 2001c; Stotzky, 2004; Flores et al., 2005). As a result, soil organisms may have a longer exposure to the *Bt* toxins as they are slowly released from organic matter and soil particles over time (Zwahlen, Hilbeck, Gugerli, et al., 2003; Stotzky, 2004).

8.4 Effects of the cultivation of Bt plants on nontarget soil organisms

Soil organisms, including bacteria, fungi, protozoa, nematodes, earthworms, and microarthropods, have a tremendously important role in maintaining plant health and soil fertility through the decomposition of organic matter, nutrient mineralization, provision of protection against disease, and improved soil structure. Moreover, symbiotic soil organisms, such as nitrogen-fixing bacteria and arbuscular mycorrhizal fungi (AMF), provide nutritional benefits to plants in exchange for carbon resources and protection by the host plant. Although ubiquitous, many soil organisms are sensitive to a variety of agricultural practices, including pesticide applications, tilling, cultivation practices (e.g., monocultures vs. intercropping, compost vs. chemical fertilizer), and even the type of plant grown. Because of their close associations with plant roots, some soil organisms, such as AMF or nitrogen-fixing bacteria, may be more sensitive to changes in the physiology of the host plant or in the composition of root exudates as a result of genetic engineering than their free-living counterparts in the soil.

8.4.1 Effects of the cultivation of Bt crops on soil bacteria

Bacteria are by far the most abundant organisms in soil and are important for nutrient mineralization, decomposition of organic matter, protection against plant pathogens, degradation of chemicals/toxins in the environment, and nutrient cycling. In both natural and agroecosystems, bacterial abundance is highest in the rhizosphere (the narrow area of soil directly surrounding and influenced by plant roots). Plants support the development of microbial communities in the rhizosphere by producing root exudates that contain carbon-rich nutrients such as carbohydrates and proteins (Grayston et al., 1996; Morgan et al., 2005). While soil organisms take advantage of these carbon resources, the plants also benefit via increased nutrient availability, improved mineral uptake, and enhanced soil fertility provided by the soil microbial community (Smith and Gianinazzi-Pearson, 1988; Morgan et al., 2005; Smith and Read, 2008).

Because of their close association with plant roots and their importance to biotic soil processes, some of the earliest environmental risk assessment research was aimed at determining the nontarget effects of *Bt* crops on soil bacteria (summarized in Table 8.3). It was thought that a change in the composition or quality of root exudates from *Bt* crops or accumulation of *Bt* toxin in the rhizosphere might modify the composition or activity of soil microbes, ultimately influencing biotically driven ecosystem processes and affecting plant growth and health. To date, however, only a few studies have reported any significant effects (either positive or negative) on soil bacteria as a result of the addition of purified *Bt* proteins to soil, amendment of soil with *Bt* plant material, or cultivation with *Bt* crops (e.g., Donegan et al., 1995; Wu et al., 2004a, 2004b; Castaldini et al., 2005; Rui et al., 2005; Xue et al., 2005; Fang et al., 2007; Sun et al., 2007; Chen et al., 2011). The reported effects of the addition of *Bt* biomass or protein to the soil, as well as the effects of *Bt* crop cultivation on soil bacteria, are summarized here.

In one of the first risk assessment studies to examine the effects of *Bt* plants on soil bacteria, Donegan et al. (1995) reported that soils amended with leaves of different lines of transgenic *Bt* cotton resulted in a statistically significant, but transient, increase in culturable bacteria. However, the plant line-specific response and the lack of effects of the addition of purified *Bt* protein (Cry1Ac) on soil bacteria suggest that these effects may not have resulted from the Cry1Ac protein but rather from a pleiotropic effect (change in a single gene that affects multiple phenotypic traits) of the genetic manipulation. In a

Table 8.3 Summary of Effects of Cultivation of *Bt* Crops or Amendment with *Bt* Proteins on Soil Bacteria

Organism/activity tested	Methodology	Study location	*Bt* crop plant/ experimental variable	Protein	Effect on bacteria	Source
Bacteria	Culturing Substrate utilization Total DNA content DNA fingerprints	Laboratory	Soil amended with *Bt* and non-*Bt* cotton leaves	Cry1Ac	Significant (but transient) increase in bacteria numbers in soil of *Bt* cotton	Donegan et al. (1995)
Bacteria	Culturing Substrate utilization Total DNA content DNA fingerprints	Laboratory	Soil amended with purified protein versus unamended soil	Cry1Ab	No effect of purified *Bt* protein on bacteria	Donegan et al. (1995)
Aerobic bacteria	Culturing technique	Field	Soil with *Bt* and non-*Bt* potato	Cry3A	Few significant differences in microflora on *Bt* and non-*Bt* potato leaves and no difference in rhizosphere populations	Donegan et al. (1996)
Leaf-litter- and fecal-colonizing microorganisms	Culturing technique	Laboratory	Bacterial growth on *Bt* (X4334-EPR) and non-*Bt* maize leaves and on feces of *Porcellio scaber* fed *Bt* or non-*Bt* maize	Cry1Ab	No difference in bacterial growth on leaves of *Bt* and non-*Bt* maize; bacterial growth lower on *Bt*-fed *P. scaber* feces	Escher et al. (2000)
Bacteria (8 Gram negative, 5 Gram positive, and cyanobacteria)	Culturing technique	Laboratory (plant growth room)	Soil with *Bt* (NK4640Bt) and non-*Bt* maize Soil amended with biomass of *Bt* and non-*Bt* maize	Cry1Ab	No significant difference in culturable bacteria in soils cultivated with or amended with *Bt* or non-*Bt* maize	Saxena and Stotzky (2001a)

(continued)

Table 8.3 Summary of Effects of Cultivation of *Bt* Crops or Amendment with *Bt* Proteins on Soil Bacteria (continued)

Organism/activity tested	Methodology	Study location	*Bt* crop plant/ experimental variable	Protein	Effect on bacteria	Source
Bacteria	Culturing technique	Laboratory	Purified *Bt* protein added to pure or mixed cultures	Cry1Ab Cry3A Cry4	No effect on bacterial growth	Koskella and Stotzky (2002)
Heterotrophic bacteria and carbon-cycling microorganisms	Culturing technique	Greenhouse	Soybean grown in soil inoculated with an insecticidal crystal protein producer of *Bt* (Cry+), a mutant nonproducer of *Bt* (Cry-), or purified insecticidal crystal protein (ICP)	Cry1Ab	No difference in heterotrophic bacterial populations inoculated with *Bt*	Ferreira et al. (2003)
Soil microbial biomass, microbial activity (N mineralization potential, short-term nitrification rate, and soil respiration), and bacterial community structure	Terminal-restriction fragment length polymorphism (T-RFLP)	Field	Soil with *Bt* (Event MON863) and non-*Bt* maize	Cry3Bb1	No negative effects of *Bt* maize cultivation on microbial activity or community structure over two growing seasons	Devare et al. (2004)
Rhizosphere bacterial counts (copiotrophic, oligotrophic, and spore-forming bacteria) Rhizosphere bacterial community structure	Culturing methods Community-level catabolic profiling (CLCP) Automated ribosomal intergenic spacer analysis (ARISA)	Greenhouse	Soil with *Bt* maize (Event 176) and non-*Bt* maize	Cry1Ab	No differences in community structure detected using culturing methods or CLCP, but differences in community structure detected using ARISA suggest that changes in root exudates may influence rhizosphere community	Brusetti et al. (2004)

(continued)

Table 8.3 Summary of Effects of Cultivation of *Bt* Crops or Amendment with *Bt* Proteins on Soil Bacteria (continued)

Organism/activity tested	Methodology	Study location	*Bt* crop plant/ experimental variable	Protein	Effect on bacteria	Source
Bacterial community structure	Phospholipid fatty acid analysis (PLFA) Community-level physiological profiles (CLPP)	Growth chamber	Soil with two different *Bt* maize lines (Events *Bt* 11 and TC1507) and non-*Bt* maize isolines	Cry1Ab Cry1F	No difference in bacterial community structure in bulk soil based on PLFA profiles Small effect on rhizosphere community structure using CLPP but differences primarily based on soil type	Blackwood and Buyer (2004)
Total aerobic culturable bacteria, the number of anaerobic fermentative bacteria, denitrifying bacteria, hydrogen-producing acetogenic bacteria, and methanogenic bacteria; enzyme activity	Culturing technique	Laboratory	Flooded soils amended with *Bt* (KMD) and non-*Bt* (Xiushui 11) rice straw	Cry1Ab	No toxic effects of amendment with *Bt* rice straw on soil bacteria	Wu et al. (2004b)

(continued)

Table 8.3 Summary of Effects of Cultivation of *Bt* Crops or Amendment with *Bt* Proteins on Soil Bacteria (continued)

Organism/activity tested	Methodology	Study location	*Bt* crop plant/ experimental variable	Protein	Effect on bacteria	Source
Protease, neutral phosphatase, cellulase, and dehydrogenase activities; methanogenesis, hydrogen production, and anaerobic respiration	Enzyme activity assays	Laboratory	Flooded soils amended with *Bt* (KMD) and non-*Bt* (Xiushui 11) rice straw	Cry1Ab	Some differences in protease, neutral phosphatase, and cellulase activities detected at the early incubation stage but did not persist Differences in dehydrogenase activity, methanogenesis, hydrogen production, and anaerobic respiration between soil amended with *Bt* or non-*Bt* rice straw persisted throughout the experiment	Wu et al. (2004a)
Rhizospheric and heterotrophic bacteria	Denaturing gradient gel electrophoresis (DGGE) Respiration rates Culturing technique	Microcosm Greenhouse	Soil with *Bt* (*Bt* 11 and *Bt* 176) and non-*Bt* maize Soil amended with *Bt* or non-*Bt* maize residue	Cry1Ab	Differences in rhizospheric eubacterial communities (both total and active) and in culturable rhizospheric heterotrophic bacteria in soil cultivated with *Bt* versus non-*Bt* maize Lower soil respiration in soils amended with *Bt* maize residue; differences in soil bacteria isolated from soil amended with *Bt* versus non-*Bt* maize	Castaldini et al. (2005)

(continued)

Table 8.3 Summary of Effects of Cultivation of *Bt* Crops or Amendment with *Bt* Proteins on Soil Bacteria (continued)

Organism/activity tested	Methodology	Study location	*Bt* crop plant/ experimental variable	Protein	Effect on bacteria	Source
Bacterial community structure	Single-strand conformation polymorphism (SSCP)	Field	Soil with *Bt* (Event MON810) and non-*Bt* maize	Cry1Ab	Bacterial community structure less affected by Cry1Ab protein than by other environmental factors (i.e., ontogeny of the plants or heterogeneities within the field)	Baumgarte and Tebbe (2005)
Rhizosphere bacterial diversity	Metabolic profiling (Biolog) DGGE	Greenhouse Field	Soil with *Bt* and non-*Bt* maize	Cry1Ab	Rhizosphere bacterial diversity affected more by soil texture than by cultivation with *Bt* maize	Fang et al. (2005)
Bacteria	PLFA Culturing technique	Field	Soil with *Bt* and non-*Bt* maize	Cry1Ab	Lower ratio of gram-positive to gram-negative bacteria in soil with *Bt* maize; no difference in fungal/bacterial ratio	Xue et al. (2005)
Culturable bacteria, enzyme activities	Culturing technique	Microcosm	Soil amended with *Bt* and non-*Bt* maize biomass	Cry1Ab	No difference in culturable bacteria or enzyme activities	Flores et al. (2005)

(continued)

Table 8.3 Summary of Effects of Cultivation of *Bt* Crops or Amendment with *Bt* Proteins on Soil Bacteria (continued)

Organism/activity tested	Methodology	Study location	*Bt* crop plant/ experimental variable	Protein	Effect on bacteria	Source
Culturable functional groups (nitrogen-fixing, inorganic phosphate-dissolving, and potassium-dissolving bacteria)	Culturing technique	Field	Rhizosphere soils of *Bt* (SGK321 and Nu-COTN99B) and non-*Bt* cotton (SHIYUAN321) Addition of purified *Bt* protein	Cry1Ac	Lower numbers of culturable functional groups of bacteria in soil with *Bt* cotton Nu-COTN99B versus non-*Bt* cotton throughout most of the growing season; nitrogen-fixing bacteria reduced in soil amended with more than 500 ng/g purified *Bt* protein	Rui et al. (2005)
Microbial community structure	PLFA CLPP	Field	Soil cultivated with *Bt* (MEB307*Bt*) and non-*Bt* maize isoline (Monumental)	Cry1Ab	No difference in microbial community structure between *Bt* and non-*Bt* soil	Griffiths et al. (2005)
Functional diversity of microbial communities as determined by enzyme activity (urease, phosphatase, dehydrogenase, phenol oxidase, and protease)	Enzyme assays Biolog	Greenhouse	Rhizosphere soil cultivated with *Bt* (Sukang-103) and non-*Bt* (Sumian-12) cotton Soil amended with biomass of *Bt* and non-*Bt* cotton	Cry1Ac	No negative effect of cultivation with or amendment with *Bt* cotton on enzyme activity, functional diversity of microbial community, or species richness	Shen et al. (2006)
Soil microbial community	PLFA	Greenhouse	Soil cultivated with *Bt* (Event MON810) and non-*Bt* (Monumental) maize collected from field sites	Cry1Ab	No effect of *Bt* soil on soil microbial community	Griffiths et al. (2006)

(continued)

Table 8.3 Summary of Effects of Cultivation of *Bt* Crops or Amendment with *Bt* Proteins on Soil Bacteria (continued)

Organism/activity tested	Methodology	Study location	*Bt* crop plant/ experimental variable	Protein	Effect on bacteria	Source
Microbial biomass, microbial activity	Chloroform fumigation-extraction Nitrogen (N) mineralization potential, short-term nitrification rate, and respiration rate	Field	Soil cultivated with *Bt* (Event MON863) and non-*Bt* maize	Cry3Bb1	No adverse effects of *Bt* maize cultivation detected in bulk or rhizosphere soil	Devare et al. (2007)
Microbial communities	Biolog DGGE	Microcosm Field	Soil amended with *Bt* (Merschman M-0012*Bt*) and non-*Bt* maize (Merschman M-00110)	Not specified	*Bt* residue in soil significantly affected the structure of microbial communities	Fang et al. (2007)
Soil microbial community structure	PLFA	Greenhouse	Soil with 8 paired *Bt* and non-*Bt* maize lines Addition of purified *Bt* protein to soil	Cry1Ab	No effect of *Bt* maize on microbial community structure No effect of amendment with purified Cry1Ab protein on microbial community structure	Griffiths, Heckmann, et al. (2007)
Nitrogen-fixing bacteria	DNA extraction, cloning, and sequencing	Field	*Bt* white spruce and non-*Bt* spruce	Cry1Ab	No negative effect of *Bt* cultivar	Lamarche and Hamelin (2007)

(continued)

Table 8.3 Summary of Effects of Cultivation of *Bt* Crops or Amendment with *Bt* Proteins on Soil Bacteria (continued)

Organism/activity tested	Methodology	Study location	*Bt* crop plant/ experimental variable	Protein	Effect on bacteria	Source
Enzyme activities (urease, acid phosphomonoesterase, arylsulfatase, invertase, cellulase) and *Bt* protein persistence in soil	ELISA to quantify *Bt* protein Enzyme activity tests	Laboratory	Field collected soil amended with leaves and stems of *Bt* cotton (*Bt*-GK and *Bt*-ZK) and a non-*Bt* isoline (non-*Bt* ZM)	Cry1Ac	*Bt* cotton stems and leaves had a positive effect on urease, acid phosphomonoesterase, invertase, and cellulase activities, but a negative effect on arylsulfatase activity; after 56 days, *Bt* protein detected in soil with 41% and 60% of the introduced amounts remaining from *Bt*-ZK and *Bt*-GK, respectively	Sun et al. (2007)
Microbial diversity (total culturable, gram-negative, chitin-utilizing, cellulose-utilizing, ammonium- and nitrite-oxidizing, and nitrate-reducing and denitrifying organisms), enzyme activities (arylsulfatase, acid and alkaline phosphatases, dehydrogenases, and proteases)	Dilution plating Most probable number (MPN) counts DGGE Enzyme assays	Field	Soil cultivated with *Bt* (Bt 11, MON810, MON863) and non-*Bt* maize	Cry1Ab Cry3Bb1	No consistent differences in microbial diversity or enzyme activities	Icoz et al. (2008)

(continued)

Table 8.3 Summary of Effects of Cultivation of *Bt* Crops or Amendment with *Bt* Proteins on Soil Bacteria (continued)

Organism/activity tested	Methodology	Study location	*Bt* crop plant/ experimental variable	Protein	Effect on bacteria	Source
Rhizosphere microbial community composition Soil enzyme activities (phosphatase activity, dehydrogenase activity, respiration, and methanogenesis)	T-RFLP DGGE	Field	*Bt* rice, non-*Bt* rice, and non-*Bt* rice treated with the insecticide Triazophos	Cry1Ab	No negative effect of *Bt* rice on microbial community structure or soil enzyme activities in the rhizosphere	Liu et al. (2008)
Aerobic bacteria, enzyme activity (dehydrogenase and nitrogenase enzymes), and ATP content	Culturing techniques	Field	Rhizosphere and nonrhizosphere soil cultivated with *Bt* (Events 176 and MON810) and non-*Bt* maize	Cry1Ab	No negative effect of *Bt* cultivar	Oliveira et al. (2008)
Rhizosphere bacteria	Culturing techniques	Field	Soil with *Bt* and non-*Bt* cotton	Cry1A, CpTI	No negative effect of *Bt* cotton on rhizosphere bacteria	Hu et al. (2009)
Microbial decomposition rates as determined by ash-free mass remaining (AFMR), total carbon (TC) and total nitrogen (TN) content after decomposition, and cellulose and lignin contents of the original rice residues	AFMR: calculation of ash weight TC and TN: combustion in an element autoanalyzer	Field	Litterbags containing roots or straw from *Bt* (KMD) and non-*Bt* (XiuShui 11) rice	Cry1Ab	*Bt* rice roots decomposed faster than non-*Bt* rice roots in buried litterbags in the first 200 days No difference in decomposition rate between *Bt* and non-*Bt* rice straw buried in litterbags	Wu et al. (2009)

(continued)

Table 8.3 Summary of Effects of Cultivation of *Bt* Crops or Amendment with *Bt* Proteins on Soil Bacteria (continued)

Organism/activity tested	Methodology	Study location	*Bt* crop plant/ experimental variable	Protein	Effect on bacteria	Source
Soil microbial community	T-RFLP	Field	Litterbags containing straw or roots from *Bt* (KMD) and non-*Bt* (XiuShui 11) rice	Cry1Ab	No negative effect of *Bt* rice residues on soil bacteria community composition	Lu et al. (2010a, 2010b)
Rhizosphere bacterial community structure	PCR-amplified 16S rRNA genes	Field	Soil with *Bt* (Event MON88017) and three non-*Bt* maize lines	Cry3Bb1	No negative effect of *Bt* maize on rhizosphere bacterial community	Miethling-Graff et al. (2010)
Soil microbial community structure	DGGE Sequences of the 16S rRNA genes	Greenhouse	Soil with two *Bt* (Events MON810 and Nongda 1246*1428) and their non-*Bt* maize isolines (Pioneer 34B23 and Nongda 3138) Soil amended with *Bt* and non-*Bt* maize leaves and stalks	Cry1Ab, Cry1A	Neither actively growing *Bt* maize nor *Bt* maize residue had a negative effect on bacterial community structure	Tan et al. (2010)
Soil bacteria: *Azotobacter* Denitrifying bacteria Ammonia-oxidizing bacteria Microbial diversity	Selective plating MPN	Field	Soil with *Bt* cotton and non-*Bt* cotton	Cry1Ac and/or CpTI protein	No significant difference in number of each microbial population or diversity indices between *Bt* or non-*Bt* cotton	Li et al. (2011)

(continued)

Table 8.3 Summary of Effects of Cultivation of *Bt* Crops or Amendment with *Bt* Proteins on Soil Bacteria (continued)

Organism/activity tested	Methodology	Study location	*Bt* crop plant/ experimental variable	Protein	Effect on bacteria	Source
Bacterial decomposer communities	T-RFLP	Field: litterbag studies	Litterbags containing plant material from *Bt* maize (Event MON863) and non-*Bt* maize	Cry3Bb	No effect of *Bt* maize biomass on bacterial decomposer communities; no effect of Cry3Bb on carbon residence time or turnover in soils amended with *Bt* residues	Xue et al. (2011)
Microbial biomass carbon Microbial activities Soil enzyme activities (catalase, nitrate reductase, acid phosphomonoesterase, arylsulfatase, and b-glucosidase)	Fumigation-extraction technique Fluorocine diacetate Enzyme assays	Greenhouse	One *Bt* (ZM30) and two stacked *Bt* and cowpea trypsin inhibitor (*Bt* + CpTI) cottons (ZM41, sGK321) and their non-*Bt* isolines (ZM16, ZM23, Shiyuan) cultivated in greenhouse pots for 4 years	Cry1Ac CpTI	Negative correlation between Cry1Ac content in soil and most microbial properties and enzyme activities; CryAc and CpTI proteins persisted in soil	Chen et al. (2011)

Source: Adapted and updated with permission from Icoz and Stotzky, 2008b.

greenhouse study, Castaldini et al. (2005) observed differences in soil bacteria isolated from soil amended with *Bt* versus non-*Bt* maize residue, with members of a subgroup of the genus *Bacillus* isolated from soil amended with *Bt* biomass, while growth-promoting rhizobacteria were isolated from soil amended with non-*Bt* maize biomass. When plant residues of *Bt* and non-*Bt* maize were kept mixed with soil for up to 4 months, soil respiration was reduced by 10% in the *Bt* maize treatments (Castaldini et al., 2005). Fang et al. (2007) reported that soil amended with *Bt* maize biomass had a significantly different microbial community structure than soil amended with non-*Bt* maize biomass as determined by substrate utilization profiles and denaturing gradient gel electrophoresis (DGGE) patterns and linked the alterations in the structure of soil microbial communities to the higher lignin content detected in the *Bt* maize plants. In soil amended with *Bt* rice straw versus non-*Bt* rice straw, differences in biological activities (dehydrogenase activity, methanogenesis, hydrogen production, and anaerobic respiration) were detected, but there was no direct toxic effect of Cry1Ab protein (Wu et al., 2004a, 2004b). In a decomposition study in which litterbags containing *Bt* rice and non-*Bt* rice roots or *Bt* and non-*Bt* rice straw were buried in the field and sampled over 2 years, Wu et al. (2009) found that *Bt* rice roots decomposed relatively faster than non-*Bt* rice roots in the first 200 days but found no difference in microbial decomposition rates between *Bt* and non-*Bt* rice straw as determined by changes in ash-free mass remaining and changes in total carbon and total nitrogen content after decomposition. The incorporation of *Bt* cotton stems and leaves into soil microcosms had a positive effect on soil urease, acid phosphomonoesterase, invertase, and cellulose activities but a negative effect on arylsulfatase activity (Sun et al., 2007). The addition of cotton tissue to soil most likely stimulated microbial activity (and thus influenced soil enzyme activity), potentially masking any negative effect of *Bt* protein on soil microbial or soil enzyme activity (Sun et al., 2007).

When soil was cultivated with *Bt* maize, Xue et al. (2005) reported a lower ratio of gram-positive to gram-negative bacteria compared to soil cultivated with non-*Bt* maize. In microcosm and greenhouse experiments, Castaldini et al. (2005) reported differences in rhizospheric eubacterial communities and in culturable rhizospheric heterotrophic bacteria in soil cultivated with *Bt* maize versus non-*Bt* maize as determined by DGGE analysis of 16S ribosomal RNA (rRNA) genes and culturing methods. In soil cultivated with *Bt* and non-*Bt* cotton, Rui et al. (2005) found that the numbers of colony-forming units (CFUs) of three different bacterial functional groups (nitrogen fixing, inorganic phosphate dissolving, and potassium dissolving) were lower in soil collected from the rhizosphere of *Bt* cotton at the early and middle stages of plant growth than in soil cultivated with non-*Bt* cotton. However, addition of purified Cry1Ac protein directly to soil had no effect on the CFU number of any of the three bacterial functional groups (except for a decrease in the nitrogen-fixing bacteria when concentrations of *Bt* protein were greater than 500 ng/g) (Rui et al., 2005). In soil consecutively cultivated for 4 years with *Bt* cotton, *Bt* and cowpea trypsin inhibitor (*Bt* + *CpTI*) cotton, and non-*Bt* cotton lines, there was a decrease in microbial biomass carbon (MBC), soil microbial activity as determined by catalase activity and fluorescein diacetate hydrolysis, and some enzyme activities (nitrate reductase, acid phosphomonoesterase, arylsulfatase, b-glucosidase, and protease) in the *Bt* and *Bt* plus *CpTI* cotton lines compared to soil cultivated with non-*Bt* cotton in greenhouse trials (Chen et al., 2011). The decreases in MBC, microbial activity, and enzyme activities were correlated with increasing CryAc protein content (Chen et al., 2011).

Most studies, however, have reported no negative effects of purified *Bt* proteins, *Bt* plant biomass, or the cultivation of *Bt* crops on soil bacteria (Donegan et al., 1995; Escher et al., 2000; Saxena and Stotzky, 2001a; Koskella and Stotzky, 2002; Ferreira et al., 2003;

Blackwood and Buyer, 2004; Brusetti et al., 2004; Devare et al., 2004; Wu et al., 2004a, 2004b; Baumgarte and Tebbe, 2005; Fang et al., 2005; Flores et al., 2005; Griffiths et al., 2005, 2006; Rui et al., 2005; Shen et al., 2006; Devare et al., 2007; Griffiths, Heckmann, et al., 2007; Lamarche and Hamelin, 2007; Icoz and Stotzky, 2008a; Liu et al., 2008; Oliveira et al., 2008; Hu et al., 2009; Lu et al., 2010a, 2010b; Miethling-Graff et al., 2010; Tan et al., 2010; Li et al., 2011; Xue et al., 2011). When purified *Bt* proteins were added to soil, there was no effect of Cry1Ab on culturable bacteria (Donegan et al., 1995), on the population size of culturable heterotrophic bacteria (Ferreira et al., 2003), or on microbial community structure as determined by phospholipid fatty acid (PLFA) analysis (Griffiths, Heckmann, et al., 2007). There was also no effect of the addition of purified Cry1Ac protein on culturable inorganic phosphate-dissolving or potassium-dissolving bacteria (Rui et al., 2005). Similarly, there was no effect of purified Cry1Ab, Cry3A, or Cry4 protein on growth of selected bacteria (eight gram-negative five gram-positive bacteria, and a cyanobacterium) in vitro, in pure and mixed cultures, using dilution, disk diffusion, and sporulation assays (Koskella and Stotzky, 2002).

In studies in which biomass of *Bt* plants was added to soil, there were generally no negative effects on the numbers of culturable bacteria (Saxena and Stotzky, 2001a; Flores et al., 2005) or on soil microbial community structure as determined by DGGE (Tan et al., 2010). In decomposition studies, there was no difference in bacterial growth on *Bt* maize versus non-*Bt* maize leaves; however, bacterial growth was lower on the feces of wood lice (*Porcellio scaber*) that were fed *Bt* maize leaves (Escher et al., 2000). There were no negative effects on bacterial decomposer communities of litterbags containing *Bt* maize biomass (cobs, roots, or stems plus leaves) (Xue et al., 2011) or *Bt* rice biomass (straw or roots) (Lu et al., 2010a; Lu et al., 2010b) as demonstrated through terminal restriction fragment length polymorphism (T-RFLP) analysis or a direct toxic effect on culturable bacteria in soil amended with *Bt* rice straw (Wu et al., 2004b). When ground, dried roots and shoots of *Bt* cotton or non-*Bt* cotton were mixed with field-collected soil and incubated for 2 months, there was no significant difference in urease, phosphatase, dehydrogenase, phenol oxidase, or protease activities between the *Bt* and non-*Bt* biomass treatments; however, enzyme activity was stimulated by the addition of both *Bt* and non-*Bt* cotton biomass (Shen et al., 2006).

In soil cultivated with *Bt* or non-*Bt* maize in the laboratory, Saxena and Stotzky (2001a) detected no significant difference in the CFUs of culturable bacteria (including actinomycetes) between rhizosphere soil of *Bt* and non-*Bt* maize after 45 days or between soil amended with *Bt* and non-*Bt* maize biomass (Saxena and Stotzky, 2001a). When *Bt* maize and non-*Bt* maize were cultivated in a growth chamber, there was no difference in bacterial community structure detected between bulk soil of *Bt* and non-*Bt* maize using PLFA analysis (Blackwood and Buyer, 2004). Although a small difference in rhizosphere bacterial community structure was detected in the *Bt* versus non-*Bt* maize cultivated soil (as determined by community-level physiological profiles [CLPPs]), the differences in the rhizosphere microbial community between the *Bt* and non-*Bt* maize cultivated soil were most influenced by soil type and texture (Blackwood and Buyer, 2004).

In greenhouse studies, there was also no difference in rhizosphere bacterial community structure in soil cultivated with *Bt* maize and non-*Bt* maize as determined by community-level catabolic profiling (CLCP) (Brusetti et al., 2004). However, when soil was treated with a root exudate solution collected from *Bt* or non-*Bt* maize plants grown hydroponically, differences in bacterial community structure were detected by automated ribosomal intergenic spacer analysis (ARISA), suggesting that changes in root exudates may influence the rhizosphere bacterial community more than the expression of *Bt* protein (Brusetti et al., 2004). Using DGGE analysis and 16S rRNA gene sequences, Tan et al. (2010) detected no difference in microbial community structure between soils cultivated with two different lines of *Bt* maize and corresponding non-*Bt* maize lines. Similarly, there was no difference in microbial community structure as determined by PLFA between *Bt* maize and non-*Bt* maize grown in greenhouse pots using soil collected from field plots that had been cultivated with the same *Bt* and non-*Bt* cultivars (Griffiths et al., 2006). When the effects of eight different *Bt* maize lines (expressing Cry1Ab) and their corresponding non-*Bt* maize isolines on soil microbes were evaluated using PLFA analysis, Griffiths, Heckmann, et al. (2007) reported that although soil microbial community structure was significantly affected by the growth stage of the plant, it was not affected by the *Bt* gene insertion. There were also no negative effects of the cultivation of *Bt* cotton on species richness or functional diversity of rhizosphere microbial communities as determined by Biolog assays or on enzyme activities (urease, phosphatase, dehydrogenase, phenol oxidase, and protease) when compared to soil cultivated with non-*Bt* cotton (Shen et al., 2006).

Most field studies have also demonstrated no significant negative effects of *Bt* crop cultivation on soil microbes. In a 2-year field experiment, there were no effects of *Bt* maize cultivation on microbial activity (N mineralization potential, short-term nitrification rate, and soil respiration) or bacterial community structure detected by T-RFLP analysis when compared to soil cultivated with a non-*Bt* maize isoline (Devare et al., 2004). After 3 years, there were still no significant differences in microbial biomass or microbial activity detected in the *Bt* versus non-*Bt* maize plots, although there were seasonal variations in microbial biomass and activity (Devare et al., 2007).

In rhizosphere soils collected from *Bt* maize and three different non-*Bt* maize cultivars over 3 years, there were no differences in rhizosphere bacterial community structure as determined by polymerase chain reaction (PCR) amplification of 16S rRNA genes (Miethling-Graff et al., 2010), and there was no difference in bacterial community structure between rhizosphere soils of *Bt* maize and two non-*Bt* maize lines grown for 2 years at three different European field sites (Denmark, eastern France, southwestern France) as determined through PLFA analysis and CLPP (Griffiths et al., 2005). There were also no differences in the culturable aerobic bacteria (including actinomycetes), dehydrogenase and nitrogenase activities, or ATP (adenosine triphosphate) content, in rhizosphere soils of *Bt* maize versus non-*Bt* maize over a 2-year field study, although there were seasonal variations as well as significant differences between rhizosphere and bulk soil samples (Oliveira et al., 2008). When soils were cultivated for 4 years with four different lines of *Bt* maize (two lines of *Bt* 11; Cry1Ab [sweet corn and field corn], Event MON810; Cry1Ab, and Event MON863; Cry3Bb1) and non-*Bt* maize, there was no difference in microbial diversity detected through dilution plating and DGGE or in enzyme activities (arylsulfatases, acid and alkaline phosphatases, dehydrogenases, and proteases) (Icoz et al., 2008).

Using metabolic profiling and molecular analysis of 16S rRNA genes, Fang et al. (2005) determined that rhizosphere bacterial diversity was affected more by soil texture than by cultivation with *Bt* maize in both greenhouse and field studies. Similarly, in soil cultivated with *Bt* maize and non-*Bt* maize, the rhizosphere bacterial community structure was more affected by environmental factors such as the ontogeny of the plants or heterogeneities within the field soil than by the Cry1Ab protein expressed in the *Bt* maize plants as determined by SSCP (single-strand conformation polymorphism) of PCR-amplified 16S rRNA genes (Baumgarte and Tebbe, 2005). When soil was cultivated with *Bt* cotton, Hu et al. (2009) found no negative effects of the *Bt* cultivar on rhizosphere bacteria, and Li et al. (2011) reported no differences in *Azotobacter*, denitrifying bacteria, ammonia-oxidizing bacteria, or microbial diversity between soil cultivated with *Bt* and non-*Bt* cotton over a 3-year field study as determined by selective plating and most probable number (MPN) assays, although there were seasonal variations not related to the *Bt* or non-*Bt* cultivars. When soil enzyme activities (phosphatase activity, dehydrogenase activity, respiration, and methanogenesis) and microbial community composition were compared in the rhizosphere of *Bt* rice, non-*Bt* parental rice, and non-*Bt* parental rice treated with the insecticide Triazophos at multiple sampling times, there were generally no significant negative effects detected on soil enzyme activity or microbial community structure as determined by DGGE and T-RFLP (Liu et al., 2008). There were, however, seasonal variations in the selected enzyme activities and microbial community composition in the rhizosphere over the course of the 2-year experiment (Liu et al., 2008).

There was no negative effect of 4 years of cultivation with *Bt* spruce (engineered to express Cry1Ab) on nitrogen-fixing bacteria compared with non-*Bt* white spruce trees as determined by molecular sequencing of a region of the nitrogenase reductase gene from genomic DNA extracted from rhizosphere soil (Lamarche and Hamelin, 2007). There were also minimal differences in culturable aerobic bacteria in rhizosphere soil cultivated with *Bt* potato, non-*Bt* russet potato treated with insecticide (Di-Syston), and non-*Bt* russet potato treated with microbial *Bt* (M-Trak) (Donegan et al., 1996). When the microflora colonizing the leaves of these potato plants were compared over multiple time points (0, 21, 42, 63, and 98 days), Donegan et al. (1996) found few significant differences across potato cultivars.

These, and other, results indicate that, in general, the insecticidal *Bt* proteins, either purified or expressed in transgenic *Bt* plants, have no significant negative effects on most soil bacteria. However, in the few studies in which effects of cultivation of *Bt* plants on soil microbes were observed (e.g., Donegan et al., 1995; Wu et al., 2004a, 2004b; Castaldini et al., 2005; Rui et al., 2005; Xue et al., 2005; Fang et al., 2007; Sun et al., 2007; Chen et al., 2011), differences in physiological properties within plants resulting from the genetic insertion may be implicated (e.g., Donegan et al., 1995; Rui et al., 2005). Genetic alterations, as a result of the insertion of *Bt* genes, that produce a change in plant root exudates or quality of plant material, for example, may influence microbial growth and species composition in the rhizosphere or affect the degradation time or quality of *Bt* plant litter. In this way, microbial communities could be affected by the cultivation of transgenic *Bt* crops without being negatively affected by *Bt* proteins directly. Fluxes in microbial community structure, however, can also be influenced by soil type, temperature, season, plant type, and other biotic and abiotic factors (e.g., Griffiths, 2000; Lottman, 2000; Kowalchuk, 2002; Dunfield and Germida, 2003; Zwahlen, Hilbeck, Gugerli, et al., 2003; Blackwood and Buyer, 2004; Icoz and Stotzky, 2008b). Thus, where an impact of the cultivation of a *Bt* crop on soil bacteria has been detected, the ecological significance has often been difficult to assess.

8.4.1.1 Research recommendations: Effects of the cultivation of Bt crops on soil bacteria

Many of the studies evaluating nontarget effects of *Bt* crops or *Bt* proteins on soil bacteria have examined effects on culturable bacteria (Table 8.3). Given the fact that less than 1% of bacterial taxa are thought to be culturable (e.g., Handelsman and Tiedje, 2007), this methodology could influence the results of many of these studies. Thus, differences in laboratory techniques may also have a role in the different outcomes of similar studies evaluating the effects of *Bt* crops on microbial communities (e.g., plating vs. DGGE vs. metabolic analysis). To evaluate the nontarget effects of the cultivation of *Bt* crops on soil bacteria, multiple detection methods should be employed as most microbes are not culturable and could be better identified, quantified, or characterized using a combination of molecular and metabolic tools. Risk assessment studies should also be conducted at multiple levels (laboratory, greenhouse, and field), and under different experimental and environmental conditions, as it is well known that bacteria and other soil organisms can be easily influenced by a multitude of biotic and abiotic factors that may not be associated with GM plants. As relatively few significant negative impacts of *Bt* crops have been reported for soil bacteria, it may be more useful to focus future research efforts on other soil organisms that are more closely associated with roots (e.g., mycorrhizal fungi) or narrowing the scope of bacterial research to investigate the effects of transgenic crops on nitrogen-fixing bacteria, many of which can form symbiotic relationships with plant roots and have clear benefits for plant health and ecosystem function by providing nutrients to plants and protection against plant pathogens (e.g., reviewed in Sessitsch et al., 2002; Dobbelaere et al., 2003; Hayat et al., 2010).

8.4.2 Effects of the cultivation of Bt crops on soil fungi

Soil fungi, including saprotrophic, parasitic/pathogenic, and mycorrhizal fungi, are another group of soil organisms that may be affected by cultivation of transgenic *Bt* crops (Table 8.4). Fungi have important roles in the soil ecosystem as decomposers, nutrient recyclers, plant symbionts, and plant pathogens. Saprotrophic fungi (also called saprophytic fungi) are the primary group of soil organisms that degrade organic material in agricultural fields, whereas parasitic/pathogenic fungi and mycorrhizal fungi are found in close association with living plant roots in the rhizosphere.

Saprotrophic fungi are free-living soil organisms that obtain their nutrients from dead organic material, such as leaves, wood, and other plant and animal materials, and are responsible for recycling a significant amount of the carbon in the soil ecosystem. To date, most studies have determined that there is no significant negative effect of purified *Bt* proteins, *Bt* crop cultivation, or *Bt* plant material on saprotrophic fungi or culturable fungi (Table 8.4) (e.g., Saxena and Stotzky, 2001a; Koskella and Stotzky, 2002; Ferreira et al., 2003; Icoz et al., 2008; Oliveira et al., 2008). For example, when soil was amended with purified Cry1Ab or Cry1Ac protein, there was no difference in the population levels of culturable fungi using selective plating compared with control soil (Donegan et al., 1995). There was also no effect of the addition of purified *Bt* proteins (Cry1Ab, Cry3A, or Cry4) on the growth of representative culturable fungi (*Cunninghamela elegans, Rhizopus nigricans, Aspergillus niger, Fusarium solani, Penicillium* sp., *Saccharomyces cerevisiae,* and *Candida albicans*) in pure and mixed cultures (Koskella and Stotzky, 2002).

Where significant effects have been reported, they have often been minor or transient. When soybean was grown in soil inoculated with a strain of *Bacillus thuringiensis* bacteria that expressed Cry1Ab protein (Cry^+), a *Bt* mutant strain that did not express Cry protein (Cry^-), purified insecticidal crystal protein (ICP), or no treatment (control), there was no significant difference in the culturable fungal populations between rhizosphere soils; however, there was a transient increase in some functional groups (saprophytic, amylolytic, cellulolytic, and proteolytic fungi) between the treatments as determined by selective plating, at the beginning of the experiment (Ferreira et al., 2003). Donegan et al. (1995) also reported a transient increase in culturable fungi in soil amended with *Bt* cotton based on selective plating methods. However, Flores et al. (2005) and Saxena and Stotzky (2001a) found no difference in culturable fungi between soils amended with *Bt* and non-*Bt* maize in soil microcosms. There were also no negative effects on the numbers of culturable fungi in flooded soils amended with *Bt* versus non-*Bt* rice straw in laboratory experiments (Wu et al., 2004b). In litterbag decomposition field studies, there was no difference in fungal

Table 8.4 Summary of Effects of Cultivation of *Bt* Crops or Amendment with *Bt* Proteins on Soil Fungi

Organism	Methods	Study location	*Bt* crop/experimental variable	Protein	Effect on fungi	Source
Fungi	Selective plating	Laboratory	Soil amended with *Bt* cotton leaves	Cry1Ac	Transient increase in culturable fungi in soil amended with *Bt* cotton	Donegan et al. (1995)
Fungi	Culturing	Laboratory	Soil amended with *Bt* protein	Cry1Ac Cry1Ab	No effect of *Bt* protein on soil fungi	Donegan et al. (1995)
Fusarium sp., *Pythium* sp., *Verticillium dahlia*	Culturing	Field	Soil cultivated with *Bt* and non-*Bt* potato	Cry3A	Minimal difference in culturable fungi between *Bt* and chemically or microbially treated potato	Donegan et al. (1996)
Zygomycetes Ascomycete Deuteromycetes Yeasts	Culturing	Plant growth room	Soil cultivated with *Bt* and non-*Bt* maize Soil amended with biomass of *Bt* and non-*Bt* maize	Cry1Ab	No difference in numbers between soils amended with biomass of *Bt* and non-*Bt* maize or in rhizosphere soil of *Bt* and non-*Bt* maize	Saxena and Stotzky (2001a)
Cunninghamela elegans, *Rhizopus nigricans*, *Aspergillus niger*, *Fusarium solani*, *Penicillium* sp., *Saccharomyces cerevisiae*, and *Candida albicans*	Culturing, dilution, disk-diffusion, and sporulation assays	Laboratory	Purified protein added to pure and mixed cultures	Cry1Ab Cry3A Cry4	No effect of *Bt* protein on growth of fungi in vitro	Koskella and Stotzky (2002)

(continued)

Table 8.4 Summary of Effects of Cultivation of *Bt* Crops or Amendment with *Bt* Proteins on Soil Fungi (continued)

Organism	Methods	Study location	*Bt* crop/experimental variable	Protein	Effect on fungi	Source
Culturable amylolytic, cellulolytic, proteolytic, saprotrophic fungi and AMF	Selective plating Grid-line intersect method	Greenhouse	Soybean grown in soil inoculated with bacterial strains of *Bt* (Cry^+), a *Bt* Cry^- mutant, purified insecticidal crystal protein (ICP), or no treatment	Cry1Ab	No difference in fungal population size; some transient differences in numbers when compared with control soil No effect of ICP on AMF; Cry^+ and Cry^- *Bt* strains inhibited AMF compared to control	Ferreira et al. (2003)
Fungi/eukaryotes	PLFA	Growth chamber	Soil cultivated with *Bt* or non-*Bt* maize	Cry1Ab Cry1F	Expression of Cry protein reduced the presence of eukaryotic PLFA in bulk soils, although it was not clear which groups of eukaryotes were affected	Blackwood and Buyer (2004)
AMF species *Glomus mosseae*	Sandwich assay Trypan blue staining	Laboratory	Effects of root exudates of *Bt* maize (Events *Bt* 11 and 176) on AMF and fungal recognition	Cry1Ab	Root exudates of *Bt* 176 reduced presymbiotic hyphal growth and reduced development of appressoria	Turrini et al. (2004)
Fungi	Culturing	Laboratory	Flooded soils amended with *Bt* and non-*Bt* rice straw	Cry1Ab	No effects on the numbers of culturable fungi	Wu et al. (2004b)

(continued)

Table 8.4 Summary of Effects of Cultivation of *Bt* Crops or Amendment with *Bt* Proteins on Soil Fungi (continued)

Organism	Methods	Study location	*Bt* crop/experimental variable	Protein	Effect on fungi	Source
AMF	Sandwich assay Trypan blue staining	Microcosm Greenhouse	Soil cultivated with or amended with residue of *Bt* maize (Events *Bt* 11 and 176) or non-*Bt* maize	Cry1Ab	Significantly lower level of AMF colonization in roots of *Bt* maize Lower AMF colonization in *Medicago sativa* grown in soil amended with *Bt* maize	Castaldini et al. (2005)
Fungi	PLFA Culturing technique	Field	Soil cultivated with *Bt* and non-*Bt* maize	Cry1Ab	No difference in fungal-to-bacterial ratio	Xue et al. (2005)
Fungi	Culturing technique	Microcosm	Soil amended with *Bt* and non-*Bt* maize	Cry1Ab	No difference in culturable fungi between soils amended with *Bt* and non-*Bt* maize	Flores et al. (2005)
Fusarium graminearum *Trichoderma atroviride*	Culturing Fungal growth on maize leaf tissue Microsatellite-based PCR, chemical fingerprints	Laboratory	Purified *Bt* protein Soil amended with *Bt* maize (Events 176, MON810) and non-*Bt* maize residue	Cry1Ab	No effect of Cry1Ab on fungal growth Some *Bt* maize hybrids differed in volatile organic composition compared to their non-*Bt* isoline	Naef et al. (2006)
AMF	Trypan blue staining	Microcosm	Soil cultivated with *Bt* maize MON810 *Bt*-variety (MEB307Bt) and non-*Bt* maize (Monumental)	Cry1Ab	No difference in frequency or intensity of root colonization between *Bt* and non-*Bt* maize	de Vaufleury et al. (2007)

(continued)

Table 8.4 Summary of Effects of Cultivation of *Bt* Crops or Amendment with *Bt* Proteins on Soil Fungi (continued)

Organism	Methods	Study location	*Bt* crop/experimental variable	Protein	Effect on fungi	Source
AMF	Trypan blue staining Grid-line intersect method	Field	Soil cultivated with *Bt* and non-*Bt* cotton	Cry1Ac Cry2Ab	No difference in root colonization (arbuscules) between *Bt* and non-*Bt* cotton	Knox et al. (2008)
Culturable fungi	Culturing	Field	Soil cultivated with four varieties of *Bt* maize (Events *Bt* 11 [field and sweet corn], MON810, MON863) and non-*Bt* maize isolines	Cry1Ab Cry3Bb1	No consistent effect of *Bt* maize on culturable fungi	Icoz et al. (2008)
Culturable fungi	Culturing	Field	Soil cultivated with *Bt* maize (Events 176, MON810) and non-*Bt* maize	Cry1Ab	No effect of *Bt* maize on culturable fungi	Oliveira et al. (2008)
Fungal communities as affected by *Bt* rice residue decomposition	T-RFLP	Field	Litterbags containing biomass (straw or roots) of *Bt* and non-*Bt* rice	Cry1Ab	No difference in fungal community composition in litterbags containing rice straw Some differences in fungal community composition at the early stage of root decomposition	Lu et al. (2010a)

(continued)

Table 8.4 Summary of Effects of Cultivation of *Bt* Crops or Amendment with *Bt* Proteins on Soil Fungi (continued)

Organism	Methods	Study location	*Bt* crop/experimental variable	Protein	Effect on fungi	Source
Fungal community structure	PCR-DGGE Sequences of 18S rRNA genes	Greenhouse Growth chamber	Soil cultivated with *Bt* maize (Events MON810 and Nongda 1246*1482) and non-*Bt* maize Soil amended with *Bt* and non-*Bt* maize	Cry1Ab Cry1A	Neither actively growing *Bt* maize nor soil amended with *Bt* maize had a consistent effect on fungal community structure in soil	Tan et al. (2010)
AMF: *Glomus mosseae*	Trypan blue staining Slide-intersect method	Greenhouse	Soil cultivated with *Bt* maize (Event *Bt* 11), non-*Bt* maize (Providence)	Cry1Ab	AMF colonization was reduced in *Bt* maize when fertilizer limited and spore density high	Cheeke et al. (2011)
Culturable fungi	Colony-forming units (CFU)	Field	Soil cultivated with *Bt* and non-*Bt* cotton	Cry1Ac or CpTI protein	Seasonal variation in numbers of CFUs but no difference between *Bt* or non-*Bt* cotton	Li et al. (2011)
Fungal decomposer communities	T-RFLP	Field	Litterbags containing biomass of *Bt* maize (Event MON863) or non-*Bt* maize	Cry3Bb	Minor effect of *Bt* maize on fungal communities but differences mostly due to environmental factors	Xue et al. (2011)
AMF	Trypan blue staining Slide-intersect method	Greenhouse	Soil cultivated with 9 paired lines of *Bt* maize and non-*Bt* maize	Cry1Ab Cry34/35Ab1 Cry1Ab+Cry3Bb1 Cry1F Cry3Bb1	Lower levels of AMF colonization in the *Bt* maize lines	Cheeke et al. (2012)

Source: Adapted and updated with permission from Icoz and Stotzky, 2008b.

community composition between *Bt* and non-*Bt* rice straw samples as determined by T-RFLP, although there were some differences in fungal community composition at the early stage of rice root decomposition (Lu et al., 2010a). In a greenhouse study, Tan et al. (2010) reported that neither actively growing *Bt* maize nor the incorporation of *Bt* maize biomass (leaves and straw) had a negative effect on fungal community structure in soil as determined by PCR-DGGE and sequences of 18S rRNA genes. Although Xue et al. (2011) found a minor effect (1 of 16 comparisons) of *Bt* maize biomass buried in litterbags on fungal decomposer communities as determined by T-RFLP, the differences were mostly due to environmental factors (i.e., litterbag placement, recovery year, and plot history) and were not a result of Cry3Bb protein in the *Bt* maize.

In growth chamber experiments, Saxena and Stotzky (2001a) found no difference in the numbers of selected culturable Zygomycetes, Ascomycetes, Deuteromycetes, and yeasts in rhizosphere soils cultivated with *Bt* and non-*Bt* maize. However, Blackwood and Buyer (2004) reported that soils cultivated with *Bt* maize expressing Cry1Ab and Cry1F reduced the presence of eukaryotic PLFA in bulk soils compared with soils cultivated with non-*Bt* maize, although it was not clear which groups of eukaryotes were affected. In field soils cultivated with *Bt* and non-*Bt* maize, Icoz et al. (2008) and Oliveira et al. (2008) found no consistent effect of *Bt* maize cultivation on culturable fungi. Similarly, there was no difference in fungal-to-bacterial ratio reported between field soils cultivated with *Bt* maize and non-*Bt* maize as determined by PLFA and culturing methods (Xue et al., 2005). Li et al. (2011) reported a seasonal variation in numbers of CFUs of culturable fungi in soils cultivated with *Bt* and non-*Bt* cotton over a 3-year field study, but there was no negative effect of *Bt* cotton cultivation on soil fungi.

Although it is not surprising that *Bt* proteins in transgenic plant material have little or no direct effect on saprotrophic fungi, it was hypothesized that the higher lignin content reported in the biomass of some *Bt* cultivars (Saxena and Stotzky, 2001b; Stotzky, 2004; Flores et al., 2005; Poerschmann et al., 2005) might take longer for fungi to degrade, thus leading to accumulation of *Bt* plant residue in the soil over time. This has turned out not to be true most of the time. Although one study showed that some *Bt* plants, including maize, canola, potato, rice, and tobacco, decomposed less in soil and linked this effect to the higher (although not always significantly higher) lignin content in each of the *Bt* cultivars tested (Flores et al., 2005), several subsequent studies have reported that *Bt* plant residue does not generally decompose more slowly than non-*Bt* plant material (Lehman et al., 2008; Tarkalson et al., 2008; Kravchenko et al., 2009; Wu et al., 2009). Moreover, some studies have not even been able to detect a difference in lignin content between *Bt* and non-*Bt* cultivars (Jung and Sheaffer, 2004; Mungai et al., 2005; Lang et al., 2006). These contrasting reports may be the result of differences in age of the plants, detection techniques, or cultivar type, highlighting the importance of a plant line-specific, multidetection, multiple sampling time approach to determine more accurately the effects of the composition of *Bt* crop material on organisms in the soil ecosystem.

Parasitic and pathogenic fungi are also prevalent in soil, and although not desirable in agricultural systems, they may also be affected by changes in the physiology of crop plants. However, during the plant selection process when GM plants are developed and tested, it is unlikely that genotypes that are more susceptible to disease would be released for commercial application. One study that examined the effects of *Bt* crops on fungal pathogens found that genetically engineered *Bt* potato had no negative effect on soilborne pathogens in the rhizosphere, including *Fusarium* sp., *Pythium* sp., *Verticillium dahliae*, potato leaf roll virus, and potato virus Y, under field conditions (Donegan et al., 1996). When fungal growth and survival of the plant pathogen *Fusarium graminearum* and its antagonist *Trichoderma atroviride* were evaluated in flask experiments using pulverized leaf tissue from four different *Bt* maize hybrids (and their corresponding non-*Bt* isolines), both fungal species degraded the Cry1Ab protein in the *Bt* maize tissue, and there was no consistent difference in fungal growth on leaf tissue from *Bt* versus non-*Bt* maize plants as determined by PCR quantification of fungal biomass (Naef et al., 2006). There was also no effect of purified Cry1Ab protein on fungal growth in agar plate assays (Naef et al., 2006). Interestingly, this study also found using mass spectrometry that some of the paired *Bt* and non-*Bt* maize lines differed more in volatile organic compound composition than could be accounted for by just the presence of the Cry protein alone (Naef et al., 2006). This study provided an example of a pleiotropic effect that can occur in GM plants that could alter the quality or composition of transgenic plant tissue, potentially affecting degradation time or the structure of fungal communities inhabiting leaf material.

Of all the different types of fungi in the soil ecosystem, AMF may be at the most risk for nontarget effects of transgenic *Bt* crop cultivation because of their close association with plant roots and their obligately biotrophic nature. Mycorrhizal fungi form symbiotic relationships with 92% of terrestrial plant families (Wang and Qiu, 2006) and are generally considered to be beneficial to plant growth and ecosystem health (Smith and Read, 2008). The mycorrhizal symbiosis is a result of bidirectional exchange: Plants supply sugars to the mycorrhizal fungi, and the fungi benefit plants by improving nutrient and water acquisition (Smith and Read, 2008). Of particular importance to plant health is the ability of AMF to obtain essential minerals such as phosphorus and zinc, which are often found in soil in forms that are unavailable to plants (Vance et al., 2003; Rillig, 2004; Morgan et al., 2005; Smith and Read, 2008). AMF are also important for nutrient acquisition in low-nutrient environments or in the absence of synthetic chemical fertilizers (Gosling et al., 2006; Lekberg et al., 2008; Sheng et al., 2008). Because AMF are obligate symbionts that require a plant host for nutrition and reproduction, they may be uniquely sensitive to alterations in plant physiology as a result of a genetic insertion, *Bt* protein accumulation in the root zone, or alterations in root exudates or plant tissue composition.

Although AMF have an important role in maintaining plant and ecosystem health, few environmental risk assessment studies have examined nontarget effects of *Bt* plants on AMF (Table 8.4; reviewed by Icoz and Stotzky, 2008b; Liu and Du, 2008; Liu, 2010). In one of the first studies that evaluated the impact of *Bt* crop plants on AMF, Turrini et al. (2004) found that root exudates of *Bt* maize (Event *Bt* 176) significantly reduced presymbiotic hyphal growth of the AMF species *Glomus mosseae,* and that 36% of the appressoria (swollen, flattened fungal filaments that enable fungal invasion) failed to produce viable infection pegs in the roots of the *Bt* maize cultivar. The fungal symbionts were able to locate their host, but they were unable to establish the same amount of colonization as in the non-*Bt* isoline, suggesting that the host recognition mechanisms of the fungus were not disrupted, but that something at the plant/fungal interface may have limited colonization. In soil microcosms, *Bt* maize (Event *Bt* 176) again exhibited a lower level of early mycorrhizal colonization and arbuscule development when compared with another line of *Bt* maize (Event *Bt* 11) and a non-*Bt* parental isoline, although both *Bt* cultivars had less arbuscule formation than the parental maize line (Castaldini et al., 2005). In a greenhouse study, the total infected root length of the *Bt* maize plants (Events *Bt* 11 and *Bt* 176) was 50% less than that of the parental maize line (Castaldini et al., 2005). Conversely, de Vaufleury et al. (2007) found no difference in frequency or intensity of root colonization by AMF between *Bt* maize (Event MON810) and its non-*Bt* isoline (Monumental) in soil microcosms, despite the fact that Event MON810 expresses the same type of Cry1Ab protein as Events *Bt* 11 and *Bt* 176. In field studies, Knox et al. (2008) also found no difference in colonization by

AMF arbuscules between *Bt* cotton (Cry1Ac, Cry2Ab) and non-*Bt* cotton. As each of these studies differed in experimental conditions (soil type; plant genotype; species of AMF; fertilizer level; spore number; location of study, i.e., microcosm, greenhouse, or field; AMF detection technique; etc.), it has been difficult to identify the primary factors influencing the patterns of AMF colonization in the different *Bt* cultivars.

When the environmental and experimental factors that may influence the symbiotic relationship between *Bt* maize and AMF were evaluated in a greenhouse study, Cheeke et al. (2011) found that significant differences in AMF colonization were only observed between *Bt* maize (Event *Bt* 11) and its non-*Bt* isoline (Providence) when fertilizer levels were limited and AMF spore density was high (80 spores of *G. mosseae*/pot). Under these experimental conditions, *Bt* maize roots had less than 50% of the AMF colonization of the non-*Bt* isoline (Figure 8.1). This study demonstrated that under circumstances where AMF would be most likely to colonize and benefit the host plant (high spore density, low fertilizer conditions), mycorrhizal colonization was limited in the *Bt* maize. This study also showed that plant response to AMF is dynamic, and AMF colonization levels by *G. mosseae* can vary, even within the same *Bt* or non-*Bt* maize cultivar, depending on the level of fertilizer application or number of spores added to soil (Figure 8.1) (Cheeke et al., 2011). In a follow-up greenhouse study in which AMF colonization levels by fungi from field-collected soil were evaluated in nine paired lines of *Bt* and non-*Bt* maize, *Bt* maize plants overall had lower levels of AMF colonization in roots than the non-*Bt* maize plants when fertilizer was limited (Cheeke et al., 2012).

The effect of *Bt* plant material incorporated into soil on AMF has also been investigated. In a greenhouse study, Castaldini et al. (2005) observed that 4 months after *Bt* maize biomass (Event *Bt* 11) was incorporated into soil, the percentage of root colonization by AMF in *Medicago sativa* (alfalfa) was significantly lower in soil containing *Bt* maize residue

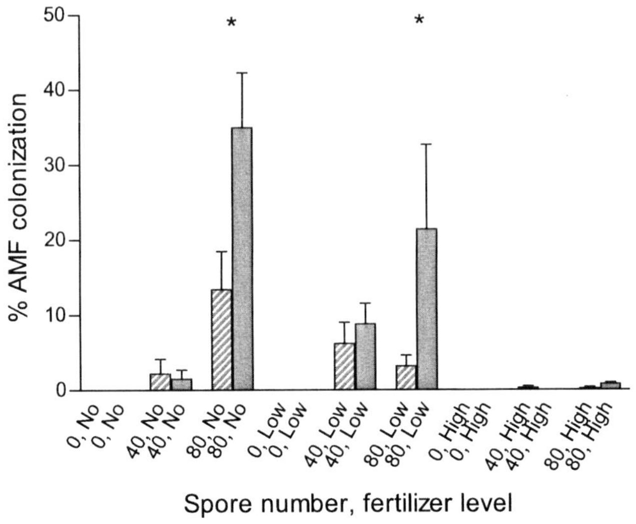

Figure 8.1 Percentage colonization by the AMF species *Glomus mosseae* in *Bt* maize (Event *Bt* 11) and non-*Bt* maize (parental isoline, Providence) inoculated with 0, 40, or 80 spores of *G. mosseae* and grown for 60 days in a greenhouse with weekly treatments of no (0 g/L), low (0.23 g/L), or high (1.87 g/L) complete fertilizer. Striped bars represent the means (±SEM) of *Bt* 11 plants, and solid bars represent the means (±SEM) of non-*Bt* maize plants. $^*P \leq 0.05$, $n = 5$ for each bar.

than in *M. sativa* plants grown in soil amended with non-*Bt* maize biomass (Castaldini et al., 2005). Although no mechanism for these results was identified, a change in nutrient composition of *Bt* plant biomass, lignin content, or time required for degradation of transgenic plant material could potentially contribute to indirect effects on AMF in the soil. Interestingly, when soybean was grown in soil inoculated with a strain of *Bacillus thuringiensis* that expressed Cry1Ab protein (Cry$^+$), a *Bt* mutant bacterial strain that did not express Cry protein (Cry$^-$), purified ICP, or no treatment (control), Ferreira et al. (2003) found that there was no effect of the addition of purified ICP on AMF, but that Cry$^+$ and Cry$^-$ bacterial strains inhibited AMF compared to the control treatment.

8.4.2.1 Research recommendations: Effects of the cultivation of Bt crops on soil fungi

In general, there appear to be no negative effects of *Bt* plants, *Bt* proteins, or the cultivation of *Bt* crops on most free-living fungal species, including saprophytic and pathogenic fungi. However, AMF appear to be sensitive to some *Bt* plant lines, perhaps because of their symbiotic relationship with host plants. As the effect does not appear to be protein specific, each *Bt* plant line containing a different insertion event should be screened for its ability to form mycorrhizal relationships. Currently, there are at least 15 different lines of *Bt* maize genetically engineered to include single, stacked, or pyramided resistance traits, yet few of these *Bt* cultivars have been evaluated for their ability to form symbioses with AMF. Because of the initial indications that certain cultivars of *Bt* maize are poorly colonized by AMF (Turrini et al., 2004; Castaldini et al., 2005; Cheeke et al., 2011; Cheeke et al., 2012), and that each *Bt* line may have a different response to mycorrhizal fungi, it is important to test the ability of each *Bt* line to form associations with AMF under the same experimental conditions as it may not be possible to generalize about the effects of *Bt* maize on AMF, especially if the effect is not protein specific. As most agriculturally important crops are mycorrhizal (excluding species of *Brassica*), even small impacts of *Bt* crops on their fungal symbionts should be closely monitored, particularly in low-input farming systems where reliance on a healthy soil community for plant health and nutrition is key. More research, including screening of multiple *Bt* cultivars for AMF responsiveness and molecular identification of the taxa of AMF colonizing *Bt* and non-*Bt* plant roots, should be conducted to determine whether *Bt* plants with lower levels of AMF colonization also have reduced diversity of AMF in their roots. If this is the case, it would then be pertinent to evaluate changes in spore abundance and diversity of AMF over time in soils in regions where *Bt* crops have been cultivated for multiple years.

8.4.3 Effects of the cultivation of Bt crops on protozoa and nematodes

Protozoa (amoebae, flagellates, and ciliates) and nematodes have an important role in the soil ecosystem, primarily as grazers. By consuming bacteria and other soil organisms, they release nutrients into the soil that can be used as a food source by other soil biota or taken up by plant roots (see Chapter 1 in this volume by Coleman et al.). Protozoa are unicellular organisms that are motile in both aquatic and soil environments and feed on bacteria, small fungi, algae, and other protozoa. Nematodes are microscopic worms that are classified into several trophic groups—bacterial feeders, fungal feeders, and root feeders—primarily based their mouth parts (e.g., whether they have a sharp stylet for piercing root or fungal tissue). Because of their importance in nutrient turnover in soil food webs and their close proximity to plant roots (reviewed by Bais et al., 2006), protozoa and nematodes are another group of soil organisms that may be impacted by cultivation of transgenic *Bt* crops.

Few studies have evaluated the impact of *Bt* plants on protozoa, and the results to date are inconsistent (Table 8.5). Some studies have reported no effect on protozoa in soil amended with purified Cry1Ab or Cry1Ac protein (Donegan et al., 1995) or with *Bt* maize biomass expressing Cry1Ab (Saxena and Stotzky, 2001a), whereas others have reported both higher (Griffiths et al., 2006; Griffiths, Caul, et al., 2007) and lower (Griffiths et al., 2005) numbers of protozoa in soil cultivated with *Bt* maize (expressing Cry1Ab) in the greenhouse and in the field (reviewed in Birch et al., 2007). There was no significant difference in protozoa numbers between soil cultivated with *Bt* maize (Cry1Ab) and non-*Bt* maize in a plant growth room (Saxena and Stotzky, 2001a) or in soil cultivated with *Bt* (Cry1Ab and Cry3Bb1) and non-*Bt* maize in the field (Icoz et al., 2008). This variation in results, even in studies conducted by the same researchers, could be the result of differences in experimental or environmental factors, such as sampling time or different ecological conditions in the greenhouse and the field (e.g., Griffiths, Caul, et al., 2007). Moreover, because amoebae and flagellates tend to be more sensitive to their environmental conditions (e.g., soil compaction, tillage, anoxic conditions) than ciliates, for example, it is important to evaluate each class of protozoa separately as each might have a different response to changes in agricultural practices resulting from the cultivation of transgenic crops or differences in *Bt* root exudates or *Bt* plant biomass.

Research on the nontarget impact of *Bt* crops on nematodes indicates that negative effects on this group of soil organisms may indeed be *Bt* protein specific (Table 8.6). Some *Bt* proteins (Cry5, Cry6, Cry12, Cry13, Cry14, Cry21) have been shown to have a direct toxicity to nematodes (e.g., Meadows et al., 1989a, 1989b, 1990; Marroquin et al., 2000; Kotze et al., 2005; Hu et al., 2010; Hoess et al., 2011), and some are used as biological control agents (i.e., Cry5, Cry6) to protect plants from root-feeding nematodes (reviewed in Bravo et al., 2007; Li, Tan, et al., 2007; Li, Wei, et al., 2007; Li et al., 2008; Khan et al., 2010). *Caenorhabditis elegans,* a common model nematode that is found in a variety of natural and agroecosystems, appears to be negatively impacted by some *Bt* proteins, including Cry1Ab, Cry3Bb1, Cry5A, and Cry5B. In field studies, a lower abundance of *C. elegans* was detected in soils cultivated with *Bt* maize expressing Cry1Ab than in soils with non-*Bt* maize (Manachini and Lozzia, 2003), and there was a negative effect of the Cry1Ab protein from *Bt* maize on nematode growth, number of eggs, and reproduction of *C. elegans* (Lang et al., 2006; Hoess et al., 2008). When *C. elegans* was fed purified Cry5A and Cry5B proteins, individuals showed substantial gut damage, decreased fertility, and increased mortality, consistent with what would be expected in target insect populations (Marroquin et al., 2000). Hoess et al. (2011) found a dose-dependent negative response on the growth and reproduction of *C. elegans* to purified Cry3Bb1 using a bioassay, and gene expression analysis demonstrated that Cry-protein-specific defense genes were upregulated in the presence of Cry1Ab or Cry3Bb1 protein. However, in field studies with *Bt* maize that expressed Cry3Bb1, there was no negative effect on *C. elegans* (Al-Deeb et al., 2003; Hoess et al., 2011), probably because of the low level of Cry3Bb1 protein in the rhizosphere (Hoess et al., 2011). Nematode abundance and diversity were also not different in the rhizospheres of *Bt* and non-*Bt* maize; however, a shift in nematode genus composition occurred in two of the three *Bt* maize plots at the end of the field season, but it did not affect functional diversity (Hoess et al., 2011).

The toxic effects of Cry1Ab observed in *C. elegans* have not been detected in other nematode species, but in some cases, nematode communities have been affected by the cultivation of *Bt* maize. Greenhouse experiments demonstrated that populations of

Table 8.5 Effects of the Cultivation of *Bt* Crops or Amendment with *Bt* Protein on Protozoa

Organism/activity tested	Methodology	Study location	*Bt* crop plant/ experimental variable	Protein	Effect on organism(s)	Source
Protozoa (amoebae, ciliates, and flagellates)	Culturing	Laboratory	Soil amended with purified protein versus unamended soil	Cry1Ab Cry1Ac	No difference in the number of protozoa in soil between *Bt* and non-*Bt* treatments	Donegan et al. (1995)
Protozoa	Culturing	Plant growth room	Soil cultivated with *Bt* and non-*Bt* maize Soil amended with biomass of *Bt* and non-*Bt* maize	Cry1Ab	No significant difference in protozoa numbers in soils cultivated with or amended with *Bt* or non-*Bt* maize	Saxena and Stotzky (2001a)
Protozoa populations	PLFA CLPP	Field	Soil cultivated with *Bt* (MEB307*Bt*) and non-*Bt* maize (Monumental)	Cry1Ab	Reduced protozoan population under *Bt* maize compared to non-*Bt* maize at two sampling times	Griffiths et al. (2005)
Protozoa numbers (active and encysted) Protozoa community profile	MPN CLPP	Greenhouse	Soil cultivated with *Bt* (MEB307*Bt*) and non-*Bt* maize (Monumental) collected from field sites	Cry1Ab	*Bt* trait resulted in more protozoa (amoebae)	Griffiths et al. (2006)
Protozoa	PLFA CLPP	Field	Soil cultivated with *Bt* maize and non-*Bt* maize	Cry1Ab	Periodic increases in protozoan abundance under *Bt*-maize but effects were not persistent	Griffiths, Caul, et al. (2007)
Protozoa	MPN	Field	Soil cultivated with *Bt* and non-*Bt* maize	Cry1Ab Cry3Bb1	No significant differences in the numbers of protozoa between soils cultivated with *Bt* and non-*Bt* maize	Icoz et al. (2008)

Source: Adapted and updated with permission from Icoz and Stotzky, 2008b.

Table 8.6 Effects of the Cultivation of *Bt* Crops or Amendment with *Bt* Protein on Nematodes

Organism/activity tested	Methodology	Study location	*Bt* crop plant/ experimental variable	Protein	Effect on organism(s)	Source
Trichostrongylus colubriformis	Lethal dose 50 (LD_{50}) tests	Laboratory	Spore-crystal preparation of *B.t. morrisoni*	Not specified	*Bt* protein lethal *in vitro* to eggs and larvae of *T. colubriformis*	Meadows et al. (1989a)
T. colubriformis	LD_{50} tests	Laboratory	A crystal-rich preparation of *B.t. kurstaki*	Not specified	*Bt* protein lethal to first- and second-stage larvae of *T. colubriformi*	Meadows et al. (1989b)
Natural populations of *Turbatrix aceti*	LD_{50} tests	Laboratory	Nematodes fed purified *Bt* protein: *B.t. israelensis, B.t. kurstaki*, and *B.t. morrisoni*	Cry1Ab	Negative effects on nematode eggs and juveniles	Meadows et al. (1990)
Caenorhabditis elegans	LC_{50} tests	Laboratory	Crystal-spore toxin lysates	Cry5B Cry6A	When fed *Bt* toxin, *C. elegans* hermaphrodites exhibited extensive gut damage, decreased fertility, and death	Marroquin et al. (2000)
Natural populations	Nematodes extracted from soil by the Baermann technique and counted	Laboratory	Soil planted in *Bt* and non-*Bt* maize Soil amended with *Bt* maize biomass	Cry1Ab	No significant differences in nematode numbers between soil with *Bt* and non-*Bt* maize or soil amended with *Bt* maize biomass	Saxena and Stotzky (2001a)

(continued)

Table 8.6 Effects of the Cultivation of *Bt* Crops or Amendment with *Bt* Protein on Nematodes (continued)

Organism/activity tested	Methodology	Study location	*Bt* crop plant/ experimental variable	Protein	Effect on organism(s)	Source
Natural populations	Assessed general composition, trophic structure, and biodiversity	Field	Cultivation of *Bt* (Event 176) and non-*Bt* maize	Cry1Ab	No effect on nematode communities or biodiversity between soils planted with *Bt* and non-*Bt* maize; however, fungal feeders were more numerous in the *Bt* maize soil	Manachini and Lozzia (2002)
C. elegans	Nematodes extracted using a centrifugal-flotation method	Field	Cultivation of *Bt* and non-*Bt* maize	Cry3Bb1	No negative effects on nematode numbers in soil with *Bt* maize compared with non-*Bt* maize	Al-Deeb et al. (2003)
C. elegans	Soil extraction	Field	Cultivation of *Bt* and non-*Bt* maize	Cry1Ab	Lower *C. elegans* abundance in soils with *Bt* maize than with non-*Bt* maize	Manachini and Lozzia (2003)
Natural populations	Soil extraction	Field	Cultivation of *Bt* and non-*Bt* eggplant	Cry3Bb1	No effect on nematode community structure	Manachini et al. (2003)
Natural populations	Soil extraction	Field	Cultivation of *Bt* and non-*Bt* canola	Cry1Ac	A distinct shift in nematode community structure with *Bt* canola when compared to non-*Bt* isoline	Manachini et al. (2004)
Natural populations	Whitehead and Hemming tray technique	Field	Cultivation of *Bt* (Event MON810) and non-*Bt* maize	Cry1Ab	Significantly lower abundance in soils with *Bt* maize than with non-*Bt* maize	Griffiths et al. (2005)

(continued)

Table 8.6 Effects of the Cultivation of *Bt* Crops or Amendment with *Bt* Protein on Nematodes (continued)

Organism/activity tested	Methodology	Study location	*Bt* crop plant/ experimental variable	Protein	Effect on organism(s)	Source
Acrobeloides spp. *Pratylenchus* spp.	Modified Whitehead & Hemming soil extraction	Greenhouse	Cultivation of *Bt* maize (Events MEB307Bt, MON810) and non-*Bt* (Monumental) maize in pots	Cry1Ab	Significantly higher populations under *Bt* maize than non-*Bt* maize	Griffiths et al. (2006)
C. elegans	Bioassays	Field	Cultivation of *Bt* and non-*Bt* maize	Cry1Ab	Negative effect of Cry1Ab on growth, egg number, and reproduction of *C. elegans*	Lang et al. (2006)
Pratylenchus spp.	Bioassays	Field	Cultivation of *Bt* and non-*Bt* maize	Cry1Ab	No effect of *Bt* maize cultivation on *Pratylenchus* spp.	Lang et al. (2006)
Meloidogyne incognita	Bioassays	Laboratory	Tomato transformed to express Cry6A and control transformed with empty vector or green fluorescent protein	Cry6A	Fourfold decrease in *M. incognita* progeny production in roots of *Bt* tomato plants	Li, Wei, et al. (2007)
M. incognita	Bioassays	Laboratory	Tomato transformed to express Cry5B and control containing empty vector	Cry5B	Cry5B expression reduced the number of root galls and led to a threefold reduction in progeny production	Li et al. (2008)

(continued)

Table 8.6 Effects of the Cultivation of *Bt* Crops or Amendment with *Bt* Protein on Nematodes (continued)

Organism/activity tested	Methodology	Study location	*Bt* crop plant/ experimental variable	Protein	Effect on organism(s)	Source
C. elegans	Bioassay	Laboratory	Rhizosphere and bulk soil cultivated with *Bt* maize (MON810) Trypsinized CrylAb protein expressed in *Escherichia coli*	Cry1Ab	Reproduction and growth reduced in rhizosphere and bulk soil of *Bt* maize; reductions correlated with concentrations of the CrylAb protein	Hoess et al. (2008)
Heligmosomoides bakeri	*Bt* protein injected into mice infected with *H. bakeri*	Laboratory	Spore crystal lysates	Cry5B	98% reduction in nematode egg production *in vivo*	Hu et al. (2010)
Meloidogyne javanica	Eggs and juveniles exposed to different *Bt* isolates at 50% concentration of cell-free filtrate in the lab Seed treatment with cell suspension of *Bt* isolates in greenhouse trials	Laboratory Greenhouse	10 isolates of *B. thuringiensis* isolated from rhizosphere of okra, brinjal, tomato, cotton, cabbage, onion, and watermelon	Not specified	*Bt* protein inhibited egg hatching and killed 2nd stage juveniles in the lab study *Bt* cell suspension reduced number of galls, egg masses, eggs/egg mass, and nematode populations in the greenhouse study	Khan et al. (2010)

(continued)

Table 8.6 Effects of the Cultivation of *Bt* Crops or Amendment with *Bt* Protein on Nematodes (continued)

Organism/activity tested	Methodology	Study location	*Bt* crop plant/ experimental variable	Protein	Effect on organism(s)	Source
C. elegans	Bioassay Gene expression analysis	Laboratory Field	Purified *Bt* protein Soil cultivated with *Bt* (Event MON88017) and non-*Bt* maize	Cry3Bb1	Dose-dependent negative effect of Cry3Bb1 on the growth and reproduction of *C. elegans* No significant effects of the cultivation of *Bt* maize on *C. elegans* in rhizosphere soil	Hoess et al. (2011)

Source: Adapted and updated with permission from Icoz and Stotzky, 2008b.

Acrobeloides sp. and *Pratylenchus* sp. were significantly higher in soil under cultivation with *Bt* maize (Cry1Ab) than with non-*Bt* maize (Griffiths et al., 2006). However, in field trials, Lang et al. (2006) found no effect of *Bt* maize (Cry1Ab) cultivation on *Pratylenchus* sp. When nematode communities were evaluated in the field, cultivation of *Bt* maize expressing Cry1Ab significantly reduced numbers of nematodes in the soil, although the effect was small and within the normal variation observed in many agricultural systems (Griffiths et al., 2005). Cultivation of *Bt* canola expressing Cry1Ac was associated with a shift in nematode community structure when compared with community structure in soil cultivated with non-*Bt* canola (Manachini et al., 2004). Other studies have shown that *Bt* maize expressing Cry1Ab has no negative effect on natural populations of nematodes in soil microcosms (Saxena and Stotzky, 2001a) or in the field (Manachini and Lozzia, 2002), and that *Bt* eggplant expressing Cry3Bb1 has no negative effect on nematode community structure in the field (Manachini et al., 2003).

8.4.3.1 *Research recommendations: Effects of the cultivation of Bt crops on protozoa and nematodes*

Because few studies have evaluated the effects of the cultivation of *Bt* crops on protozoa, the results reported to date are difficult to assess. However, a review of the literature indicates that there appear to be no consistent negative effects of purified *Bt* proteins (Cry1Ab and Cry1Ac) or the cultivation of *Bt* maize (expressing Cry1Ab or Cry3Bb1) on protozoa in greenhouse or in field studies. Future studies would benefit from evaluating a greater variety of *Bt* crops for nontarget effects on protozoa in the soil and under a range of experimental conditions. Moreover, data should be reported separately for the different classes of protozoa (amoebae, flagellates, and ciliates) as each group may have a different response to *Bt* proteins, *Bt* residue in the soil, or the cultivation of *Bt* crops.

Although nematode community structure, biodiversity, and number of individuals in natural populations do not appear to be affected by cultivation of *Bt* crops, individual species, such as *C. elegans*, appear to be sensitive to some Cry proteins, including the Cry1Ab protein that is expressed in most lines of transgenic maize. As *Bt* maize is one of the most commonly cultivated transgenic crops worldwide, and as nematodes are key indicators of soil quality (Blair et al., 1996), more research on the impacts of Cry1Ab and other *Bt* proteins should be conducted to evaluate nontarget effects at both the individual and population levels of nematodes. Insomuch as nematodes are one of the only groups of soil organisms that have displayed direct toxicity of certain Cry proteins produced in transgenic *Bt* crops, they should be carefully evaluated for nontarget effects of *Bt* crop cultivation under a variety of environmental and experimental scenarios.

8.4.4 *Effects of the cultivation of Bt crops on earthworms*

Earthworms are common soil invertebrates responsible for much of the degradation of large pieces of plant material, incorporation of organic matter into lower strata of soil, and increasing aeration of soil by creating large pores in soil as they burrow. Impacts of *Bt* plant biomass, *Bt* proteins, and the cultivation of *Bt* crops on a variety of earthworm species have been studied in numerous laboratory, greenhouse, and field experiments, and most studies have shown that there are few or no effects on this group of soil organisms (Table 8.7; reviewed by Stotzky, 2004; O'Callaghan et al., 2005; Birch et al., 2007; Icoz and Stotzky, 2008b). In laboratory studies, there was no negative effect of soil amended with biomass of *Bt* maize expressing Cry1Ab on *Eisenia fetida* mortality and weight (Ahl Goy et al., 1995), and there were no deleterious effects on survival and reproduction of *E. fetida* fed leaves of *Bt* maize expressing Cry1Ab (Clark and Coats, 2006). Similarly, Saxena and Stotzky (2001a) found no significant difference in mortality and weight of *Lumbricus terrestris* grown for 40 days in soil cultivated with *Bt* maize (Cry1Ab) or non-*Bt* maize and no effect on mortality and weight of *L. terrestris* grown in soil amended with ground, dried *Bt* maize biomass for 45 days. No lethal effects of *Bt* maize (Cry1Ab) residues on adult or immature earthworms were found in field studies (Zwahlen, Hilbeck, Howald, et al., 2003; Lang et al., 2006; Zwahlen et al., 2007). However, adult *L. terrestris* had a significant loss in weight in a laboratory experiment when worms were grown in soil amended with *Bt* (Cry1Ab) versus non-*Bt* maize biomass for 200 days (Zwahlen, Hilbeck, Howald, et al., 2003).

Laboratory studies with *Aporrectodea caliginosa* reported no negative effect on survival, growth, development, or reproduction of individuals grown in soil amended with leaves of *Bt* maize (Cry1Ab) or grown in pots cultivated with *Bt* maize (Vercesi et al., 2006), and there was no effect on survival and reproduction of *Enchytraeus albidus* that were fed diets with *Bt* maize (Cry3Bb1) versus non-*Bt* maize leaf biomass (Honemann and Nentwig, 2009). However, *E. albidus* fed *Bt* leaves (Cry1Ab) had a lower reproduction rate but a higher survival rate when compared with worms fed leaves of the non-*Bt* maize isoline (Honemann and Nentwig, 2009). When litterbags containing biomass of nine different *Bt* maize (Cry1Ab and Cry3Bb1) and non-*Bt* maize cultivars were buried in the field, there was no difference in the numbers of Enchytraeidae or Lumbricidae found in litterbags after 9 months or a difference in the degradation rate between the different maize varieties (Honemann et al., 2008). Interestingly, when maize residues were added to soil microcosms, consumption of *Bt* plant material by *L. terrestris* and *A. caliginosa* lowered the concentration of immunoreactive Cry1Ab protein in soil to less than 10% of the original concentration after 5 weeks without causing any detectable harm on the earthworms (Schrader et al., 2008). *E. fetida* fed *Bt* cotton leaves expressing Cry1Ac in the laboratory also displayed no toxic effects and even resulted in slight increases in growth and reproduction compared with earthworms fed non-*Bt* cotton leaves, although this difference was not statistically significant (Liu, Cui, et al., 2009; Liu, Wang, et al., 2009).

In other studies, there have been no negative effects reported of the cultivation of *Bt* maize expressing Cry3Bb1 on the abundance of Oligochaeta in the field (Bhatti et al., 2005) or soil planted with or amended with *Bt* maize (Cry3Bb1) on weight and mortality of *L. terrestris* in the greenhouse (Ahmad et al., 2006). There was also no difference in the biomass of juveniles or adults of *A. caliginosa, Aporrectodea trapezoides, Aporrectodea tuberculata,* and *L. terrestris* in soil cultivated with *Bt* maize expressing Cry1Ab and Cry3Bb1 compared to soil cultivated with non-*Bt* maize over a 4-year field study (Zeilinger et al., 2010). When effects of the cultivation of *Bt* maize (Cry1Ab) and HT maize were tested on natural populations of earthworms (including *A. caliginosa, Aporrectodea longa, Aporrectodea rosea, L. terrestris, Allolobophora chlorotica, Prosellodrilus amplisetosus,* and *Allolobophora cupulifera*) over two consecutive growing seasons at two different field sites, earthworm populations were reduced only in the plots with the HT crop, likely due to reduced tillage (RT) practices associated with HT crops (Krogh et al., 2007). Plots with the HT maize allowed the practice of RT; plants can be sprayed with herbicides for weed control rather than using tillage measures. In the HT, RT plots, earthworm populations were significantly reduced to about half of earthworm populations in the HT plots that received the conventional tillage treatment. The authors speculated that the reduction in earthworm populations in the HT plots was likely because of exposure to the herbicide Basta®, as the HT maize plants were the same across treatments (Krogh et al., 2007).

Table 8.7 Summary of the Effects of the Cultivation of *Bt* Crops or Amendment with *Bt* Protein on Earthworms

Organism/activity tested	Study location	*Bt* crop plant/experimental variable	Protein	Effect on organism(s)	Source
Eisenia fetida	Laboratory	Soil amended with biomass of *Bt* and non-*Bt* maize	Cry1Ab	No effect on mortality and weight	Ahl Goy et al. (1995)
Lumbricus terrestris	Laboratory	Soil amended with biomass of *Bt* and non-*Bt* maize Soil cultivated with *Bt* and non-*Bt* maize	Cry1Ab	No difference in mortality and weight in soil amended with or cultivated with *Bt* maize	Saxena and Stotzky (2001a)
L. terrestris	Laboratory Field	Soil amended with biomass of *Bt* and non-*Bt* maize	Cry1Ab	Significant weight loss of worms fed *Bt* biomass in laboratory trials No difference in mortality and weight in the field	Zwahlen, Hilbeck, Howald, et al. (2003)
Oligochaeta	Field	Split plots cultivated with *Bt* maize (Event MON 863) and non-*Bt* maize (RX670)	Cry3Bb1	*Bt* maize had no consistent negative impact on abundance	Bhatti et al. (2005)
L. terrestris	Greenhouse	Soil containing *Bt* and non-*Bt* maize roots or dried biomass	Cry3Bb1	No difference in weight or mortality of *L. terrestris* in soil planted with or amended with *Bt* and non-*Bt* maize	Ahmad et al. (2006)
Lumbriciadae community	Field	Cultivation of *Bt* and non-*Bt* maize; cultivation of maize treated with insecticide and untreated maize	Cry1Ab	No effect of Cry protein on worm numbers in soil planted with *Bt* and non-*Bt* maize or in soil with maize treated with insecticide	Lang et al. (2006)
E. fetida	Laboratory	Fed leaves of *Bt* and non-*Bt* maize	Cry1Ab	No negative effects on survival or reproduction	Clark and Coats (2006)
Aporrectodea caliginosa	Laboratory	Soil amended with leaves of *Bt* and non-*Bt* maize Soil cultivated with *Bt* maize	Cry1Ab	No effect on survival, development, growth, or reproduction	Vercesi et al. (2006)

(continued)

Table 8.7 Summary of the Effects of the Cultivation of *Bt* Crops or Amendment with *Bt* Protein on Earthworms (continued)

Organism/activity tested	Study location	*Bt* crop plant/experimental variable	Protein	Effect on organism(s)	Source
Natural earthworm populations including *A. caliginosa* *Aporrectodea longa* *Aporrectodea rosea* *L. terrestris* *Allolobophora chlorotica* *Prosellodrilus amplisetosus* *Allolobophora cupulifera*	Field	*Bt* maize (Event MON810) with conventional tillage (CT) or reduced tillage (RT)	Cry1Ab	RT of *Bt* maize reduced earthworm numbers, probably due to herbicide applications	Krogh et al. (2007)
Enchytraeidae	Field	Litterbags containing *Bt* or non-*Bt* maize	Cry1Ab	Higher numbers of Enchytraeidae extracted from non-*Bt* litterbags than *Bt* litterbags	Zwahlen et al. (2007)
Enchytraeidae Lumbricidae	Field	Litterbags with plant material from 9 different *Bt* or non-*Bt* maize	Cry1Ab Cry3Bb1	No difference in earthworm numbers or degradation rate between the 9 maize varieties	Honemann et al. (2008)
L. terrestris *A. caliginosa*	Soil microcosm	Leaves and roots of *Bt* maize (Event MON810) and non-*Bt* maize	Cry1Ab	Earthworms reduced the immunoreactive properties of Cry1Ab from maize residues	Schrader et al. (2008)
Enchytraeus albidus	Laboratory	Fed *Bt* (N4640*Bt* and DKC5143*Bt*) and non-*Bt* maize leaves	Cry1Ab Cry3Bb1	No effect of Cry3Bb1 on survival and reproduction; higher survival but lower reproduction in the Cry1Ab treatment compared with control	Honemann and Nentwig (2009)
E. fetida	Laboratory	Fed leaves of *Bt* (GK19) and non-*Bt* cotton	Cry1Ac	No negative effects of consuming *Bt* cotton leaves	Liu et al. (2009b)
E. fetida	Laboratory	Fed leaves of *Bt* + CpTI cotton and non-*Bt* cotton	Cry1Ac	Leaves of *Bt* cotton enhanced growth and reproduction	Liu et al. (2009a)
A. caliginosa, *Aporrectodea trapezoides*, *Aporrectodea tuberculata*, and *L. terrestris*	Field	Fields planted in *Bt* and non-*Bt* maize	Cry1Ab Cry3Bb1	No difference in the biomass of juveniles or adults of any species tested	Zeilinger et al. (2010)

Source: Adapted and updated with permission from Icoz and Stotzky, 2008b.

8.4.4.1 Research recommendations: Effects of the cultivation of Bt crops on earthworms

The laboratory, greenhouse, and field studies outlined above indicate that cultivation of *Bt* crops, including maize expressing Cry1Ab or Cry3Bb1 and cotton expressing Cry1Ac, have no deleterious effects in soil on numbers or populations of earthworms, and that differences in agricultural practices (e.g., herbicide applications) associated with the cultivation of GM crops may be more important to earthworm health and ecosystem function than the *Bt* proteins associated with GM crops. Nevertheless, continued monitoring for effects of *Bt* plants on earthworms would be useful, especially with the ever-increasing hectarage dedicated to the cultivation of *Bt* crops across the globe and the continued development of new and different types of *Bt* plants. Because earthworms do not appear to be affected by most *Bt* plants examined thus far, in future studies, the research focus should be narrowed to evaluate earthworm species based on their association with a particular *Bt* crop or their importance to ecosystem processes in a given area (Zeilinger et al., 2010).

8.4.5 Effects of the cultivation of Bt crops on microarthropods

Microarthropods, including isopods (pillbugs, woodlice), collembolans (springtails), and mites, are other organisms in the soil ecosystem that may be affected by the cultivation of transgenic *Bt* crops (Table 8.8). Their important role in nutrient cycling and degradation of plant material may be compromised by changes in the structure, chemistry, and other properties of soil by Cry proteins deposited in soil from transgenic plant biomass or root exudates. Despite being more closely related to target insect pest populations (e.g., Lepidoptera, Coleoptera, Diptera) than the other soil organisms already discussed, most microarthropods, including the woodlouse *Porcellio scaber* (Escher et al., 2000; Pont and Nentwig, 2005), the pillbugs *Armadillidium nastum* and *Trachelipus rathkii* (Clark et al., 2006), the collembolans *Folsomia candida* (Sims and Martin, 1997; EPA, 2001b; Bakonyi et al., 2006; Clark and Coats, 2006; Bakonyi et al., 2011), *Xenylla griesea* (Sims and Martin, 1997), and *Protaphorura armata* (Heckmann et al., 2006), as well as natural populations of collembolans (Lang et al., 2006; de Vaufleury et al., 2007; Priestley and Brownbridge, 2009), have shown no adverse effects in soil from the cultivation of *Bt* Cry1Ab maize, consumption of *Bt* Cry1Ab maize plant material, or consumption of purified Cry1Ab protein in studies in microcosms and in the field. Similarly, there was no negative effect of transgenic *Bt* cotton expressing Cry1Ab or Cry1Ac or *Bt* potato expressing Cry3A on *F. candida* (Yu et al., 1997; EPA, 2001b) and also no effect of *Bt* potato on the number of eggs and body length of the mite *Oppia nitens* (Yu et al., 1997). In laboratory experiments in which the orbatid mite *Scheloribates praeincisus* was grown in rearing chambers and fed *Bt* cotton leaves (Bollgard; Cry1Ac), non-*Bt* cotton leaves, and the *Bt* biopesticide Dipel® (containing spores of *B. thuringiensis* var. *kurstaki,* HD-1, coding for the expression of the insecticidal d-endotoxin Cry1Ab), there was no effect on survival or development of adult or immature mites or an effect on food consumption (Oliveira et al., 2007). Cultivation of *Bt* maize expressing Cry3Bb1 showed no deleterious effects on the numbers of collembolans or mites in natural populations of field soil (Al-Deeb et al., 2003). In a field experiment in which litterbags containing plant material from nine different *Bt* (Cry1Ab, Cry3Bb1) and non-*Bt* maize lines were buried in soil, there was no difference in decomposer communities (Collembola, Acari, and 12 taxa of other arthropods) or degradation rate between the different cultivars (Honemann et al., 2008). When the effects of individual purified proteins were tested (Cry1Ab, Cry2A, Cry1Ac, Cry3A) in a feeding study over 21 days in Petri dish microcosms, there were no

Table 8.8 Summary of the Effects of Cultivation with *Bt* Crops or Amendment with *Bt* Proteins on Microarthropods

Organism/activity tested	Species	Study location	*Bt* plant/experimental variable	Protein	Effect on organism(s)	Source
Collembola	*Folsomia candida* *Xenylla grisea*	Laboratory (Petri dish microcosms)	Added four purified proteins to diet	Cry1Ab Cry1Ac Cry2A Cry3A	No effect on survival or reproduction over 21 days	Sims and Martin (1997)
Woodlouse	*Porcellio scaber*	Laboratory	Fed purified protein	Cry2A	No toxic effect	Sims (1997)
Collembola Mite	*Folsomia candida* *Oppia nitens*	Laboratory	Fed leaves of *Bt* and non-*Bt* cotton or leaves of *Bt* and non-*Bt* potato	Cry1Ab/Ac Cry3A	No effects on oviposition, egg number, or body length	Yu et al. (1997)
Woodlouse	*Porcellio scaber*	Laboratory	Fed biomass of *Bt* and non-*Bt* maize	Cry1Ab	No negative effect of *Bt* maize litter on consumption, reproduction, and growth	Escher et al. (2000)
Collembola	*Folsomia candida*	Field	Cultivation of *Bt* and non-*Bt* maize or *Bt* and non-*Bt* cotton	Cry1Ab Cry1Ac	No effect on *F. candida* numbers	EPA (2001b)
Woodlouse	*Porcellio scaber*	Laboratory	Fed biomass of *Bt* and non-*Bt* maize	Cry1Ab	Fed less on *Bt* maize than on non-*Bt* maize during a 20-day feeding period	Wandeler et al. (2002)
Collembola Mites	Natural populations	Field	Cultivation of *Bt* and non-*Bt* maize	Cry3Bb1	No negative effects on numbers of collembolans or mites	Al-Deeb et al. (2003)
Woodlouse	*Porcellio scaber*	Laboratory	Fed *Bt* and non-*Bt* maize leaves	Cry1Ab	No toxic effect	Pont and Nentwig (2005)

(continued)

Table 8.8 Summary of the Effects of Cultivation with *Bt* Crops or Amendment with *Bt* Proteins on Microarthropods (continued)

Organism/activity tested	Species	Study location	*Bt* plant/experimental variable	Protein	Effect on organism(s)	Source
Collembola	*Folsomia candida* *Heteromurus nitidus* *Sinella coeca*	Laboratory	Fed dried leaves of *Bt* and non-*Bt* maize	Cry1Ab	No effect on survival; species-specific effects in distributions and feeding preferences	Bakonyi et al. (2006)
Collembola	*Folsomia candida*	Soil microcosm	Fed leaves of *Bt* maize (Events *Bt* 11 and MON810) and non-*Bt* isolines	Cry1Ab	No significant effect on survival and reproduction	Clark and Coats (2006)
Isopods	*Armadillidium nasatum* *Trachelipus rathkii*	Laboratory	Fed purified protein or leaves of *Bt* (Events *Bt* 11 and MON810) and non-*Bt* maize	Cry1Ab	No adverse effects of *Bt* on survival and growth	Clark et al. (2006)
Collembola Mites	Natural populations	Laboratory	Cultivation of *Bt* and non-*Bt* maize in pots	Cry1Ab	Lower collembolan abundance and higher mite populations under *Bt* maize	Griffiths et al. (2006)
Collembola	Natural populations	Field	Cultivation of *Bt* and non-*Bt* maize	Cry1Ab	No differences in population density	Lang et al. (2006)
Collembola	*Protaphorura armata*	Laboratory	Fed purified protein or biomass of *Bt* and non-*Bt* maize	Cry1Ab	No effect of purified protein or biomass of *Bt* maize on growth and reproduction	Heckmann et al. (2006)
Mites Collembola	Mite suborder: *Actinedida, Gamasida, Oribatida, Acaridida* Collembola not identified to species	Field	Cultivation of *Bt* and non-*Bt* maize	Cry1Ab	Minor negative effect of *Bt* maize on microarthropod abundance in high-clay soils	Cortet et al. (2007)

(continued)

Table 8.8 Summary of the Effects of Cultivation with *Bt* Crops or Amendment with *Bt* Proteins on Microarthropods (continued)

Organism/activity tested	Species	Study location	*Bt* plant/experimental variable	Protein	Effect on organism(s)	Source
Collembola	Functional groups: Euedaphic, eu-hemiedaphic, hemiedaphic, hemiepiedaphic, epiedaphic	Field	Cultivation of *Bt* and non-*Bt* maize	Cry1Ab	Lower abundance of Collembola in *Bt* soil, but only at one site in early fall; no effect of *Bt* maize on functional groups	Debeljak et al. (2007)
Collembola, Actinedida, Acaridida, Gamasida, Oribatida	*Entomobrya* sp., *Entomobrya multifasciata*, *Orchesella* sp., *Lepidocyrtus lanuginosus*, *Pseudosinella alba*, *Folsomia fimetaria*, *Paraisotoma notabilis*, *Protaphorura armata*, *Sminthurinus aureus*	Soil microcosm	Cultivation of *Bt* maize (Event MON810) and non-*Bt* maize (Monumental)	Cry1Ab	No difference in abundance and diversity in *Bt* and non-*Bt* soil	de Vaufleury et al. (2007)
Oribatid mite	*Scheloribates praeincisus*	Laboratory	Fed *Bt* cotton (Bollgard) leaves, non-*Bt* cotton leaves, and *Bt* biopesticide (Dipel)	Cry1Ac	No effect on adult and immature survivorship, food consumption, or development	Oliveira et al. (2007)

(continued)

Table 8.8 Summary of the Effects of Cultivation with *Bt* Crops or Amendment with *Bt* Proteins on Microarthropods (continued)

Organism/activity tested	Species	Study location	*Bt* plant/experimental variable	Protein	Effect on organism(s)	Source
Collembola, Acari, and 12 taxa of other arthropods	Collembolan families: *Isotomidae*, *Hypogastruridae*, *Entomobryidae*, and *Sminthuridae* Mites: *Gamasina*, *Uropodina*, and *Cryptostigmata*	Field	Litterbags with plant material from 9 different *Bt* or non-*Bt* maize cultivars	Cry1Ab Cry3Bb1	No difference in decomposer communities or degradation speed between the 9 maize varieties	Honemann et al. (2008)
Collembola	Surface-dwelling and soil-dwelling Collembola species	Field	Cultivation of *Bt* and non-*Bt* maize	Cry1Ab	No negative effect of *Bt* maize on abundance or diversity	Priestley and Brownbridge (2009)
Collembola	*Folsomia candida*	Laboratory	Fed leaves of *Bt* maize (Event MON810) and non-*Bt* maize	Cry1Ab	No negative effect on *F. candida* in long-term feeding studies	Bakonyi et al. (2011)

Source: Adapted and updated with permission from Icoz and Stotzky, 2008b.

toxic effects detected on the survival or reproduction of *F. candida* or *X. griesea* (Sims and Martin, 1997) and no effects of purified Cry2A on *P. saber* (Sims, 1997).

Only in a few studies have negative effects of *Bt* proteins on microarthropods been reported: Wandeler et al. (2002) found that *P. scaber* fed significantly less on *Bt* maize expressing Cry1Ab than on non-*Bt* maize during a 20-day feeding trial in the laboratory in soil microcosms; Griffiths et al. (2006) reported lower collembolan abundance and higher mite populations in soil cultivated with *Bt* maize expressing Cry1Ab than in soil cultivated with non-*Bt* maize in the laboratory; and Bakonyi et al. (2006) found that there were species-specific effects in feeding preference when the collembolans *F. candida, Heteromurus nitidus,* and *Sinella coeca* were fed dried leaves of *Bt* (Cry1Ab) or non-*Bt* maize in laboratory feeding tests—*F. candida* preferred non-*Bt* maize as a food source over *Bt* maize, but there was no difference in feeding preference for *H. nitidus* and *S. coeca*. Cortet et al. (2007) detected a minor negative effect of *Bt* maize (Cry1Ab) cultivation on microarthropod abundance (mites and collembolans) in high-clay soils in field trials; however, agricultural practices had a comparable or greater effect on microarthropod abundance than the *Bt* crop. Debeljak et al. (2007) also reported a lower abundance of collembolans in field soil cultivated in *Bt* maize (Cry1Ab) but only at one site and only in early fall; no difference in functional groups of Collembola was detected.

8.4.5.1 Research recommendations: Effects of the cultivation of Bt crops on microarthropods

In general, cultivation of transgenic *Bt* crops, *Bt* plant biomass, or purified *Bt* proteins, including Cry1Ab, Cry1Ac, and Cry3Bb1, has had little to no effect on most microarthropods tested. When effects have been reported, they were minor, and the microarthropods were often more affected by temporal differences in sampling time, agricultural practices, plant varietal differences, or other biotic and abiotic factors not related to a particular *Bt* protein. Although most studies have not shown a consistent pattern of *Bt* effects on microarthropods, longer-term field experiments would improve the current understanding *Bt* effects on the abundance and diversity of microarthropods in the soil environment (e.g., Theissen and Russell, 2009). Future research studies should focus on multitrophic-level interactions and continue to evaluate the effects of new and different types of *Bt* crops on nontarget microarthropods in the soil.

8.4.6 Effects of the cultivation of Bt crops on nontarget Lepidopteran larvae and aquatic insects

Although most insects are not generally classified as soil organisms per se, many insect orders (e.g., Lepidoptera, Coleoptera, and Hymenoptera) have a larval stage that lives in or feeds in close proximity to *Bt* plant leaves and other residues, *Bt* pollen, or Cry proteins in soil. Thus, it is possible for nontarget insects, including pollinators, prey and predator species, and biocontrol agents, to be affected by the cultivation of *Bt* crops, even if they do not feed on the GM crop directly. As the number of studies evaluating the effects of *Bt* crops on nontarget insects are too numerous to review in this chapter, only the nontarget effects of the cultivation of *Bt* crops on selected species of Lepidoptera and aquatic insects are summarized here (Table 8.9). More information on the nontarget effects of the cultivation of *Bt* crops on a variety of insect species can be found in several reviews (including those of Romeis et al., 2006; Marvier et al., 2007; Thies and Devare, 2007; Duan et al., 2008;

Wolfenbarger et al., 2008; Lovei et al., 2009; Lundgren et al., 2009; Lang and Otto, 2010; Gatehouse et al., 2011).

Many of the studies evaluating the effects of the cultivation of *Bt* crops on nontarget insects have been conducted on Lepidoptera, primarily the larvae of monarch (*Danaus plexippus*) and swallowtail (*Papilio* sp.) butterflies. Larvae of the monarch butterfly, for example, do not feed on *Bt* crop plants, but their primary food source (milkweed) often grows in or near agricultural fields where *Bt* pollen deposition on milkweed leaves could occur (Lang et al., 2004). Losey et al. (1999) reported that monarch caterpillars fed pollen from *Bt* maize (N4640, Cry1Ab) in laboratory studies ate less, grew more slowly, and had higher mortality than larvae fed non-*Bt* maize pollen and cautioned that *Bt* crop cultivation may have serious unintended consequences to nontarget Lepidoptera in the environment. However, critics pointed out that the authors neglected to record the amount of *Bt* corn pollen deposited on the milkweed leaves fed to the monarch caterpillars and argued that the results reported in laboratory studies were unlikely to be observed under field conditions (reviewed and discussed in Shelton and Sears, 2001). Since then, additional studies have evaluated the effects of *Bt* pollen on nontarget Lepidoptera under both laboratory and field conditions (e.g., Jesse and Obrycki, 2000; Wraight et al., 2000; Hellmich et al., 2001; Sears et al., 2001; Stanley-Horn et al., 2001; Tschenn et al., 2001; Zangerl et al., 2001; reviewed in Gatehouse et al., 2002; Anderson et al., 2004, 2005; Candolfi et al., 2004; Dively et al., 2004; Gathmann et al., 2006; Lang and Vojtech, 2006; Prasifka et al., 2007) and reviewed by others (Sears et al., 2001; Shelton and Sears, 2001; Sears, 2004; Lang and Otto, 2010).

Wind-pollinated crops, such as *Bt* maize, have a greater potential for nontarget effects of pollen deposition on Lepidopteran larvae (and other insects) than non-wind-pollinated GM crops such as *Bt* potato, *Bt* cotton, or *Bt* canola and thus have been the focus of many risk assessment studies. In an analysis of 20 peer-reviewed publications (16 laboratory feeding studies with purified *Bt* toxin, *Bt* maize pollen, or *Bt* maize anthers in an artificial diet or on leaf disks, whole leaves, or whole plants and 7 field-based studies using unfenced host plants, field cages, or natural conditions), Lang and Otto (2010) found adverse effects of *Bt* maize or *Bt* protein on nontarget Lepidopteran larvae (*Danaus plexippus, Papilio polyxenes, Papilio machaon, Pieris rapae, Pieris brassicae, Pseudozizeeria maha, Inachis io, Euchaetes pernyi, Plutella xylostella,* or *Galleria mellonella*) in 52% of laboratory-based and in 21% of field-based studies. The majority of these studies were based in the United States and focused on nontarget effects of *Bt* maize pollen (primarily from Events MON810, *Bt* 176, *Bt* 11) on monarch butterfly larvae. The effects most often studied in these experiments were on larval body mass, survival, and developmental time, whereas effects on adult Lepidoptera were rarely examined (Lang and Otto, 2010). The results of each study appeared to depend primarily on the methodology used, the *Bt* cultivar, and the amount of Cry protein expressed in the pollen of each *Bt* cultivar. Although several field-based studies found no effect of *Bt* crop cultivation on nontarget Lepidoptera (black swallowtail and monarch) (e.g., Wraight et al., 2000; reviewed in Sears et al., 2001; Tschenn et al., 2001; Anderson et al., 2004), the results of Losey et al. (1999) were supported by those of other studies that demonstrated significant adverse effects on swallowtail or monarch caterpillars in the presence of *Bt* pollen in the field (e.g., Jesse and Obrycki, 2000; Stanley-Horn et al., 2001; Zangerl et al., 2001; Dively et al., 2004) and in laboratory feeding studies (e.g., Hellmich et al., 2001; Anderson et al., 2004, 2005; Lang and Vojtech, 2006; Prasifka et al., 2007). Most studies have shown that pollen from *Bt* maize Event 176 had the most lethal and sublethal effects (e.g., mortality, growth rate, body weight, development time) on larvae of swallowtail and monarch butterflies, whereas *Bt* maize Events *Bt* 11 and MON810 had negligible effects on the larvae. *Bt* maize Event 176 was

Table 8.9 Summary of the Effects of the Cultivation of *Bt* Crops or Amendment with *Bt* Protein on Selected Nontarget Lepidopteran and Aquatic Insect Larvae

Organism(s)	Species/ classification	Study location	*Bt* crop plant/ experimental variable	Protein	Effect on organism(s)	Source
Monarch butterfly larvae	*Danaus plexippus*	Laboratory	Fed milkweed leaves dusted with pollen from *Bt* maize (Event N4640), pollen from non-*Bt* maize, or no pollen	Cry1Ab	*D. plexippus* reared on milkweed leaves dusted with *Bt* pollen, ate less, grew more slowly, and suffered higher mortality	Losey et al. (1999)
Monarch butterfly larvae	*Danaus plexippus*	Field	*Bt* maize (Event 176, *Bt*11) and non-*Bt* maize pollen naturally deposited on *Asclepias syriaca* (common milkweed)	Cry1Ab	Caused significant mortality of *D. plexippus* larvae after 48 hours and 120 hours	Jesse and Obrycki (2000)
Black swallowtail butterfly larvae	*Papilio polyxenes*	Field	Potted host plants with larvae placed along fields of *Bt* maize (Event MON810)	Cry1Ab	No effect of proximity to the *Bt* maize field or *Bt* pollen deposition on host plants	Wraight et al. (2000)
Monarch butterfly larvae	*Danaus plexippus*	Laboratory	Larvae fed purified *Bt* toxins, pollen from *Bt* maize applied directly to milkweed leaf disks, and *Bt* pollen with tassel material applied directly to milkweed leaf disks	Cry1Ab Cry1Ac Cry9C Cry1F	Purified Cry9C and Cry1F had minimal effect; first instars were sensitive to Cry1Ab and Cry1Ac *Bt* 176 maize pollen had a consistent negative effect on larvae	Hellmich et al. (2001)

(continued)

Table 8.9 Summary of the Effects of the Cultivation of *Bt* Crops or Amendment with *Bt* Protein on Selected Nontarget Lepidopteran and Aquatic Insect Larvae (continued)

Organism(s)	Species/ classification	Study location	*Bt* crop plant/ experimental variable	Protein	Effect on organism(s)	Source
Monarch butterfly larvae	*Danaus plexippus*	Field	Survival and growth of larvae after exposure to pollen from *Bt* maize (Events *Bt*11, *Bt*176, and MON810) or to an insecticide	Cry1Ab	Negative effects of *Bt* 176 on larvae but negligible effects of *Bt* 11 and MON810	Stanley-Horn et al. (2001)
Monarch butterfly larvae	*Danaus plexippus*	Cage and flight chambers	Oviposition behavior of monarch on milkweed plants dusted with *Bt* or non-*Bt* maize pollen, gravel dust, or undusted	Cry1Ab	No difference in oviposition behavior between treatments	Tschenn et al. (2001)
Black swallowtail butterfly larvae	*Papilio polyxenes*	Field	*Bt* maize, Event 176	Cry1Ab	Significant reduction in growth rates of black swallowtail larvae	Zangerl et al. (2001)
Monarch butterfly larvae	*Danaus plexippus*	Laboratory Field	Larvae fed *Bt* maize anthers Measured anther density on milkweed in the field and evaluated effects on monarch larvae	Cry1Ab	Negative effects on larvae that increased with *Bt* anther density in the lab No negative effect in the field	Anderson et al. (2004)

(continued)

Table 8.9 Summary of the Effects of the Cultivation of *Bt* Crops or Amendment with *Bt* Protein on Selected Nontarget Lepidopteran and Aquatic Insect Larvae (continued)

Organism(s)	Species/ classification	Study location	*Bt* crop plant/ experimental variable	Protein	Effect on organism(s)	Source
Natural arthropod communities	Various	Field	Cultivation of *Bt* maize (Event 176) and non-*Bt* maize with chemical insecticide (Karate Xpress), biopesticide (Delfin), or no treatment	Cry1Ab	Lower abundance of adult Lepidoptera, Lonchopteridae, Mycetophilidae, Syrphidae, and Ceraphronidae in *Bt* maize plots but few effects on most other arthropods	Candolfi et al. (2004)
Monarch butterfly larvae	*Danaus plexippus*	Laboratory Field	Natural deposits of *Bt* and non-*Bt* maize pollen on milkweed	Cry1Ab	~24% fewer larvae reached the adult stage when exposed to *Bt* pollen; 0.6% mortality after long-term exposure	Dively et al. (2004)
Monarch butterfly larvae	*Danaus plexippus*	Laboratory Cage study	*Bt* and non-*Bt* maize pollen deposited on milkweed	Cry1Ab	Additive effects of *Bt* anthers and pollen led to behavioral and feeding changes in the lab Larvae exposed to *Bt* anthers and pollen took longer to develop and pupae weighed less in a cage study	Anderson et al. (2005)

(continued)

Table 8.9 Summary of the Effects of the Cultivation of *Bt* Crops or Amendment with *Bt* Protein on Selected Nontarget Lepidopteran and Aquatic Insect Larvae (continued)

Organism(s)	Species/ classification	Study location	*Bt* crop plant/ experimental variable	Protein	Effect on organism(s)	Source
Common swallowtail butterfly	*Papilio machaon*	Laboratory	Different densities of pollen from *Bt* maize (Event 176) applied to leaf disks	Cry1Ab	Consumption of *Bt* pollen led to reduced appetite, lower body weight, and longer larval development time	Lang and Vojtech (2006)
Monarch butterfly larvae	*Danaus plexippus*	Laboratory	Larvae exposed to milkweed leaves with no anthers, *Bt* maize anthers, or non-*Bt* anthers	Cry1Ab	Larvae fed less and weighed less when exposed to *Bt* anthers, likely due to behavioral changes—larvae tended to move off the milkweed leaves	Prasifka et al. (2007)
Caddis flies	*Lepidostoma liba*	Laboratory	Fed *Bt* and non-*Bt* maize leaves	Cry1Ab	*Bt* maize residue reduced growth and increased mortality of nontarget stream insects	Rosi-Marshall et al. (2007)
Aquatic insect larvae	*Chironomus dilutus* (midge)	Laboratory	Fed *Bt* protein from maize root extracts	Cry3Bb1	After 10 days, a significant decrease in *C. dilutus* survival at low *Bt* concentrations, but no effect on growth among surviving larvae	Prihoda and Coats (2008)

(continued)

Table 8.9 Summary of the Effects of the Cultivation of *Bt* Crops or Amendment with *Bt* Protein on Selected Nontarget Lepidopteran and Aquatic Insect Larvae (continued)

Organism(s)	Species/ classification	Study location	*Bt* crop plant/ experimental variable	Protein	Effect on organism(s)	Source
Decomposition rate, aquatic invertebrate abundance, and community composition	10 different invertebrate taxa	Field	*Bt* and non-*Bt* maize tissue in litterbags in 9 streams	Cry1Ab Cry1Ab + Cry3Bb1	Minimal effects; most differences due to site-to-site variation and difference in agricultural practices	Swan et al. (2009)
Aquatic invertebrate abundance, diversity, biomass, and functional structure	Trichopteran Amphipod Snail Natural populations in the field	Laboratory Field	Laboratory feeding trials Benthic cores and litterbags in 12 streams adjacent to *Bt* or non-*Bt* maize fields	Cry protein not specified; *Bt* maize lines Crow4635 and Agventure	*Lepidostoma liba* grew slower when fed *Bt* maize in the lab No difference in abundance or diversity between *Bt* and non-*Bt* fields	Chambers et al. (2010)
Aquatic invertebrates (sedge, caddis fly, crane fly, and an aquatic isopod)	*Lepidostoma* spp., *Pycnopsyche* cf. *scabripennis*, *Tipula (Nippotipula)* cf. *abdominalis*, *Caecidotia communis*	Laboratory	Fed leaves of *Bt* or non-*Bt* maize	Cry1Ab Cry1Ab + Cry3Bb1	*Bt* maize negatively affected growth and survivorship of crane fly and *C. communis;* effects were more due to differences in plant tissue content than Cry protein	Jensen et al. (2010)

shown to have more than 50 times the level of Cry1Ab found in *Bt* maize Events MON810 and *Bt* 11 (EPA, 2001a). When the EPA registration of *Bt* maize Event 176 expired in 2001, it was not renewed by the producing seed companies, primarily as a result of concerns about pest resistance, and is thus no longer grown in the United States (EPA, 2011).

Insect larvae in aquatic ecosystems may also be impacted by the cultivation of *Bt* crops as a result of the transport and accumulation of transgenic *Bt* biomass in rivers and streams (Douville et al., 2007; Rosi-Marshall et al., 2007; Prihoda and Coats, 2008; Swan et al., 2009; Chambers et al., 2010; Jensen et al., 2010; Tank et al., 2010; Wolt and Peterson, 2010; reviewed by Viktorov, 2011). In the Midwestern part of the United States, where most of the nation's corn crop is grown, *Bt* plant biomass is often deposited into aquatic ecosystems by wind and transported downstream, where it can accumulate in the bends or edges of the waterways (Tank et al., 2010) and in sediments (Douville et al., 2007). Invertebrate consumers are abundant in agricultural streams (e.g., Moore and Palmer, 2005; Menninger and Palmer, 2007) and are likely to feed directly on decaying plant material, including *Bt* maize, in waterways. Tank et al. (2010) reported that 86% of 217 stream sites surveyed in Indiana, United States, in 2007—when 75% of all maize cultivated in the United States was engineered to express single or stacked combinations of insecticidal or herbicide resistance traits (USDA, 2010)—contained plant material from maize, including leaves, cobs, husks, and stalks. Thus, it was not surprising that Cry1Ab protein was detected in the water column at 23% of these sites. What was unexpected, however, was that Cry1Ab was detected in streams even in areas where no maize fields were present within 500 m, indicating that *Bt* toxins may be more widely distributed in aquatic ecosystems than previously realized. As these stream sites were sampled 6 months after the corn fields were harvested, levels of Cry1Ab (and other *Bt* proteins) would probably be even higher immediately following harvest when plant biomass is fresh. Douville et al. (2007) reported that Cry1Ab gene from *Bt* maize cultivation could be detected in surface water and sediments for more than 21 and 40 days, respectively. The Cry1Ab protein was found in higher concentrations in the clay and sand-rich sediments than in surface water and could still be detected at least 82 km downstream from the *Bt* maize plot, although the *Bt* concentration tended to decrease the farther away from the plot it was detected.

Some of the first evidence of the potential sensitivity of some aquatic insects to *Bt* biomass was reported in a laboratory experiment; caddisfly larvae fed *Bt* maize residue grew less and had higher mortality than larvae fed non-*Bt* maize residue (Rosi-Marshall et al., 2007). Caddisflies are closely related to target pests, and the larvae are an important food source for fish and other marine life. Chambers et al. (2010) reported that the aquatic leaf shredder *Lepidostoma liba* grew more slowly when fed *Bt* maize residue in laboratory trials. However, no negative effects were detected on the abundance or diversity of natural populations of nontarget aquatic invertebrates, including Coleoptera, Diptera, Ephemeroptera, Hemiptera, Odonata, Plecoptera, and Trichoptera in a field study, possibly because the *Bt* plant biomass was already highly degraded at the time of sampling (Chambers et al., 2010). In a laboratory feeding study, differences in the composition of plant tissue from different maize cultivars, rather than a direct toxic effect of Cry protein in the *Bt* maize varieties, were found to affect the growth and survivorship of nontarget aquatic detritivores (e.g., leaf-chewing shredders), including crane fly larvae and the aquatic isopod *Caecidotia communis* (Jensen et al., 2010). Until recently, most risk assessment research has focused on the nontarget effects of *Bt* crops in the terrestrial environment. Thus, it is largely unknown what long-term effects, if any, could be expected on nontarget aquatic invertebrates and detritivores that are exposed to *Bt* plant material, which appears to be widely dispersed throughout aquatic ecosystems.

8.4.6.1 *Research recommendations: Effects of the cultivation of Bt crops on nontarget Lepidopteran larvae and aquatic insects*

Although it has been demonstrated that some nontarget insects (e.g., Lepidoptera) can be negatively impacted by certain *Bt* cultivars, particularly those that express high levels of the Cry proteins in pollen, root exudates, and plant biomass, further studies are needed to determine the long-term health and reproductive success of nontarget insect larvae that feed on naturally deposited *Bt* pollen and other plant residues in the field. Because most *Bt* crops developed thus far are not wind pollinated and would therefore have minimal impacts of pollen drift on nontarget insect larvae, in future studies, it will be important to evaluate the impact of *Bt* plants that are wind pollinated or are cultivated in close proximity to the primary food sources of nontarget insects. Moreover, the scope of future biotech risk assessment research should be broadened to assess the impacts of *Bt* crops on a variety of insect species in different parts of the world, as many risk assessment studies to date have focused primarily on effects on nontarget insect larvae in the United States.

Continuing to monitor and test for effects of the cultivation of *Bt* crops on nontarget aquatic insects is also important, particularly as the deposition of *Bt* plant biomass into streams and rivers is common throughout *Bt* crop-growing regions of the United States. Because Cry proteins from *Bt* maize have been detected in streams at least up to 82 km downstream, nontarget insects such as caddisfly larvae and other aquatic invertebrates that serve as an important food source for fish should be monitored for nontarget effects of different types of *Bt* proteins and over multiple distances away from the source. In some cases, there may even be a continual source of *Bt* residue deposition into aquatic ecosystems, as more and different types of *Bt* crops are cultivated near waterways and *Bt* plant debris enters aquatic ecosystems at different times throughout the growing season. If further research demonstrates negative effects of *Bt* proteins, the cultivation of *Bt* crops, or *Bt* biomass on nontarget insect larvae in aquatic or terrestrial ecosystems, it may be possible to use these data to develop new conditions that can be imposed on registration of *Bt* products and to establish new scientific evaluation protocols that would minimize the environmental impacts of different types of *Bt* crops in both types of systems.

8.5 *Gene escape/introgression*

Gene flow and introgression of *Bt* crops has been a concern in both the agriculture and scientific communities since transgenic crops were commercially released, so much so that the planting of GM PIP crops in close proximity to SCWR in the United States is regulated by the EPA (for more details, see the section on how *Bt* and other GM plants are regulated in the United States in this chapter). Although the regulatory agencies in the United States have a fairly rigorous oversight role in protecting the interests of farmers, consumers, seed companies, and the environment, other countries, particularly those with small economies, may not be able to enforce the types of regulations that the United States has designed to minimize environmental impact, especially with regard to containment or separation of GM products (produce, grain, seed), setting aside refuge areas to minimize the development of pest resistance, creating buffers to limit gene flow, or monitoring for nontarget effects in the environment. This potential lack of enforcement or regulations to monitor *Bt* (and other transgenic) crops is of concern as almost half of the global hectarage of GM crops is grown in developing nations where this level of oversight may not be feasible. It is therefore imperative that GM crops be carefully evaluated for nontarget effects and the potential for gene flow under a variety of environmental and experimental scenarios, even after they have been approved for commercial use in the United States.

Examples of gene flow from transgenic crops to SCWR are becoming more prevalent, and as more and different types of transgenic crops are introduced each year, gene flow between them will become inevitable (Snow, 2002). In one of the first studies to examine "transgene escape," Quist and Chapela (2001) reported that genes from *Bt* maize had introgressed into ancient landraces of traditional maize in Oaxaca, Mexico. This was despite a 6-year moratorium, implemented in 1998, on the cultivation of GM maize in Mexico. A follow-up study, conducted in 2005, found no traces of the genetically engineered traits (specifically the cauliflower mosaic virus promoter) in any of the Mexican maize samples tested, but rather than attempting to disprove the findings of Quist and Chapela, Ortiz-Garcia et al. (2005) concluded that the moratorium, as well as increased education among the farmers, likely led to the lack of genetic contamination detected in their study. Pineyro-Nelson et al. (2009) employed new molecular techniques to confirm the presence of transgenes in 3 of 23 localities sampled in Oaxaca in 2001 and demonstrated that the persistence or reintroduction of transgenes from *Bt* maize continued until at least 2004. It is not clear how changes in plant physiology due to flow of transgenes might influence rhizosphere ecology, if at all, but as new and different types of GM crops are developed, it is important to consider the potential impacts of gene flow on nontarget soil organisms in different agricultural and nonagricultural environments.

Escape of transgenes has also been reported in other GM crops. In Oregon, for example, EPA researchers discovered gene escape (CP4 EPSPS protein and the corresponding transgene) from glyphosate-resistant creeping bentgrass (*Agrostis stolonifera*) grown in USDA-approved Monsanto-Scotts test plots (Reichman et al., 2006). The researchers found the transgenes incorporated into resident populations of compatible *Agrostis* species up to 3.8 km away from the USDA-APHIS permitted field test site, supporting a previous study in which gene flow between glyphosate-resistant creeping bentgrass and sentinel and resident plants occurred up to 21 and 14 km away, respectively, from the perimeter of the GM bentgrass test plot (Watrud et al., 2004). These studies provide evidence that novel traits can spread to wild-type plants and related species over much greater distances than previously realized.

Gene flow and introgression of some GM traits may also affect plant population or plant-soil dynamics and may indirectly aid in the spread of invasive species via the alteration of the soil community. When the effect of glyphosate drift (10% application rate of Roundup Ready®) was evaluated on mixed-species mesocosm communities consisting of GM glyphosate resistant *Brassica*, two sexually compatible *Brassica* relatives, and a variety of annual weeds (*Digitaria sanguinalis, Panicum capillare,* and *Lapsana communis*), Watrud et al. (2011) found that crabgrass was the dominant weed in the control treatments, and that Brassica dominated in the glyphosate treatments, increasing the incidence of the Roundup resistance gene in the plant community. When *Trifolium incarnatum* (crimson clover) was planted in soil mesocosms that had received glyphosate drift treatments, shoot biomass and AMF colonization were reduced compared with plants grown in mesocosms that received no herbicide treatment (Watrud et al., 2011). These results suggest that glyphosate drift associated with HT GM crops could contribute to the persistence and spread of certain invasive species, such as *Brassica* sp., which are nonmycorrhizal, and thus could facilitate the process of invasion by altering the mycorrhizal community in the soil over time. Introgression of a *Bt* gene that had a negative impact on AMF (or any other group of soil organisms) might be expected to have similar effects on the soil ecosystem, but this remains to be seen.

8.6 Can Bt crops have a role in sustainable agroecosystems?

It has been proposed that *Bt* crops could contribute to agricultural sustainability by reducing the amounts of chemical insecticides that are usually applied in conventional agricultural systems, improving yield in areas where insect control measures are limited and minimizing the negative effects to soil structure associated with tillage, as *Bt* genes are often also stacked with HT traits. By contributing to reduced insecticide usage, *Bt* crops could confer benefits to farmworkers and the environment, and because many *Bt* cultivars also include herbicide tolerance traits, conservation tillage measures could be employed to reduce the effects of tillage on soil organisms and help minimize loss of soil from erosion. The potential environmental benefits of *Bt* crops may perhaps be best illustrated by *Bt* cotton. Historically, 25% of all insecticides used in agriculture were applied to cotton—more than to any other crop (James, 2010). By planting *Bt* cotton that contains its own insecticide targeted against the cotton bollworm, the environment has been spared from pollution with thousands of pounds of broad-spectrum insecticides each year. The EPA reported that *Bt* cotton reduced insecticide use by nearly 1 million gallons in 1999 alone and saved farmers nearly $500/acre in chemical costs. The high level of confidence that many farmers have in this type of crop biotechnology is reflected in the large amount of land dedicated to GM crop production in the United States: In 2010, 93% of the cotton crop and 86% of the corn crop in the United States was genetically engineered (USDA, 2010). *Bt* corn has also been shown to offer communal benefits; non-*Bt* plants grown in close proximity to *Bt* fields also benefit from reduced pest damage through a "halo effect" on the target pest population (Alstad and Andow, 1996; Hutchison et al., 2010). Moreover, because the *Bt* protein expressed in crop plants has high specificity to certain insect groups (i.e., Lepidoptera, Coleoptera), the Cry proteins are not likely to have direct toxic effects on nontarget organisms (with a few exceptions). There is also the potential of increased yield of *Bt* crops (when compared with crops without insect protection), which may help to reduce land area required for agricultural production. It remains to be seen if this, indeed, will be the case, however, as high-yielding crop varieties can also be developed through conventional breeding methods, and the use of cultivars adapted to particular agricultural regions can also improve yields.

The relatively rapid and widespread commercialization of transgenic crop technology, however, has contributed to a certain level of mistrust and suspicion by the general public, particularly in countries outside the United States. The public perception is that GM crops are being rapidly adopted by farmers throughout the world without a complete understanding of the long-term environmental impacts. Even in the United States, 100 times more resources are invested in developing transgenic crops than are spent on risk assessment and monitoring for nontarget effects after their commercial release (Thies and Devare, 2007). While *Bt* crops may help to improve the sustainability of conventional agricultural systems by reducing insecticide usage, they may not be of benefit in agricultural systems in which long-term sustainability is the goal (i.e., organic farming systems or low-input farming systems) as they offer few, if any, benefits to enhancing soil fertility and may even have negative effects on fungal symbionts that are essential for nutrient uptake in low-input systems. Moreover, caution should be exercised when cultivating certain types of GM crops outside the United States, where the precautions necessary to prevent gene flow and insect resistance may not be practical. Gene flow may be of particular concern in tropical areas where the wild relatives of many agricultural crops are endemic and the hectarage dedicated to the cultivation of GM crops is increasing.

8.7 Conclusions and future directions

In general, the results of risk assessments of GM crops indicate, to date, that there are few to no consistent negative effects of *Bt* protein, the cultivation of *Bt* plants, or *Bt* plant residues on most of the soil bacteria, saprotrophic and pathogenic fungi, protozoa, earthworms, and microarthropods evaluated. When significant effects have been detected, they have often been minimal or transitory, making their ecological significance difficult to assess. While the research conducted thus far is by no means exhaustive, it is encouraging to note that few negative effects have been detected on soil organisms, despite the widespread and long-term cultivation of *Bt* crops. The soil organisms that appear to be most sensitive to *Bt* crop cultivation include AMF, nematodes, and nontarget insect larvae that live or feed in soil in close proximity to areas cultivated with *Bt* crops.

As nontarget effects cannot always be attributed to a particular Cry protein, risk assessment research should include investigations to identify genetic alterations that produce a change in the physiology or tissue composition of the plant. A genetic insertion that results in a change in plant root exudates, for example, could influence species composition in the rhizosphere without being an effect of the *Bt* protein itself. Internationally, more research funding should be directed toward risk assessment of GM plants in the areas of fungal ecology, nematode abundance and diversity, and insect ecology in both terrestrial and aquatic ecosystems. Moreover, each new GM plant line should be tested for effects on nontarget organisms under a variety of environmental and experimental conditions, as results of previous studies varied depending on biotic and abiotic factors that may or may not be associated with the expression of *Bt* proteins. Researchers should also continue to investigate nontarget effects on symbionts that improve plant performance (e.g., nitrogen-fixing bacteria, AMF), as these are soil organisms that may be most sensitive to *Bt* proteins or to cultivation of *Bt* crops because of their dependence on a plant host. A focus on long-term field experiments and collaborative research efforts between soil ecologists, agroecologists, microbial ecologists, and others will help to understand better the long-term effects of the cultivation of *Bt* crops on multitrophic-level interactions (e.g., Hilbeck et al., 1999; Groot and Dicke, 2002; Guo et al., 2008) and ecosystem functioning. Future investigations should incorporate the use of molecular tools for the identification and quantification of different trophic groups of soil microorganisms, as plate counts and other culturing methods are generally not sufficiently specific. As more and different types of *Bt* crops are developed, it will be important to evaluate the effects of each genetic insertion event on a variety of nontarget organisms under different environmental and experimental scenarios. Allocating sufficient resources to postrelease monitoring will also be important for mitigating any potential negative effects of the cultivation of GM crops on nontarget organisms in both terrestrial and aquatic ecosystems.

Acknowledgments

The preparation of this review and the *Bt* maize-AMF research herein by T. E. Cheeke was supported by the Charles A. and Anne Morrow Lindbergh Foundation, the U.S. Environmental Protection Agency (EPA) under the Science to Achieve Results (STAR)

Graduate Fellowship Program, a Sigma Delta Epsilon Graduate Women in Science Vessa Notchev grant, a Forbes-Lea Foundation grant, a Miller Grant for Sustainability, a Botanical Society of America Graduate Research Award, a National Science Foundation Student (DEB-1011525), and a Sigma Xi Grant-in-Aid of Research. The funding agencies have not officially endorsed this chapter, and the views expressed herein may not reflect the views of the EPA or any of the other funding agencies. The topics covered in this review have benefited from many discussions with M. B. Cruzan, L. Weasel, and T. N. Rosenstiel at Portland State University and the Bever-Schultz lab at Indiana University. This chapter has also benefited from the valuable feedback provided by Guenther Stotzky, Chris Wozniak, Tom Platt, M. B. Cruzan and members of the Cruzan lab, and an anonymous reviewer. Permission to use or expand tables was granted by the International Service for the Acquisition of Agri-biotech Application (ISAAA) and Icoz and Stotzky. I would like to thank W. J. Landesman and the Soil Ecology, Microbial Ecology, and Agroecology section of the Ecological Society of America (ESA) for helping to organize the 2009 ESA symposium from which this volume was derived.

References

Abel, C. A., and J. J. Adamczyk. 2004. Relative concentration of Cry1A in maize leaves and cotton bolls with diverse chlorophyll content and corresponding larval development of fall armyworm (Lepidoptera:Noctuidae) and southwestern corn borer (Lepidoptera:Crambidae) on maize whorl leaf profiles. *Journal of Economic Entomology* 97:1737–1744.

Adamczyk, J. J., and D. D. Hardee. 2002. Insect-resistant transgenic crops. *Crop Biotechnology* 829:23–37.

Ahl Goy, P., G. Warren, J. White, L. Pivalle, P. L. Fearing, and D. Vlachos. 1995. Interaction of insect tolerant maize with organisms in the ecosystem. *Mitteilungen des Biologischen Bundesamts fur Forst- und Landwirschaft* 309:50–53.

Ahmad, A., G. E. Wilde, and K. Y. Zhu. 2006. Evaluation of effects of Coleopteran-specific Cry3Bb1 protein on earthworms exposed to soil containing corn roots or biomass. *Environmental Entomology* 35:976–985.

Al-Deeb, M. A., G. E. Wilde, J. M. Blair, and T. C. Todd. 2003. Effect of *Bt* corn for corn rootworm control on nontarget soil microarthropods and nematodes. *Environmental Entomology* 32:859–865.

Alstad, D. N., and D. A. Andow. 1996. Implementing management of insect resistance to transgenic crops. *AgBiotech News and Information* 8:177–181.

Anderson, P. L., R. L. Hellmich, J. R. Prasifka, and L. C. Lewis. 2005. Effects on fitness and behavior of monarch butterfly larvae exposed to a combination of Cry1Ab-expressing corn anthers and pollen. *Environmental Entomology* 34:944–952.

Anderson, P. L., R. L. Hellmich, M. K. Sears, D. V. Sumerford, and L. C. Lewis. 2004. Effects of Cry1Ab-expressing corn anthers on monarch butterfly larvae. *Environmental Entomology* 33:1109–1115.

Bais, H. P., T. L. Weir, L. G. Perry, S. Gilroy, and J. M. Vivanco. 2006. The role of root exudates in rhizosphere interations with plants and other organisms. *Annual Review of Plant Biology* 57:233–266.

Bakonyi, G., A. Dolezsai, N. Matrai, and A. Szekacs. 2011. Effects of consumption of *Bt*-maize (MON 810) on the Collembolan *Folsomia candida*, over multiple generations: a laboratory study. *Insects* 2:243–252.

Bakonyi, G., F. Szira, I. Kiss, I. Villanyi, A. Seres, and A. Szekacs. 2006. Preference tests with collembolas on isogenic and *Bt*-maize. *European Journal of Soil Biology* 42:S132–S135.

Baumgarte, S., and C. C. Tebbe. 2005. Field studies on the environmental fate of the Cry1Ab *Bt*-toxin produced by transgenic maize (MON810) and its effect on bacterial communities in the maize rhizosphere. *Molecular Ecology* 14:2539–2551.

Beegle, C. C., and T. Yamamoto. 1992. Invitation paper (CP-Alexander-Fund)—history of *Bacillus-thuringiensis* Berliner Research-and-Development. *Canadian Entomologist* 124:587–616.

Bhatti, M. A., J. Duan, G. Head, C. J. Jiang, M. J. McKee, T. E. Nickson, C. L. Pilcher, and C. D. Pilcher. 2005. Field evaluation of the impact of corn rootworm (Coleoptera:Chrysomelidae)-protected *Bt* corn on ground-dwelling invertebrates. *Environmental Entomology* 34:1325–1335.

Birch, A. N. E., B. S. Griffiths, S. Caul, J. Thompson, L. H. Heckmann, P. H. Krogh, and J. Cortet. 2007. The role of laboratory, glasshouse and field scale experiments in understanding the interactions between genetically modified crops and soil ecosystems: a review of the ECOGEN project. *Pedobiologia* 51:251–260.

Blackwood, C. B., and J. S. Buyer. 2004. Soil microbial communities associated with *Bt* and non-*Bt* corn in three soils. *Journal of Environmental Quality* 33:832–836.

Blair, J. M., P. J. Bohlen, and D. W. Freckman. 1996. Soil invertebrates as indicators of soil quality. In *Methods for Assessing Soil Quality*, eds. J. W. Doran and A. J. Jones, 283–301. Soil Science Society of America, Madison, WI.

Bravo, A., S. S. Gill, and M. Soberon. 2007. Mode of action of *Bacillus thuringiensis* Cry and Cyt toxins and their potential for insect control. *Toxicon* 49:423–435.

Broderick, N. A., K. F. Raffa, and J. Handelsman. 2006. Midgut bacteria required for *Bacillus thuringiensis* insecticidal activity. *Proceedings of the National Academy of Sciences of the United States of America* 103:15196–15199.

Broderick, N. A., C. J. Robinson, M. D. McMahon, J. Holt, J. Handelsman, and K. F. Raffa. 2009. Contributions of gut bacteria to *Bacillus thuringiensis*-induced mortality vary across a range of Lepidoptera. *BMC Biology* 7:11.

Brusetti, L., P. Francia, C. Bertolini, A. Pagliuca, S. Borin, C. Sorlini, A. Abruzzese, G. Sacchi, C. Viti, L. Giovannetti, E. Giuntini, M. Bazzicalupo, and D. Daffonchio. 2004. Bacterial communities associated with the rhizosphere of transgenic *Bt* 176 maize (*Zea mays*) and its non transgenic counterpart. *Plant and Soil* 266:11–21.

Candolfi, M. P., K. Brown, C. Grimm, B. Reber, and H. Schmidli. 2004. A faunistic approach to assess potential side-effects of genetically modified *Bt*-corn on non-target arthropods under field conditions. *Biocontrol Science and Technology* 14:129–170.

Castaldini, M., A. Turrini, C. Sbrana, A. Benedetti, M. Marchionni, S. Mocali, A. Fabiani, S. Landi, F. Santomassimo, B. Pietrangeli, M. P. Nuti, N. Miclaus, and M. Giovannetti. 2005. Impact of *Bt* corn on rhizospheric and soil eubacterial communities and on beneficial mycorrhizal symbiosis in experimental microcosms. *Applied and Environmental Microbiology* 71:6719–6729.

Chambers, C. P., M. R. Whiles, E. J. Rosi-Marshall, J. L. Tank, T. V. Royer, N. A. Griffiths, M. A. Evans-White, and A. R. Stojak. 2010. Responses of stream macroinvertebrates to *Bt* maize leaf detritus. *Ecological Applications* 20:1949–1960.

Cheeke, T. E., B. A. Pace, T. N. Rosenstiel, and M. B. Cruzan. 2011. The influence of fertilizer level and spore density on arbuscular mycorrhizal colonization of transgenic *Bt* 11 maize (*Zea mays*) in experimental microcosms. *FEMS Microbiology Ecology* 75:304–312.

Cheeke, T. E., T. N. Rosenstiel, and M. B. Cruzan. 2012. Evidence of reduced arbuscular mycorrhizal fungal colonization in multiple lines of *Bt* maize. *American Journal of Botany* 99(4):700–707.

Chen, Z. H. C. Z. H., L. J. Chen, Y. L. Zhang, and Z. J. Wu. 2011. Microbial properties, enzyme activities and the persistence of exogenous proteins in soil under consecutive cultivation of transgenic cottons (*Gossypium hirsutum* L.). *Plant Soil and Environment* 57:67–74.

Clark, B. W., and J. R. Coats. 2006. Subacute effects of CrylAb *Bt* corn litter on the earthworm *Eisenia fetida* and the springtail *Folsomia candida*. *Environmental Entomology* 35:1121–1129.

Clark, B. W., T. A. Phillips, and J. R. Coats. 2005. Environmental fate and effects of *Bacillus thuringiensis* (*Bt*) proteins from transgenic crops: a review. *Journal of Agricultural and Food Chemistry* 53:4643–4653.

Clark, B. W., K. R. Prihoda, and J. R. Coats. 2006. Subacute effects of transgenic Cry1Ab *Bacillus thuringiensis* corn litter on the isopods *Trachelipus rathkii* and *Armadillidium nasatum*. *Environmental Toxicology and Chemistry* 25:2653–2661.

Cortet, J., B. S. Griffiths, M. Bohanec, D. Demsar, M. N. Andersen, S. Caul, A. N. E. Birch, C. Pernin, E. Tabone, A. de Vaufleury, X. Ke, and P. H. Krogh. 2007. Evaluation of effects of transgenic *Bt* maize on microarthropods in a European multi-site experiment. *Pedobiologia* 51:207–218.

Crecchio, C., and G. Stotzky. 1998. Insecticidal activity and biodegradation of the toxin from *Bacillus thuringiensis* subsp. *kurstaki* bound to humic acids from soil. *Soil Biology and Biochemistry* 30:463–470.

Debeljak, M., J. Cortet, D. Demsar, P. H. Krogh, and S. Dzeroski. 2007. Hierarchical classification of environmental factors and agricultural practices affecting soil fauna under cropping systems using *Bt* maize. *Pedobiologia* 51:229–238.

Devare, M. H., C. M. Jones, and J. E. Thies. 2004. Effect of Cry3Bb transgenic corn and tefluthrin on the soil microbial community: biomass, activity, and diversity. *Journal of Environmental Quality* 33:837–843.

Devare, M., L. M. Londoño-R, and J. E. Thies. 2007. Neither transgenic *Bt* maize (MON863) nor tefluthrin insecticide adversely affect soil microbial activity or biomass: a 3-year field analysis. *Soil Biology and Biochemistry* 39:2038–2047.

de Vaufleury, A., P. E. Kramarz, P. Binet, J. Cortet, S. Caul, M. N. Andersen, E. Plumey, M. Coeurdassier, and P. H. Krogh. 2007. Exposure and effects assessments of *Bt*-maize on non-target organisms (gastropods, microarthropods, mycorrhizal fungi) in microcosms. *Pedobiologia* 51:185–194.

Dively, G. P., R. Rose, M. K. Sears, R. L. Hellmich, D. E. Stanley-Horn, D. D. Calvin, J. M. Russo, and P. L. Anderson. 2004. Effects on monarch butterfly larvae (Lepidoptera:Danaidae) after continuous exposure to Cry1Ab-expressing corn during anthesis. *Environmental Entomology* 33:1116–1125.

Dobbelaere, S., J. Vanderleyden, and Y. Okon. 2003. Plant growth-promoting effects of diazotrophs in the rhizosphere. *Critical Reviews in Plant Sciences* 22:107–149.

Donegan, K. K., C. J. Palm, V. J. Fieland, L. A. Porteous, L. M. Ganio, D. L. Schaller, L. Q. Bucao, and R. J. Seidler. 1995. Changes in levels, species and DNA fingerprints of soil-microorganisms associated with cotton expressing the *Bacillus thuringiensis* var *kurstaki* endotoxin. *Applied Soil Ecology* 2:111–124.

Donegan, K. K., D. L. Schaller, J. K. Stone, L. M. Ganio, G. Reed, P. B. Hamm, and R. J. Seidler. 1996. Microbial populations, fungal species diversity and plant pathogen levels in field plots of potato plants expressing the *Bacillus thuringiensis* var *tenebrionis* endotoxin. *Transgenic Research* 5:25–35.

Douville, M., F. Gagne, C. Blaise, and C. Andre. 2007. Occurrence and persistence of *Bacillus thuringiensis* (*Bt*) and transgenic *Bt* corn Cry1Ab gene from an aquatic environment. *Ecotoxicology and Environmental Safety* 66:195–203.

Duan, J. J., M. Marvier, J. Huesing, G. Dively, and Z. Y. Huang. 2008. A meta-analysis of effects of *Bt* crops on honey bees (Hymenoptera:Apidae). *PloS One* 3:e1415.

Dunfield, K. E., and J. J. Germida. 2003. Seasonal changes in the rhizosphere microbial communities associated with field-grown genetically modified canola (*Brassica napus*). *Applied and Environmental Microbiology* 69:7310–7318.

Dunfield, K. E., and J. J. Germida. 2004. Impact of genetically modified crops on soil- and plant-associated microbial communities. *Journal of Environmental Quality* 33:806–815.

Escher, N., B. Käch, and W. Nentwig. 2000. Decomposition of transgenic *Bacillus thuringiensis* maize by microorganisms and woodlice *Porcellio scaber* (Crustacea:Isopoda). *Basic and Applied Ecology* 1:161–169.

Fang, M., R. J. Kremer, P. P. Motavalli, and G. Davis. 2005. Bacterial diversity in rhizospheres of non-transgenic and transgenic corn. *Applied and Environmental Microbiology* 71:4132–4136.

Fang, M., P. P. Motavalli, R. J. Kremer, and K. A. Nelson. 2007. Assessing changes in soil microbial communities and carbon mineralization in *Bt* and non-*Bt* corn residue-amended soils. *Applied Soil Ecology* 37:150–160.

Federici, B. A. 2002. Case study: *Bt* crops, a novel mode of insect control. In *Genetically Modified Crops: Assessing Safety*, ed. K. T. Atherton, 164–200. Taylor and Francis, New York.

Ferreira, L., J. C. Molina, C. Brasil, and G. Andrade. 2003. Evaluation of *Bacillus thuringiensis* bioinsecticidal protein effects on soil microorganisms. *Plant and Soil* 256:161–168.

Flores, S., D. Saxena, and G. Stotzky. 2005. Transgenic *Bt* plants decompose less in soil than non-*Bt* plants. *Soil Biology and Biochemistry* 37:1073–1082.

Gao, Y. L., J. J. An, C. X. Liu, and K. M. Wu. 2010. Monitoring of *Bt* resistance from field-collected *Helicoverpa armigera* populations in northern China. *Southwestern Entomologist* 35:399–402.

Gatehouse, A. M. R., N. Ferry, M. G. Edwards, and H. A. Bell. 2011. Insect-resistant biotech crops and their impacts on beneficial arthropods. *Philosophical Transactions of the Royal Society B–Biological Sciences* 366:1438–1452.

Gatehouse, A. M. R., N. Ferry, and R. J. M. Raemaekers. 2002. The case of the monarch butterfly: a verdict is returned. *Trends in Genetics* 18:249–251.

Gathmann, A., L. Wirooks, L. A. Hothorn, D. Bartsch, and I. Schuphan. 2006. Impact of *Bt* maize pollen (MON810) on Lepidopteran larvae living on accompanying weeds. *Molecular Ecology* 15:2677–2685.

Giovannetti, M. 2003. The ecological risks of transgenic plants. *Rivista Di Biologia-Biology Forum* 96:207–223.

Giovannetti, M., C. Sbrana, and A. Turrini. 2005. The impact of genetically modified crops on soil microbial communities. *Rivista Di Biologia–Biology Forum* 98:393–417.

Gosling, P., A. Hodge, G. Goodlass, and G. D. Bending. 2006. Arbuscular mycorrhizal fungi and organic farming. *Agriculture Ecosystems and Environment* 113:17–35.

Gould, F., N. Blair, M. Reid, T. L. Rennie, J. Lopez, and S. Micinski. 2002. *Bacillus thuringiensis*-toxin resistance management: stable isotope assessment of alternate host use by *Helicoverpa zea*. *Proceedings of the National Academy of Sciences of the United States of America* 99:16581–16586.

Grayston, S. J., D. Vaughan, and D. Jones. 1996. Rhizosphere carbon flow in trees, in comparison with annual plants; the importance of root exudation and its impact on microbial activity and nutrient availability. *Applied Soil Ecology* 5:29–56.

Griego, V. M., and K. D. Spence. 1978. Inactivation of *Bacillus thuringiensis* spores by ultraviolet and visible-light. *Applied and Environmental Microbiology* 35:906–910.

Griffiths, B. S., S. Caul, J. Thompson, A. N. E. Birch, J. Cortet, M. N. Andersen, and P. H. Krogh. 2007a. Microbial and microfaunal community structure in cropping systems with genetically modified plants. *Pedobiologia* 51:195–206.

Griffiths, B. S., S. Caul, J. Thompson, A. N. E. Birch, C. Scrimgeour, M. N. Andersen, J. Cortet, A. Messean, C. Sausse, B. Lacroix, and P. H. Krogh. 2005. A comparison of soil microbial community structure, protozoa and nematodes in field plots of conventional and genetically modified maize expressing the *Bacillus thuringiensis* Cry1Ab toxin. *Plant and Soil* 275:135–146.

Griffiths, B. S., S. Caul, J. Thompson, A. N. E. Birch, C. Scrimgeour, J. Cortet, A. Foggo, C. A. Hackett, and P. H. Krogh. 2006. Soil microbial and faunal community responses to *Bt* maize and insecticide in two soils. *Journal of Environmental Quality* 35:734–741.

Griffiths, B. S., S. Caul, J. Thompson, C. A. Hackett, J. Cortet, C. Pernin, and P. H. Krogh. 2008. Soil microbial and faunal responses to herbicide tolerant maize and herbicide in two soils. *Plant and Soil* 308:93–103.

Griffiths, B. S., I. E. Geoghegan, and W. M. Robertson. 2000. Testing genetically engineered potato, producing the lectins GNA and Con A, on non-target soil organisms and processes. *Journal of Applied Ecology* 37:159–170.

Griffiths, B. S., L. H. Heckmann, S. Caul, J. Thompson, C. Scrimgeour, and P. H. Krogh. 2007b. Varietal effects of eight paired lines of transgenic *Bt* maize and near-isogenic non-*Bt* maize on soil microbial and nematode community structure. *Plant Biotechnology Journal* 5:60–68.

Groot, A. T., and M. Dicke. 2002. Insect-resistant transgenic plants in a multi-trophic context. *Plant Journal* 31:387–406.

Guo, J. Y., F. H. Wan, L. Dong, G. L. Lovei, and Z. J. Han. 2008. Tri-trophic interactions between *Bt* cotton, the herbivore *Aphis gossypii glover* (Homoptera:Aphididae), and the predator *Chrysopa pallens* (Rambur) (Neuroptera:Chrysopidae). *Environmental Entomology* 37:263–270.

Handelsman, J., and J. Tiedje. 2007. *The New Science of Metagenomics*. National Academies Press, Washington, DC.

Hayat, R., S. Ali, U. Amara, R. Khalid, and I. Ahmed. 2010. Soil beneficial bacteria and their role in plant growth promotion: a review. *Annals of Microbiology* 60:579–598.

Heckmann, L. H., B. S. Griffiths, S. Caul, J. Thompson, M. Pusztai-Carey, W. J. Moar, M. N. Andersen, and P. H. Krogh. 2006. Consequences for *Protaphorura armata* (Collembola:Onychiuridae) following exposure to genetically modified *Bacillus thuringiensis* (*Bt*) maize and non-*Bt* maize. *Environmental Pollution* 142:212–216.

Hellmich, R. L., B. D. Siegfried, M. K. Sears, D. E. Stanley-Horn, M. J. Daniels, H. R. Mattila, T. Spencer, K. G. Bidne, and L. C. Lewis. 2001. Monarch larvae sensitivity to *Bacillus thuringiensis*-purified proteins and pollen. *Proceedings of the National Academy of Sciences of the United States of America* 98:11925–11930.

Hilbeck, A., W. J. Moar, M. Pusztai-Carey, A. Filippini, and F. Bigler. 1999. Prey-mediated effects of Cry1Ab toxin and protoxin and Cry2A protoxin on the predator *Chrysoperla carnea*. *Entomologia Experimentalis et Applicata* 91:305–316.

Hoess, S., M. Arndt, S. Baurngarte, C. C. Tebbe, H. T. Nguyen, and J. A. Jehle. 2008. Effects of transgenic corn and CrylAb protein on the nematode, *Caenorhabditis elegans*. *Ecotoxicology and Environmental Safety* 70:334–340.

Hoess, S., H. T. Nguyen, R. Menzel, S. Pagel-Wieder, R. Miethling-Graf, C. C. Tebbe, J. A. Jehle, and W. Traunspurger. 2011. Assessing the risk posed to free-living soil nematodes by a genetically modified maize expressing the insecticidal Cry3Bb1 protein. *Science of the Total Environment* 409:2674–2684.

Hofte, H., and H. R. Whiteley. 1989. Insecticidal crystal proteins of *Bacillus thuringiensis*. *Microbiological Reviews* 53:242–255.

Honemann, L., and W. Nentwig. 2009. Are survival and reproduction of *Enchytraeus albidus* (Annelida: Enchytraeidae) at risk by feeding on *Bt*-maize litter? *European Journal of Soil Biology* 45:351–355.

Honemann, L., C. Zurbrugg, and W. Nentwig. 2008. Effects of *Bt*-corn decomposition on the composition of the soil meso- and macrofauna. *Applied Soil Ecology* 40:203–209.

Hopkins, D. W., and E. G. Gregorich. 2003. Detection and decay of the *Bt* endotoxin in soil from a field trial with genetically modified maize. *European Journal of Soil Science* 54:793–800.

Hu, H. Y., X. X. Liu, Z. W. Zhao, J. G. Sun, Q. W. Zhang, X. Z. Liu, and Y. Yu. 2009. Effects of repeated cultivation of transgenic *Bt* cotton on functional bacterial populations in rhizosphere soil. *World Journal of Microbiology and Biotechnology* 25:357–366.

Hu, Y., S. B. Georghiou, A. J. Kelleher, and R. V. Aroian. 2010. *Bacillus thuringiensis* Cry5B protein is highly efficacious as a single-dose therapy against an intestinal roundworm infection in mice. *PloS Neglected Tropical Diseases* 4(3):e614. DOI:10.1371/journal.pntd.0000614.

Huang, F. N., B. R. Leonard, D. R. Cook, D. R. Lee, D. A. Andow, J. L. Baldwin, K. V. Tindall, and X. Y. Wu. 2007. Frequency of alleles conferring resistance to *Bacillus thuringiensis* maize in Louisiana populations of the southwestern corn borer. *Entomologia Experimentalis et Applicata* 122:53–58.

Hutchison, W. D., E. C. Burkness, P. D. Mitchell, R. D. Moon, T. W. Leslie, S. J. Fleischer, M. Abrahamson, K. L. Hamilton, K. L. Steffey, M. E. Gray, R. L. Hellmich, L. V. Kaster, T. E. Hunt, R. J. Wright, K. Pecinovsky, T. L. Rabaey, B. R. Flood, and E. S. Raun. 2010. Areawide suppression of European corn borer with *Bt* maize reaps savings to non-*Bt* maize growers. *Science* 330:222–225.

Icoz, I., D. Saxena, D. A. Andow, C. Zwahlen, and G. Stotzky. 2008. Microbial populations and enzyme activities in soil in situ under transgenic corn expressing Cry proteins from *Bacillus thuringiensis*. *Journal of Environmental Quality* 37:647–662.

Icoz, I., and G. Stotzky. 2008a. Cry3Bb1 protein from *Bacillus thuringiensis* in root exudates and biomass of transgenic corn does not persist in soil. *Transgenic Research* 17:609–620.

Icoz, I., and G. Stotzky. 2008b. Fate and effects of insect-resistant *Bt* crops in soil ecosystems. *Soil Biology and Biochemistry* 40:559–586.

Ito, A., Y. Sasaguri, S. Kitada, Y. Kusaka, K. Kuwano, K. Masutomi, E. Mizuki, T. Akao, and M. Ohba. 2004. A *Bacillus thuringiensis* crystal protein with selective cytocidal action to human cells. *Journal of Biological Chemistry* 279:21282–21286.

James, C. 2010. *Global Status of Commercialized Biotech/GM Crops: 2010*. International Service for the Acquisition of Agri-Biotech Applications, Ithaca, NY. http://isaaa.org/resources/publications/briefs/42/executivesummary/default.asp.

Jensen, P. D., G. P. Dively, C. M. Swan, and W. O. Lamp. 2010. Exposure and nontarget effects of transgenic *Bt* corn debris in streams. *Environmental Entomology* 39:707–714.

Jesse, L. C. H., and J. J. Obrycki. 2000. Field deposition of *Bt* transgenic corn pollen: lethal effects on the monarch butterfly. *Oecologia* 125:241–248.

Jung, H. G., and C. C. Sheaffer. 2004. Lignin concentration of whole plants and stems of *Bt* corn hybrids. *Journal of Animal Science* 82:250–250.

Khan, M. Q., M. W. Abbasi, M. J. Zaki, and S. A. Khan. 2010. Evaluation of *Bacillus thuringiensis* isolates against root-knot nematodes following seed application in okra and mungbean. *Pakistan Journal of Botany* 42:2903–2910.

Knox, O. G. G., D. B. Nehl, T. Mor, G. N. Roberts, and V. Gupta. 2008. Genetically modified cotton has no effect on arbuscular mycorrhizal colonisation of roots. *Field Crops Research* 109:57–60.

Koskella, J., and G. Stotzky. 2002. Larvicidal toxins from *Bacillus thuringiensis* subspp. *kurstaki, morrisoni* (strain *tenebrionis*), and *israelensis* have no microbicidal or microbiostatic activity against selected bacteria, fungi, and algae in vitro. *Canadian Journal of Microbiology* 48:262–267.

Kotze, A. C., J. O'Grady, J. M. Gough, R. Pearson, N. H. Bagnall, D. H. Kemp, and R. J. Akhurst. 2005. Toxicity of *Bacillus thuringiensis* to parasitic and free-living life-stages of nematode parasites of livestock. *International Journal for Parasitology* 35:1013–1022.

Kowalchuk, G. A., M. Bruinsma, and J. A. van Veen. 2003. Assessing responses of soil microorganisms to GM plants. *Trends in Ecology and Evolution* 18:403–410.

Kowalchuk, G. A., D. S. Buma, W. de Boer, P. G. L. Klinkhamer, and J. A. van Veen. 2002. Effects of above-ground plant species composition and diversity on the diversity of soil-borne microorganisms. *Antonie van Leeuwenhoek* 81:509–520.

Kravchenko, A. N., X. M. Hao, and G. P. Robertson. 2009. Seven years of continuously planted *Bt* corn did not affect mineralizable and total soil C and total N in surface soil. *Plant and Soil* 318:269–274.

Krogh, P. H., B. Griffiths, D. Demsar, M. Bohanec, M. Debeljak, M. N. Andersen, C. Sausse, A. N. E. Birch, S. Caul, M. Holmstrup, L. H. Heckmann, and J. Cortet. 2007. Responses by earthworms to reduced tillage in herbicide tolerant maize and *Bt* maize cropping systems. *Pedobiologia* 51:219–227.

Lamarche, J., and R. C. Hamelin. 2007. No evidence of an impact on the rhizosphere diazotroph community by the expression of *Bacillus thuringiensis* Cry1Ab toxin by *Bt* white spruce. *Applied and Environmental Microbiology* 73:6577–6583.

Lang, A., K. Abdel-Kader, M. Arndt, J. Bauchhenss, R. Beck, U. Benker, A. Hermann, D. Mautz, M. Zellner, and G. Pommer. 2006. Monitoring the environmental impact of *Bt* maize: a research project of the Bavarian State Ministry for Health, Environment and Consumer Protection, and the Bavarian State Research Center for Agriculture. *Mitteilungen aus der Biologischen Bundesanstalt fuer Land- und Forstwirtschaft Berlin-Dahlem* 403:136–139.

Lang, A., C. Ludy, and E. Vojtech. 2004. Dispersion and deposition of *Bt* maize pollen in field margins. *Zeitschrift Fur Pflanzenkrankheiten Und Pflanzenschutz–Journal of Plant Diseases and Protection* 111:417–428.

Lang, A., and M. Otto. 2010. A synthesis of laboratory and field studies on the effects of transgenic *Bacillus thuringiensis* (*Bt*) maize on non-target Lepidoptera. *Entomologia Experimentalis et Applicata* 135:121–134.

Lang, A., and E. Vojtech. 2006. The effects of pollen consumption of transgenic *Bt* maize on the common swallowtail, *Papilio machaon* L. (Lepidoptera, Papilionidae). *Basic and Applied Ecology* 7:296–306.

Lee, L., D. Saxena, and G. Stotzky. 2003. Activity of free and clay-bound insecticidal proteins from *Bacillus thuringiensis* subsp *israelensis* against the mosquito *Culex pipiens*. *Applied and Environmental Microbiology* 69:4111–4115.

Lehman, R. M., S. L. Osborne, and K. A. Rosentrater. 2008. No differences in decomposition rates observed between *Bacillus thuringiensis* and non-*Bacillus thuringiensis* corn residue incubated in the field. *Agronomy Journal* 100:163–168.

Lekberg, Y., R. T. Koide, and S. J. Twomlow. 2008. Effect of agricultural management practices on arbuscular mycorrhizal fungal abundance in low-input cropping systems of southern Africa: a case study from Zimbabwe. *Biology and Fertility of Soils* 44:917–923.

Li, X. G., B. A. Liu, J. J. Cui, D. D. Liu, S. A. Ding, B. Gilna, J. Y. Luo, Z. X. Fang, W. Cao, and Z. M. Han. 2011. No evidence of persistent effects of continuously planted transgenic insect-resistant cotton on soil microorganisms. *Plant and Soil* 339:247–257.

Li, X. G., B. A. Liu, S. Heia, D. D. Liu, Z. M. Han, K. X. Zhou, J. J. Cui, J. Y. Luo, and Y. P. Zheng. 2009. The effect of root exudates from two transgenic insect-resistant cotton lines on the growth of *Fusarium oxysporum*. *Transgenic Research* 18:757–767.

Li, X. Q., A. Tan, M. Voegtline, S. Bekele, C. S. Chen, and R. V. Aroian. 2008. Expression of Cry5B protein from *Bacillus thuringiensis* in plant roots confers resistance to root-knot nematode. *Biological Control* 47:97–102.

Li, X., A. Tan, M. Voegtline, S. Bekele, J. Wei, and R. V. Aroian. 2007. Nematicidal *Bt* crystal proteins targeting plant endoparasitic nematodes. *Phytopathology* 97:S144–S144.

Li, X. Q., J. Z. Wei, A. Tan, and R. V. Aroian. 2007. Resistance to root-knot nematode in tomato roots expressing a nematicidal *Bacillus thuringiensis* crystal protein. *Plant Biotechnology Journal* 5:455–464.

Liu, B., J. J. Cui, J. Meng, W. J. Hu, J. Y. Luo, and Y. P. Zheng. 2009a. Effects of transgenic *Bt* plus CpTI cotton on the growth and reproduction of earthworm *Eisenia foetida*. *Frontiers in Bioscience* 14:4008–4014.

Liu, B., L. Wang, Q. Zeng, J. Meng, W. J. Hu, X. G. Li, K. X. Zhou, K. Xue, D. D. Liu, and Y. P. Zheng. 2009b. Assessing effects of transgenic Cry1Ac cotton on the earthworm *Eisenia fetida*. *Soil Biology and Biochemistry* 41:1841–1846.

Liu, B., Q. Zeng, F. M. Yan, H. G. Xu, and C. R. Xu. 2005. Effects of transgenic plants on soil microorganisms. *Plant and Soil* 271:1–13.

Liu, W., and L. F. Du. 2008. Interactions between *Bt* transgenic crops and arbuscular mycorrhizal fungi: a new urgent issue of soil ecology in agroecosystems. *Acta Agriculturae Scandinavica Section B-Soil and Plant Science* 58:187–192.

Liu, W., H. H. Lu, W. Wu, Q. K. Wei, Y. X. Chen, and J. E. Thies. 2008. Transgenic *Bt* rice does not affect enzyme activities and microbial composition in the rhizosphere during crop development. *Soil Biology and Biochemistry* 40:475–486.

Liu, W. K. 2010. Do genetically modified plants impact arbuscular mycorrhizal fungi? *Ecotoxicology* 19:229–238.

Losey, J. E., L. S. Rayor, and M. E. Carter. 1999. Transgenic pollen harms monarch larvae. *Nature* 399:214–214.

Lottman, J., H. Heier, J. de Vries, A. Mahn, K. During, W. Wackernagel, K. Smalla, and G. Berg. 2000. Establishment of introduced antagonistic bacteria in the rhizosphere of transgenic potatoes and their effect on the bacterial community. *FEMS Microbiology Ecology* 33:41–49.

Lovei, G. L., D. A. Andow, and S. Arpaia. 2009. Transgenic insecticidal crops and natural enemies: a detailed review of laboratory studies. *Environmental Entomology* 38:293–306.

Lu, H. H., W. X. Wu, Y. X. Chen, H. L. Wang, M. Devare, and J. E. Thies. 2010a. Soil microbial community responses to *Bt* transgenic rice residue decomposition in a paddy field. *Journal of Soils and Sediments* 10:1598–1605.

Lu, H. H., W. X. Wu, Y. X. Chen, X. J. Zhang, M. Devare, and J. E. Thies. 2010b. Decomposition of *Bt* transgenic rice residues and response of soil microbial community in rapeseed-rice cropping system. *Plant and Soil* 336:279–290.

Lundgren, J. G., A. J. Gassmann, J. Bernal, J. J. Duan, and J. Ruberson. 2009. Ecological compatibility of GM crops and biological control. *Crop Protection* 28:1017–1030.

Lupwayi, N. Z., K. G. Hanson, K. N. Harker, G. W. Clayton, R. E. Blackshaw, J. T. O'Donovan, E. N. Johnson, Y. Gan, R. B. Irvine, and M. A. Monreal. 2007. Soil microbial biomass, functional diversity and enzyme activity in glyphosate-resistant wheat-canola rotations under low-disturbance direct seeding and conventional tillage. *Soil Biology and Biochemistry* 39:1418–1427.

Manachini, B., M. C. Fiore, S. Landi, and S. Arpaia. 2003. Nematode species assemblage in *Bt*-expressing transgenic eggplants and their isogenic control. In *Biodiversity Implications of Genetically Modified Plants*, 31. Ascona, Switzerland.

Manachini, B., S. Landi, M. C. Fiore, M. Festa, and S. Arpaia. 2004. First investigations on the effects of *Bt*-transgenic *Brassica nupus* L. on the trophic structure of the nematofauna. IOBC/WPRS Bulletin, 103–108.

Manachini, B., and G. C. Lozzia. 2002. First investigations into the effects of *Bt* corn crop on Nematofauna. *Bollettino di Zoologia Agraria e di Bachicoltura* 34:85–96.

Manachini, B., and G. C. Lozzia. 2003. Biodiversity and structure on Nematofauna in *Bt* corn. In *Biodiversity Implications of Genetically Modified Plants*, 32. Ascona, Switzerland.

Marroquin, L. D., D. Elyassnia, J. S. Griffitts, J. S. Feitelson, and R. V. Aroian. 2000. *Bacillus thuringiensis* (*Bt*) toxin susceptibility and isolation of resistance mutants in the nematode *Caenorhabditis elegans*. *Genetics* 155:1693–1699.

Marvier, M., C. McCreedy, J. Regetz, and P. Kareiva. 2007. A meta-analysis of effects of *Bt* cotton and maize on nontarget invertebrates. *Science* 316:1475–1477.

Meadows, J., S. S. Gill, and L. W. Bone. 1989a. Factors influencing lethality of *Bacillus thuringiensis-kurstaki* toxin for eggs and larvae of *Trichostronglus columbriformis* (Nematoda). *Journal of Parasitology* 75:191–194.

Meadows, J., S. S. Gill, and L. W. Bone. 1990. *Bacillus thuringiensis* strains affect population-growth of the free-living nematode *Turbatrix aceti*. *Invertebrate Reproduction and Development* 17:73–76.

Meadows, J. R., S. S. Gill, and L. W. Bone. 1989b. Lethality of *Bacillus thuringiensis morrisoni* for eggs of *Trichostronglus colubriformis* (Nematoda). *Invertebrate Reproduction and Development* 15:159–161.

Menninger, H. L., and M. A. Palmer. 2007. Herbs and grasses as an allochthonous resource in open-canopy headwater streams. *Freshwater Biology* 52:1689–1699.

Mercer, K. L., and J. D. Wainwright. 2008. Gene flow from transgenic maize to landraces in Mexico: an analysis. *Agriculture Ecosystems and Environment* 123:109–115.

Miethling-Graff, R., S. Dockhorn, and C. C. Tebbe. 2010. Release of the recombinant Cry3Bb1 protein of *Bt* maize MON88017 into field soil and detection of effects on the diversity of rhizosphere bacteria. *European Journal of Soil Biology* 46:41–48.

Monsanto. 2011. Research and development pipeline, St. Louis, MO. http://www.monsanto.com/products/Pages/research-development-pipeline.aspx#/Overview.

Moore, A. A., and M. A. Palmer. 2005. Invertebrate biodiversity in agricultural and urban headwater streams: implications for conservation and management. *Ecological Applications* 15:1169–1177.

Morgan, J. A. W., G. D. Bending, and P. J. White. 2005. Biological costs and benefits to plant-microbe interactions in the rhizosphere. *Journal of Experimental Botany* 56:1729–1739.

Motavalli, P. P., R. J. Kremer, M. Fang, and N. E. Means. 2004. Impact of genetically modified crops and their management on soil microbially mediated plant nutrient transformations. *Journal of Environmental Quality* 33:816–824.

Mungai, N. W., P. P. Motavalli, K. A. Nelson, and R. J. Kremer. 2005. Differences in yields, residue composition and N mineralization dynamics of *Bt*- and non-*Bt*-maize. *Nutrient Cycling in Agroecosystems* 73:101–109.

Naef, A., T. Zesiger, and G. Defago. 2006. Impact of transgenic *Bt* maize residues on the mycotoxigenic plant pathogen *Fusarium graminearum* and the biocontrol agent *Trichoderma atroviride*. *Journal of Environmental Quality* 35:1001–1009.

Nagamatsu, Y., S. Okamura, H. Saitou, T. Akao, and E. Mizuki. 2010. Three Cry toxins in two types from *Bacillus thuringiensis* strain M019 preferentially kill human hepatocyte cancer and uterus cervix cancer cells. *Bioscience Biotechnology and Biochemistry* 74:494–498.

O'Callaghan, M., T. R. Glare, E. P. J. Burgess, and L. A. Malone. 2005. Effects of plants genetically modified for insect resistance on nontarget organisms. *Annual Review of Entomology* 50:271–292.

Ohba, M., E. Mizuki, and A. Uemori. 2009. Parasporin, a new anticancer protein group from *Bacillus thuringiensis*. *Anticancer Research* 29:427–433.

Oliveira, A. P., M. E. Pampulha, and J. P. Bennett. 2008. A two-year field study with transgenic *Bacillus thuringiensis* maize: effects on soil microorganisms. *Science of the Total Environment* 405:351–357.

Oliveira, A. R., T. R. Castro, D. M. F. Capalbo, and I. Delalibera. 2007. Toxicological evaluation of genetically modified cotton (Bollgard®) and Dipel®WP on the non-target soil mite *Scheloribates praeincisus* (Acari:Oribatida). *Experimental and Applied Acarology* 41:191–201.

Ortiz-Garcia, S., E. Ezcurra, B. Schoel, F. Acevedo, J. Sobernon, and A. A. Snow. 2005. Absence of detectable transgenes in local landraces of maize in Oaxaca, Mexico (2003–2004). *Proceedings of the National Academy of Sciences of the United States of America* 102:12338–12343.

Pineyro-Nelson, A., J. Van Heerwaarden, H. R. Perales, J. A. Serratos-Hernandez, A. Rangel, M. B. Hufford, P. Gepts, A. Garay-Arroyo, R. Rivera-Bustamante, and E. R. Alvarez-Buylla. 2009. Transgenes in Mexican maize: molecular evidence and methodological considerations for GMO detection in landrace populations. *Molecular Ecology* 18:750–761.

Poerschmann, J., A. Gathmann, J. Augustin, U. Langer, and T. Gorecki. 2005. Molecular composition of leaves and stems of genetically modified *Bt* and near-isogenic non-*Bt* maize—characterization of lignin patterns. *Journal of Environmental Quality* 34:1508–1518.

Pont, B., and W. Nentwig. 2005. Quantification of *Bt*-protein digestion and excretion by the primary decomposer *Porcellio scaber*, fed with two *Bt*-corn varieties. *Biocontrol Science and Technology* 15:341–352.

Poornima, K., P. Selvanayagam, and R. Shenbagarathai. 2010. Identification of native *Bacillus thuringiensis* strain from South India having specific cytocidal activity against cancer cells. *Journal of Applied Microbiology* 109:348–354.

Prasifka, P. L., R. L. Hellmich, J. R. Prasifka, and L. C. Lewis. 2007. Effects of Cry1Ab-expressing corn anthers on the movement of monarch butterfly larvae. *Environmental Entomology* 36:228–233.

Priestley, A. L., and M. Brownbridge. 2009. Field trials to evaluate effects of *Bt*-transgenic silage corn expressing the Cry1Ab insecticidal toxin on non-target soil arthropods in northern New England, USA. *Transgenic Research* 18:425–443.

Prihoda, K. R., and J. R. Coats. 2008. Aquatic fate and effects of *Bacillus thuringiensis* Cry3Bb1 protein: toward risk assessment. *Environmental Toxicology and Chemistry* 27:793–798.

Quist, D., and I. H. Chapela. 2001. Transgenic DNA introgressed into traditional maize landraces in Oaxaca, Mexico. *Nature* 414:541–543.

Reichman, J. R., L. S. Watrud, E. H. Lee, C. A. Burdick, M. A. Bollman, M. J. Storm, G. A. King, and C. Mallory-Smith. 2006. Establishment of transgenic herbicide-resistant creeping bentgrass (*Agrostis stolonifera* L.) in nonagronomic habitats. *Molecular Ecology* 15:4243–4255.

Rillig, M. C. 2004. Arbuscular mycorrhizae and terrestrial ecosystem processes. *Ecology Letters* 7:740–754.

Romeis, J., M. Meissle, and F. Bigler. 2006. Transgenic crops expressing *Bacillus thuringiensis* toxins and biological control. *Nature Biotechnology* 24:63–71.

Rosi-Marshall, E. J., J. L. Tank, T. V. Royer, M. R. Whiles, M. Evans-White, C. Chambers, N. A. Griffiths, J. Pokelsek, and M. L. Stephen. 2007. Toxins in transgenic crop byproducts may affect headwater stream ecosystems. *Proceedings of the National Academy of Sciences of the United States of America* 104:16204–16208.

Rui, Y. K., G. X. Yi, J. Zhao, B. M. Wang, Z. H. Li, Z. X. Zhai, Z. P. He, and Q. X. Li. 2005. Changes of *Bt* toxin in the rhizosphere of transgenic *Bt* cotton and its influence on soil functional bacteria. *World Journal of Microbiology and Biotechnology* 21:1279–1284.

Sanchis, V. 2011. From microbial sprays to insect-resistant transgenic plants: history of the biospesticide *Bacillus thuringiensis*. A review. *Agronomy for Sustainable Development* 31:217–231.

Saxena, D., S. Flores, and G. Stotzky. 1999. Insecticidal toxin in root exudates from *Bt* corn. *Nature* 402:480.

Saxena, D., S. Flores, and G. Stotzky. 2002a. *Bt* toxin is released in root exudates from 12 transgenic corn hybrids representing three transformation events. *Soil Biology and Biochemistry* 34:133–137.

Saxena, D., S. Flores, and G. Stotzky. 2002b. Vertical movement in soil of insecticidal Cry1Ab protein from *Bacillus thuringiensis*. *Soil Biology and Biochemistry* 34:111–120.

Saxena, D., S. Pushalkar, and G. Stotzky. 2010. Fate and effects in soil of Cry proteins from *Bacillus thuringiensis* in sprays and transgenic plants: influence of the physicochemical and biological characteristics of soil. In *Entomopathogenic Bacterial Toxins as Biological Control Agents, the Open Toxinology Journal*, eds. E. Ben-Dov and A. Zaritsky, 133–153. Bentham Science, Oak Park, IL.

Saxena, D., C. N. Stewart, I. Altosaar, Q. Y. Shu, and G. Stotzky. 2004. Larvicidal Cry proteins from *Bacillus thuringiensis* are released in root exudates of transgenic *B. thuringiensis* corn, potato, and rice but not of *B. thuringiensis* canola, cotton, and tobacco. *Plant Physiology and Biochemistry* 42:383–387.

Saxena, D., and G. Stotzky. 2000. Insecticidal toxin from *Bacillus thuringiensis* is released from roots of transgenic *Bt* corn in vitro and in situ. *FEMS Microbiology Ecology* 33:35–39.

Saxena, D., and G. Stotzky. 2001a. *Bacillus thuringiensis* (*Bt*) toxin released from root exudates and biomass of *Bt* corn has no apparent effect on earthworms, nematodes, protozoa, bacteria, and fungi in soil. *Soil Biology and Biochemistry* 33:1225–1230.

Saxena, D., and G. Stotzky. 2001b. *Bt* corn has a higher lignin content than non-*Bt* corn. *American Journal of Botany* 88:1704–1706.

Saxena, D., and G. Stotzky. 2001c. *Bt* toxin uptake from soil by plants. *Nature Biotechnology* 19:199–199.
Saxena, D., and G. Stotzky. 2002. *Bt* toxin is not taken up from soil or hydroponic culture by corn, carrot, radish, or turnip. *Plant and Soil* 239:165–172.
Saxena, D., and G. Stotzky. 2003. Fate and effects in soil of insecticidal toxins from *Bacillus thuringiensis* in transgenic plants. In *Collection of Biosafety Reviews*, 7–83. International Centre for Genetic Engineering and Biotechnology, Trieste, Italy.
Schnepf, E., N. Crickmore, J. Van Rie, D. Lereclus, J. Baum, J. Feitelson, D. R. Zeigler, and D. H. Dean. 1998. *Bacillus thuringiensis* and its pesticidal crystal proteins. *Microbiology and Molecular Biology Reviews* 62:775–806.
Schrader, S., T. Munchenberg, S. Baumgarte, and C. C. Tebbe. 2008. Earthworms of different functional groups affect the fate of the *Bt*-toxin Cry1Ab from transgenic maize in soil. *European Journal of Soil Biology* 44:283–289.
Sears, M. K. 2004. Impact of *Bacillus thuringiensis* corn pollen on monarch butterfly populations: a risk assessment. *Agricultural Biotechnology: Challenges and Prospects* 866:125–137.
Sears, M. K., R. L. Hellmich, D. E. Stanley-Horn, K. S. Oberhauser, J. M. Pleasants, H. R. Mattila, B. D. Siegfried, and G. P. Dively. 2001. Impact of *Bt* corn pollen on monarch butterfly populations: a risk assessment. *Proceedings of the National Academy of Sciences of the United States of America* 98:11937–11942.
Sessitsch, A., J. G. Howieson, X. Perret, H. Antoun, and E. Martinez-Romero. 2002. Advances in *Rhizobium* research. *Critical Reviews in Plant Sciences* 21:323–378.
Shelton, A. M., and M. K. Sears. 2001. The monarch butterfly controversy: scientific interpretations of a phenomenon. *The Plant Journal* 27:483–488.
Shen, R. F., H. Cai, and W. H. Gong. 2006. Transgenic *Bt* cotton has no apparent effect on enzymatic activities or functional diversity of microbial communities in rhizosphere soil. *Plant and Soil* 285:149–159.
Sheng, M., M. Tang, H. Chen, B. W. Yang, F. F. Zhang, and Y. H. Huang. 2008. Influence of arbuscular mycorrhizae on photosynthesis and water status of maize plants under salt stress. *Mycorrhiza* 18:287–296.
Sheveleva, E. V., S. Marquez, W. Chmara, A. Zegeer, R. G. Jensen, and H. J. Bohnert. 1998. Sorbitol-6-phosphate dehydrogenase expression in transgenic tobacco—high amounts of sorbitol lead to necrotic lesions. *Plant Physiology* 117:831–839.
Shu, Q. Y., H. R. Cui, G. Y. Ye, D. X. Wu, Y. W. Xia, M. W. Gao, and I. Altosaar. 2002. Agronomic and morphological characterization of *Agrobacterium*-transformed *Bt* rice plants. *Euphytica* 127:345–352.
Siciliano, S. D., and J. J. Germida. 1999. Taxonomic diversity of bacteria associated with the roots of field-grown transgenic *Brassica napus* cv. Quest, compared to the non-transgenic *B. napus* cv. Excel and *B. rapa* cv. Parkland. *FEMS Microbiology Ecology* 29:263–272.
Sims, S. R. 1997. Host activity spectrum of the CryIIA *Bacillus thuringiensis* subsp. *kurstaki* protein: effect on Lepidoptera, Diptera, and non-target arthropods. *Southwestern Entomologist* 22:395–404.
Sims, S. R., and J. W. Martin. 1997. Effect of the *Bacillus thuringiensis* insecticidal proteins CryIA(b), CryIA(c), CryIIA, and CryIIIA on *Folsomia candida* and *Xenylla grisea* (Insecta: Collembola). *Pedobiologia* 41:412–416.
Smith, S. E., and V. Gianinazzi-Pearson. 1988. Physiological interactions between symbionts in vesicular-arbuscular mycorrhizal plants. *Annual Review of Plant Physiology and Plant Molecular Biology* 39:221–244.
Smith, S. E., and D. J. Read. 2008. *Mycorrhizal Symbiosis*. Academic Press, London.
Snow, A. A. 2002. Transgenic crops—why gene flow matters. *Nature Biotechnology* 20:542–542.
Stanley-Horn, D. E., G. P. Dively, R. L. Hellmich, H. R. Mattila, M. K. Sears, R. Rose, L. C. H. Jesse, J. E. Losey, J. J. Obrycki, and L. Lewis. 2001. Assessing the impact of Cry1Ab-expressing corn pollen on monarch butterfly larvae in field studies. *Proceedings of the National Academy of Sciences of the United States of America* 98:11931–11936.
Stotzky, G. 2000. Persistence and biological activity in soil of insecticidal proteins from *Bacillus thuringiensis* and of bacterial DNA bound on clays and humic acids. *Journal of Environmental Quality* 29:691–705.

Stotzky, G. 2002. Release, persistence, and biological activity in soil of insecticidal proteins from *Bacillus thuringiensis*. In *Genetically Engineered Organisms: Assessing Environmental and Human Health Effects*, eds. D. K. Letourneau and B. E. Burrows, 187–222. CRC Press, Boca Raton, FL.

Stotzky, G. 2004. Persistence and biological activity in soil of the insecticidal proteins from *Bacillus thuringiensis*, especially from transgenic plants. *Plant and Soil* 266:77–89.

Stotzky, G., and D. Saxena. 2009. Is molecular "pharming" a potential hazard to the environment? In *Environmental Impact Assessments*, eds. G. T. Halley and Y. T. Fridian, 77–86. Nova Science, Hauppauge, NY.

Sun, C. X., L. J. Chen, Z. J. Wu, L. K. Zhou, and H. Shimizu. 2007. Soil persistence of *Bacillus thuringiensis (Bt)* toxin from transgenic *Bt* cotton tissues and its effect on soil enzyme activities. *Biology and Fertility of Soils* 43:617–620.

Swan, C. M., P. D. Jensen, G. P. Dively, and W. O. Lamp. 2009. Processing of transgenic crop residues in stream ecosystems. *Journal of Applied Ecology* 46:1304–1313.

Tan, F. X., J. W. Wang, Y. J. Feng, G. L. Chi, H. L. Kong, H. F. Qiu, and S. L. Wei. 2010. *Bt* corn plants and their straw have no apparent impact on soil microbial communities. *Plant and Soil* 329:349–364.

Tank, J. L., E. J. Rosi-Marshall, T. V. Royer, M. R. Whiles, N. A. Griffiths, T. C. Frauendorf, and D. J. Treering. 2010. Occurrence of maize detritus and a transgenic insecticidal protein (Cry1Ab) within the stream network of an agricultural landscape. *Proceedings of the National Academy of Sciences of the United States of America* 107:17645–17650.

Tapp, H., L. Calamai, and G. Stotzky. 1994. Adsorption and binding of the insecticidal proteins from *Bacillus thuringiensis* subsp. *kurstaki* and subsp. *tenebrionis* on clay minerals. *Soil Biology and Biochemistry* 26:663–679.

Tapp, H., and G. Stotzky. 1995. Insecticidal activity of the toxins from *Bacillus thuringiensis* subspecies *kurstaki* and *tenebrionis* adsorbed and bound on pure and soil clays. *Applied and Environmental Microbiology* 61:1786–1790.

Tapp, H., and G. Stotzky. 1998. Persistence of the insecticidal toxin from *Bacillus thuringiensis* subsp. *kurstaki* in soil. *Soil Biology and Biochemistry* 30:471–476.

Tarkalson, D. D., S. D. Kachman, J. M. N. Knops, J. E. Thies, and C. S. Wortmann. 2008. Decomposition of *Bt* and non-*Bt* corn hybrid residues in the field. *Nutrient Cycling in Agroecosystems* 80:211–222.

Tharakan, B. D., G. Devaki, H. B. Nair, and J. Padikkala. 2009. Cytotoxic effects of crystal proteins from *Bacillus thuringiensis* against breast cancer cells. *Cancer Research* 69:226S-226S.

Theissen, B., and D. J. Russell. 2009. The relevance of Collembola in monitoring soil-ecological effects of GMOs. *Gefahrstoffe Reinhaltung Der Luft* 69:391–394.

Then, C. 2010. Risk assessment of toxins derived from *Bacillus thuringiensis*—synergism, efficacy, and selectivity. *Environmental Science and Pollution Research* 17:791–797.

Thies, J. E., and M. H. Devare. 2007. An ecological assessment of transgenic crops. *Journal of Development Studies* 43:97–129.

Tschenn, J., J. E. Losey, L. H. Jesse, J. J. Obrycki, and R. Hufbauer. 2001. Effects of corn plants and corn pollen on monarch butterfly (Lepidoptera:Danaidae) oviposition behavior. *Environmental Entomology* 30:495–500.

Turrini, A., C. Sbrana, M. P. Nuti, B. M. Pietrangeli, and M. Giovanetti. 2004. Development of a model system to assess the impact of genetically modified corn and aubergine plants on arbuscular mycorrhizal fungi. *Plant and Soil* 266:69–75.

U.S. Department of Agriculture (USDA), National Agricultural Statistics Service (NASS). 2010. Adoption of genetically engineered crops in the U.S.: corn varieties. Extent of adoption, Table 1. http://www.ers.usda.gov/Data/BiotechCrops/ExtentofAdoptionTable1.htm.

U.S. Environmental Protection Agency (EPA). 2001a. Amended revised response to EPA's data call-in notice concerning the potential for adverse effects of *Bt* corn on non-target lepidopterans. http://www.epa.gov/oppbppd1/biopesticides/pips/executive_summary_and_preface.pdf.

U.S. Environmental Protection Agency (EPA). 2001b. SAP Report No 2000–07, sets of scientific issues being considered by the Environmental Protection Agency regarding: *Bt* plant pesticides risk and benefits assessments. http://www.epa.gov/scipoly/sap/meetings/2000/october/octoberfinal.pdf.

U.S. Environmental Protection Agency (EPA). 2009. SmartStax pesticide fact sheet. http://www.epa.gov/oppbppd1/biopesticides/pips/smartstax-factsheet.pdf.

U.S. Environmental Protection Agency (EPA). 2010. Introduction to biotechnology regulation for pesticides. http://www.epa.gov/oppbppd1/biopesticides/regtools/biotech-reg-prod.htm.

U.S. Environmental Protection Agency (EPA). 2011. Pesticides: regulating biopesticides, plant incorporated protectants, current and previously registered Section 3 PIP registrations. http://www.epa.gov/pesticides/biopesticides/pips/pip_list.htm.

Vance, C. P., C. Uhde-Stone, and D. L. Allan. 2003. Phosphorus acquisition and use: critical adaptations by plants for securing a non-renewable resource. *New Phytologist* 157:423–447.

Vercesi, M. L., P. H. Krogh, and M. Holmstrup. 2006. Can *Bacillus thuringiensis* (*Bt*) corn residues and *Bt*-corn plants affect life-history traits in the earthworm *Aporrectodea caliginosa*? *Applied Soil Ecology* 32:180–187.

Viktorov, A. G. 2011. Transfer of *Bt* corn byproducts from terrestrial to stream ecosystems. *Russian Journal of Plant Physiology* 58:543–548.

Wandeler, H., J. Bahylova, and W. Nentwig. 2002. Consumption of two *Bt* and six non-*Bt* corn varieties by the woodlouse *Porcellio scaber*. *Basic and Applied Ecology* 3:357–365.

Wang, B., and Y. L. Qiu. 2006. Phylogenetic distribution and evolution of mycorrhizas in land plants. *Mycorrhiza* 16:299–363.

Wang, W. X., B. Vinocur, and A. Altman. 2003. Plant responses to drought, salinity and extreme temperatures: towards genetic engineering for stress tolerance. *Planta* 218:1–14.

Watrud, L. S., G. King, J. P. Londo, R. Colasanti, B. M. Smith, R. S. Waschmann, and E. H. Lee. 2011. Changes in constructed *Brassica* communities treated with glyphosate drift. *Ecological Applications* 21:525–538.

Watrud, L. S., E. H. Lee, A. Fairbrother, C. Burdick, J. R. Reichman, M. Bollman, M. Storm, G. King, and P. K. Van de Water. 2004. Evidence for landscape-level, pollen-mediated gene flow from genetically modified creeping bentgrass with CP4 EPSPS as a marker. *Proceedings of the National Academy of Sciences of the United States of America* 101:14533–14538.

West, A. W. 1984. Fate of the insecticidal, proteinaceous parasporal crystal of *Bacillus thuringiensis* in soil. *Soil Biology and Biochemistry* 16:357–360.

West, A. W., H. D. Burges, T. J. Dixon, and C. H. Wyborn. 1985. Survival of *Bacillus thuringiensis* and *Bacillus cereus* spore inocula in soil—effects of pH, moisture, nutrient availability and indigenous microorganisms. *Soil Biology and Biochemistry* 17:657–665.

Wolfenbarger, L. L., S. E. Naranjo, J. G. Lundgren, R. J. Bitzer, and L. S. Watrud. 2008. *Bt* crop effects on functional guilds of non-target arthropods: a meta-analysis. *PloS One* 3:e2118.

Wolt, J. D., and R. K. D. Peterson. 2010. Prospective formulation of environmental risk assessments: probabilistic screening for Cry1A(b) maize risk to aquatic insects. *Ecotoxicology and Environmental Safety* 73:1182–1188.

Wong, R. S. Y., S. M. Mohamed, V. D. Nadarajah, and I. A. T. Tengku. 2010. Characterisation of the binding properties of *Bacillus thuringiensis* 18 toxin on leukaemic cells. *Journal of Experimental and Clinical Cancer Research* 29:86.

Wraight, C. L., A. R. Zangerl, M. J. Carroll, and M. R. Berenbaum. 2000. Absence of toxicity of *Bacillus thuringiensis* pollen to black swallowtails under field conditions. *Proceedings of the National Academy of Sciences of the United States of America* 97:7700–7703.

Wu, W. X., H. H. Lu, W. Liu, M. Devare, J. E. Thies, and Y. X. Chen. 2009. Decomposition of *Bacillus thuringiensis* (*Bt*) transgenic rice residues (straw and roots) in paddy fields. *Journal of Soils and Sediments* 9:457–467.

Wu, W. X., Q. F. Ye, and H. Min. 2004a. Effect of straws from *Bt*-transgenic rice on selected biological activities in water-flooded soil. *European Journal of Soil Biology* 40:15–22.

Wu, W. X., Q. F. Ye, H. Min, X. J. Duan, and W. M. Jin. 2004b. *Bt*-transgenic rice straw affects the culturable microbiota and dehydrogenase and phosphatase activities in a flooded paddy soil. *Soil Biology and Biochemistry* 36:289–295.

Xu, B. L., Y. X. Wang, X. X. Liu, F. Yuan, N. Su, Y. Chen, Y. J. Wu, and Q. W. Zhang. 2006. Effects of Cry1Ac and secondary metabolites in *Bt* transgenic cottonseed on *Lycoriella pleuroti* Yang et Zhang (Diptera:Sciaridae). *Environmental Entomology* 35:807–810.

Xue, K., H. F. Luo, H. Y. Qi, and H. X. Zhang. 2005. Changes in soil microbial community structure associated with two types of genetically engineered plants analyzing by PLFA. *Journal of Environmental Sciences*-China 17:130–134.

Xue, K., R. C. Serohijos, M. Devare, and J. E. Thies. 2011. Decomposition rates and residue-colonizing microbial communities of *Bacillus thuringiensis* insecticidal protein Cry3Bb-expressing (*Bt*) and non-*Bt* corn hybrids in the field. *Applied and Environmental Microbiology* 77:839–846.

Yamashita, S., H. Katayama, H. Saitoh, T. Akao, Y. S. Park, E. Mizuki, M. Ohba, and A. Ito. 2005. Typical three-domain cry proteins of *Bacillus thuringiensis* strain A1462 exhibit cytocidal activity on limited human cancer cells. *Journal of Biochemistry* 138:663–672.

Yu, L., R. E. Berry, and B. A. Croft. 1997. Effects of *Bacillus thuringiensis* toxins in transgenic cotton and potato on *Folsomia candida* (Collembola: Isotomidae) and *Oppia nitens* (Acari: Orbatidae). *Journal of Economic Entomology* 90:113–118.

Zangerl, A. R., D. McKenna, C. L. Wraight, M. Carroll, P. Ficarello, R. Warner, and M. R. Berenbaum. 2001. Effects of exposure to event 176 *Bacillus thuringiensis* corn pollen on monarch and black swallowtail caterpillars under field conditions. *Proceedings of the National Academy of Sciences of the United States of America* 98:11908–11912.

Zeilinger, A. R., D. A. Andow, C. Zwahlen, and G. Stotzky. 2010. Earthworm populations in a northern US Cornbelt soil are not affected by long-term cultivation of *Bt* maize expressing Cry1Ab and Cry3Bb1 proteins. *Soil Biology and Biochemistry* 42:1284–1292.

Zwahlen, C., A. Hilbeck, P. Gugerli, and W. Nentwig. 2003. Degradation of the Cry1Ab protein within transgenic *Bacillus thuringiensis* corn tissue in the field. *Molecular Ecology* 12:765–775.

Zwahlen, C., A. Hilbeck, R. Howald, and W. Nentwig. 2003. Effects of transgenic *Bt* corn litter on the earthworm *Lumbricus terrestris*. *Molecular Ecology* 12:1077–1086.

Zwahlen, C., A. Hilbeck, and W. Nentwig. 2007. Field decomposition of transgenic *Bt* maize residue and the impact on non-target soil invertebrates. *Plant and Soil* 300:245–257.

chapter 9

Maize legume relay intercrops in Malawi

Meeting short- and long-term sustainability goals

Carol Shennan
University of California, Santa Cruz

Dorothy Sirrine
University of California, Santa Cruz
and Utopia Foundation, Traverse City

Contents

9.1 Introduction

The challenges to improving food security and agricultural sustainability in Africa are great and multifaceted. The first green revolution failed in Africa, and roughly 50% of its population has remained in poverty (International Assessment of Agricultural Knowledge, Science and Technology for Development [IAASTD], 2009), despite the considerable resources invested in its implementation. This failure has been attributed to Africa's high agroecological variability, lack of infrastructure and irrigation (Toenniessen et al., 2008), and poor understanding of the important role of socioeconomic and cultural complexity (Sanginga, 2010). As Giller et al. (2010) pointed out, there are no silver bullets for improving agricultural productivity in Africa. For any set of approaches to be effective, it will need to take into account the tremendous diversity of local farming systems in the region and the variation in underlying resources and capacities within and between communities and improve livelihoods to increase nutritional self-sufficiency and reduce poverty. These approaches will also need to be sustainable, that is, to be able to maintain improvements into the foreseeable future.

Despite improvements in many other regions, sub-Saharan Africa has continued to experience a decline in food security and agricultural productivity per capita, leading to an increase in undernourishment since 1990 (Food and Agricultural Organization [FAO], 2006; Toenniessen et al., 2008). Biophysical challenges to food security include production-limiting constraints faced by resource-poor farmers, such as shrinking farm sizes and inequitable land distribution patterns, depleted soils and limited use of fertilizer and soil amendments, unreliable rainfall and lack of irrigation capacity, inadequate pest and disease control, and limited access to improved varieties and seed distribution systems (Diao et al., 2007).

These production challenges for agriculture in Africa are likely to be made even more difficult in the future by the effects of global climate change. Predicted changes for sub-Saharan Africa include increased rainfall variability, more frequent extreme events such as droughts and floods, and increased average temperatures (National Research Council [NRC], 2008; IAASTD, 2009). Only 4% of agricultural land in sub-Saharan Africa is irrigated, which means that more unpredictable rainfall will greatly impact the primarily rain-fed systems throughout the region. Biophysical production constraints are further aggravated by infrastructural issues such as poorly maintained roads and transportation systems, lack of access to regional or international markets, poor or nonexistent access to credit, problems with labor availability, unstable political systems, poor security, and warfare (Diao et al., 2007). Further, a lack of innovation networks and underinvestment by national governments and other institutions in the physical, institutional, and human capital needed to support sustainable agricultural intensification in Africa are widespread (NRC, 2010).

Against this backdrop, there have been a number of multistakeholder, international groups convened to develop strategic approaches to what has been termed either a second green revolution (InterAcademy Council, 2004; Toenniessen et al., 2008; African Green Revolution, 2009) or the sustainable intensification of African agriculture (NRC, 2010; IAASTD, 2009). Members of the African Union have created the Comprehensive Africa Agriculture Development Program (CAADP) to help African countries improve economic growth through agricultural development that "eliminates hunger, reduces poverty and food insecurity, and enables expansion of exports" (African Union Report, 2008). The road maps offered by these organizations vary but include use of improved crop varieties, increasing soil productivity, building more equitable access to input and product markets

(Toenniessen et al., 2008), and adopting new policies that encourage input use by farmers and fair prices for their produce (Sanginga, 2010). Some envisage a tripling of cereal grain yields through the use of inorganic and organic fertilizers and high-yielding crop varieties, farmer education and empowerment, and improved markets (Sanchez, 2010). Others emphasize the need for interdisciplinary and participatory system approaches that recognize the importance of socioeconomic and biophysical context, make effective use of local resources, use well-adapted crop varieties and livestock breeds, and involve judicious use of external inputs when needed (NRC, 2010; IAASTD, 2009).

9.2 *Improving soil fertility in sub-Saharan Africa*

Improving soil fertility is a central goal for sustainable agriculture in Africa (IAASTD, 2009; NRC, 2010). Although use of improved cultivars that are more productive, drought tolerant, and resistant to pests and diseases is an important intervention, it is clear that poor soil fertility is a major constraint to increased crop productivity (Sanchez, 2010). This has led some to argue that a primary focus on increasing fertilizer use, together with the use of improved crop varieties where possible, is needed to improve yields and livelihoods across many African cropping systems (Vanlauwe et al., 2010; Sanchez, 2010). Others caution that while fertilizer use may be an important component, an overreliance on purchased inputs makes resource-poor farmers vulnerable to what can be wide fluctuations in price and availability, and that high prices put fertilizer out of reach for resource-poor farmers (Denning et al., 2009; NRC, 2010; Sirrine, Shennan, Snapp, et al., 2010; Snapp et al., 2010). Systems that rely on and enhance locally available fertility resources also need to be sought. The reasons for low fertilizer use and the myriad challenges to increasing fertilizer use effectively in Africa, especially among the poorest farmers, were aptly described by Morris et al. (2007). Issues such as low value/cost ratios for fertilizer use, high price instability, high yield response variability, and high costs of or limited access to credit led the authors to suggest that low fertilizer use may in fact be a rational response to managing risk by African farmers (Morris et al., 2007).

Degraded soils are a severe problem in much of sub-Saharan Africa (Bekunda et al., 2010), and significant declines in soil organic matter (SOM) are well documented (Moebius-Clune et al., 2011; Joergensen, 2010; Lal, 2006). A review of organic matter in the tropics as a whole highlighted the lower levels of total soil carbon (C) and nitrogen (N) associated with arable farming as compared to forest and pastures, with even greater differences observed in microbial biomass C and N (Joergensen, 2010). Further, the amount of land converted from forest and savannah to cropland is still increasing in sub-Saharan Africa (Verchot, 2010), a problem exacerbated by degradation of cropland soils.

The loss of SOM affects not only nutrient availability but also other aspects of soil biological and physical functions that often cannot be rectified by use of fertilizers alone. SOM levels have an impact on soil physical structure and hence aggregate stability, water infiltration rates, and water-holding capacity. Indeed, in addition to fertility limitation, poor soil physical structure, reduced resistance to erosion, and reduced drought resistance are thought to be responsible for low crop yields observed in the region (Lal, 2006). Organic matter inputs to soil are also critical for supporting a diverse and active soil microbial community and macrofauna responsible for residue decomposition, carbon turnover, nutrient cycling, and other functions, such as disease and pest suppression (Joergensen, 2010; Akinnifesi et al., 2010; Shennan, 2008). While use of inorganic fertilizers can increase yields, when used alone they may not provide enough crop residue to affect SOM positively. Indeed, a meta-analysis concluded that organic resource additions were necessary to increase soil organic C in maize-based systems (Chivenge, Vanlauwe, and J. Six, 2011).

There is an increasing recognition that integrated use of inorganic fertilizer together with technologies to improve SOM (such as use of legumes as cover crops, relay crops, managed fallows and other agroforestry (AF) systems, and recycling of manure and composts) may hold the greatest potential for both short-term livelihood gains and longer-term soil fertility improvements (Sileshi et al., 2008; Sanchez, 2010; Snapp et al., 1998, 2010; Vanlauwe et al., 2010; Bekunda et al., 2010; Chivenge, Vanlauwe, and J. Six, 2011). Integrated soil fertility management (ISFM) is now being widely promoted and can be defined as "a set of soil fertility management practices that necessarily include the use of fertilizer, organic inputs and improved germplasm, combined with the knowledge of how to adapt these practices to local conditions aimed at maximizing agronomic use efficiency of applied nutrients and improving crop productivity" (Vanlauwe et al., 2010, p. 17). In essence, this approach encompasses the use of any of the organic matter technologies listed, in conjunction with supplemental fertilizer, depending on the specific characteristics of the area. Conservation agriculture is also being promoted as a means to improve SOM levels and crop productivity, which brings in the added dimension of reduced tillage along with maintenance of permanent soil cover (by crop residues and cover crops) and diversified crop rotations or plant associations, including legumes (Meyer, 2009). This approach is proving to be successful in many regions globally, including sub-Saharan Africa (European Technology Assessment Group, 2009; Naudin et al., 2010), but there are challenges in some areas related to soil conditions, competing uses for crop residues, labor requirements for weeding, and access to herbicides (Gowing and Palmer, 2008; Giller et al., 2009).

9.3 *Legume-based systems*

To what extent can these various integrated approaches and organic matter technologies improve SOM levels, soil structure, and soil biology and thus sustainably support higher crop yields over the long term? Here, we focus on annual cropping system diversification with legumes and highlight important findings and areas needing more research, rather than attempt an extensive review of the literature. Various types of legume systems are being promoted for improved soil fertility and crop production in different contexts. These include improved fallows, AF, rotation, green manures, and intercropping systems (Akinnifesi et al., 2010). Most reviews of legume strategies emphasize the evidence for increased crop yields relative to current farmer practice.

A recent meta-analysis of 94 studies involving legumes concluded overwhelmingly that the response to legumes is positive, often resulting in two- to threefold yield increases in low to moderate-yielding sites (Sileshi et al., 2008). Addition of 50% of the recommended fertilizer dose further increased yields by more than 25% over legumes alone, indicating that legumes can significantly reduce fertilizer requirements. In a large on-farm trial in Malawi, Snapp et al. (2010) also demonstrated the benefits of diversification of cropping systems, with legumes combined with half the recommended fertilizer rates producing equivalent yields to fully fertilized monoculture systems and with lower yield variability. Similar conclusions of improved yields and reduced fertilizer requirements were reached in a review of the effects of various systems that incorporated leguminous trees and shrubs in maize systems of eastern and southern Africa (Akinnifesi et al., 2010).

There is a variety of mechanisms by which legumes can potentially improve crop production. These include increased N provision through biological nitrogen fixation (BNF), recycling of nutrients from deeper in the soil profile, improved soil structure and physical properties due to increased organic matter inputs, and improved soil biological activity with associated increased nutrient availability and in some cases via enhanced disease or pest suppression (Akinnifesi et al., 2010; Franke et al., 2008).

BNF contributions vary considerably among species, growth habit, and by location, but amounts as high as 300 kg N ha^{-1} have been measured for some tree legumes (*Leucaena colinsii*) and in the range of 35–150 kg N ha^{-1} for shrubby species like fish bean (*Tephrosia vogellii*), common sesban or river bean (*Sesbania sesban*), and pigeon pea (*Cajanus cajan*) (Akinnifesi et al., 2010; Adu-Gyamfi et al., 2007). There is considerable variation in BNF and legume productivity from site to site, which can be related to soil nutrient limitations, especially P availability, water stress, or lack of appropriate or sufficient indigenous rhizobium inoculant (Bekunda et al., 2010). An assessment of indigenous soil rhizobia in eastern and southern Africa found that many soils had low populations and may benefit from the addition of inoculant, but inoculant use is presently low in the region (Bekunda et al., 2010). Some work suggests that inoculation with indigenous rhizobia can be more beneficial than with exotic strains (Makatiani and Odee, 2007), and that combined inoculation with arbuscular mycorrhiza (AM) (Mishra, 2008; Singh, 1996; Sekhon et al., 1992) or with endosphere bacteria (Rajendran et al., 2008) can improve nodulation and BNF compared to rhizobial inoculation alone. There clearly remains significant potential for improvement of legume growth and BNF to improve poor-quality soils through addition of appropriate fertility amendments or microbial inoculants.

While much of the crop response to legumes can be attributed to increased N supply, other effects may also be important (Franke et al., 2008; Yusuf et al., 2009). It is widely agreed that the use of legumes is an important strategy to help maintain soil quality (Lal, 2009; Akinnifesi et al., 2010; Bationo et al., 2007), but relatively few studies in the published literature have investigated changes in SOM or physical and biological properties directly. In some cases, integration of legumes was found to increase total soil C when used as intercrops (Beedy et al., 2010; Bationo and Buerkert, 2001), as improved fallows (Nyamadzawo et al., 2009; Bossio et al., 2005), or as cover crops (Barthes et al., 2004); others found that incorporation of high-quality legume biomass led to a decline in soil C, but less than when fertilizer was used alone or in the no input control (Mugwe et al., 2009). In another study, rotation with legumes did not affect total soil C after 2 years relative to continuous maize, but total soil N was increased by 23% and water-soluble carbon was increased by 79–106%, depending on the species and cultivar of legume used (Yusuf et al., 2009).

Snapp and Pound (2008) have argued that in some land-limited systems improvements in total soil C may not be observed given the relatively low levels of residue input possible with intercropping or relay intercropping, as compared with managed fallow or use of green manures. Instead, they asserted that it is more important to assess effects on nutrient availability mediated through changes in labile pools of soil C and N, microbial activity, and recycling of nutrients from deeper in the soil profile. Various measures of labile C and N pools provide indicators of readily available sources of nutrients for microbial activity and food sources for soil fauna, such as particulate organic matter (POM), microbial carbohydrates involved in aggregate formation (C extractable with hot water), nonspecific labile pools (permanaganate oxidizable), soil N supply capacity (N mineralization), and microbial biomass and activity, which provide a measure of cycling of organic matter and nutrients (Haynes, 2008). These measures can provide more sensitive indications of management impacts that relate to both nutrient cycling and improved soil physical properties.

There is some evidence that labile soil C pools are increased by legumes. For example, a gliricidia/maize intercrop increased POM more than total SOM and enriched the POM with N relative to maize alone (Beedy et al., 2010). Similarly, high-quality legume residue incorporation increased labile C (measured as potassium permanaganate oxidizable) and mineralizable N in POM in a manner equivalent to manure application (Mtambanengwe and Mapfumo, 2008). In another study, total soil C, microbial biomass C (MBC), soil C mineralization, and soil-specific respiration were all higher in legume-treated soils, although the difference in MBC was not statistically significant (Kone et al., 2008). Total microbial biomass increased with legume fallows relative to continuous maize (Nyamadzawo et al., 2009; Bossio et al., 2005), and 16S ribosomal RNA (rRNA) gene and phospholipid fatty acid (PLFA) analysis demonstrated that shifts in the microbial community structure also occurred (Bossio et al., 2005). Seasonal changes in microbial biomass C and N and crop N uptake led Sugihara et al. (2010) to conclude that soil microbes serve as an important N source for crop growth during the grain-forming stage in dry tropical cropland in Tanzania.

The effect of residue incorporation on changes in soil C and N pools is complex and depends in part on soil and residue properties. In a clayey soil, adding organic residues, irrespective of quality, increased total soil C and N as well as macroaggregates, whereas residue inputs had little effect in a loamy, sandy soil (Gentile et al., 2010). Conversely, another study found that in a sandy soil a medium-quality residue (higher polyphenol content) led to greater amounts of residue-derived N being found in coarse POM as compared to high-quality residue (Chivenge, Vanlauwe, Gentile, et al., 2011), yet residue quality had no effect on C pools in a clayey soil in the same study. The reason for these differences across soil types is unclear. Also, soil quality measured as MBC and available P improved more rapidly at a higher-fertility site with cover crop use than when cover crops were used at a lower-fertility location (Kone et al., 2008b).

Even when there is no effect of residue quality on long-term soil C accumulation, it may affect short-term C and N release dynamics and hence synchrony between N release and crop demand. Another study found that residue quality did not affect long-term soil C accumulation, but it did affect short-term C and N release dynamics and interactions with fertilizer additions (Gentile et al., 2011). The combination of fertilizer and low-quality residue immobilized more fertilizer N than with a high-quality residue. Under field conditions, this temporary immobilization actually reduced N losses and led to a positive effect on crop N uptake (Gentile et al., 2011). Nitrogen mineralization rates are known to vary among legume residues of different qualities, with polyphenol content and N content being the major properties controlling N release (Baijukya et al., 2006). Improved understanding of seasonal dynamics of mineralization, microbial populations, and changes in labile pools of C and N, associated with incorporation of legume residues of different qualities and fertilizer use, is needed to enhance synchrony between nutrient availability and crop demand.

More work is also needed to relate changes in different SOM fractions and microbial populations noted to key soil physical properties, such as improved aggregate stability and reduced crusting, that are important for reducing soil erosion. Some work has shown improvements in water infiltration rates (Nyamadzawo et al., 2007), reduced bulk density, and improved aggregate stability (Sileshi and Mafongoya, 2006) following improved legume fallows, but little information is available for other systems.

While most research has focused on C and N dynamics, some studies have shown benefits of legume incorporation on crop growth and P uptake (Pypers et al., 2007; Akinnifesi et al., 2007; Alvey et al., 2001). This may be due to increased microbial mineralization of soil organic P (Randhawa et al., 2005), solubilization of inorganic P fixed by iron (Fe) and aluminum (Al) in the soil (Mweta et al., 2007), stimulation of AM infection (Bagayoko et al., 2000), or a combination of effects (Alvey et al., 2001). Given the low P status of many African soils, the effects of long-term legume use on P fertility warrant further study.

It is interesting to note that fields classified by farmers as productive versus poor differed in many of the same measures as discussed, with productive soils having higher total C and N, labile C, microbial biomass C and N, N mineralization, and soil respiration, as well as increased effective cation exchange capacity, exchangeable cations, and extractable P (Murage et al., 2000). By preferentially applying residues, manures, or fertilizers to their most productive fields, farmers ensured that at least some of their land had good SOM and fertility levels (Mtambanengwe and Mapfumo, 2008).

Clearly, legumes can have many beneficial impacts on soil quality, in large part by impacting nutrient cycling through soil microbial populations, but legume BNF and productivity are often suboptimal due to soil fertility or other limitations. There is considerable interest in whether soil microbial populations can be altered to enhance different functions such as BNF, or increase P availability, through the addition of beneficial bacteria or fungi. A publication summarized the current state of use of microbial soil amendments or biofertilizers in the tropics (Uribe et al., 2010). Most of the work on biofertilizers has focused on Latin America and Asia, with relatively few studies in Africa. The main types of materials being developed are inoculants of symbiotic BNF bacteria (rhizobia), nonsymbiotic BNF bacteria, phosphorus-solubilizing bacteria (Yarzabal, 2010), and AM fungi. All have had success in increasing yields of various crops in experimental trials but suffered from variable performance in field situations for a variety of possible reasons.

Challenges remain in the production of high-quality, low-cost inoculants, as well as a lack of understanding of species- and site-specific effects. These can include inoculant survival issues, competition from indigenous microorganisms, soil type and fertility effects, and differential interactions based on crop species and varieties. From the discussion, however, given the importance of legumes for helping improve African soils that are degraded and with low organic matter, further research on biofertilizers is warranted. One example of a product developed in Africa is PREP PAC, an inexpensive mix of urea, rock phosphate, legume seeds, and rhizobium, which is being tested in participatory on-farm trials (Okalebo et al., 2006).

9.4 A conceptual framework for assessing relative sustainability of production systems

In all probability, increased use of fertilizers, legumes, AF, ISFM, conservation agriculture, improved seed, and crop/livestock integration will all play a role in improving food production and soil quality in Africa (NRC, 2010), but no single technology package is likely to be broadly applicable across the diversity of systems in the region. Rather, it can be argued that a systems approach with interdisciplinary research that is participatory and grounded in the local context and needs is required to develop locally appropriate and sustainable solutions (NRC, 2010; InterAcademy Council, 2004; Pretty, 2008; IAASTD, 2009; Snapp and Pound, 2008). To identify what are the most locally appropriate and sustainable systems among different options requires us to think about how we assess "appropriateness" and sustainability.

Gains in agricultural production must be sustained into the future, yet what is involved in creating sustainable systems and what makes a system sustainable are complex questions that merit critical exploration. It is not simply about increasing yields; agricultural sustainability also encompasses other considerations, such as longer-term impacts on the environment and resource base (especially soil), as well as on farmers' livelihoods. As discussed in the NRC report *Sustainable Agricultural Systems in the 21st Century* (NRC, 2010), agricultural sustainability is a complex concept that involves meeting broadly agreed-on societal goals into the future. Achieving these goals will require systems that are adequately productive; that use resources efficiently and protect or enhance the resource base; and importantly, that demonstrate robustness in the face of fluctuating conditions and unpredictable shocks (NRC, 2010).

Overarching sustainability goals revolve around the need to satisfy human food and fiber needs (in terms of amount and accessibility), while protecting or enhancing the resource base and surrounding environments and providing viable livelihoods and a good quality of life for farming communities (NRC, 2010). More specific goals are embedded within each of these general ones, and the relative importance given to each will vary depending on the system in question and the social context within which the system is embedded. For example, it can be argued that questions regarding the sustainability of contemporary agricultural systems in the United States and other developed countries have arisen in large part due to concerns about negative impacts of agricultural inputs on the environment, whereas primary questions of sustainability in sub-Saharan Africa revolve around the ability to produce sufficient and accessible food to alleviate hunger and poverty and regenerate degraded soils. Thus, while there is ultimately a need to balance each of the goals for long-term sustainability, in the more immediate term one goal may be given higher priority than another depending on the specific context.

The question of who decides the specific goals and priorities is critical since many parties have vested interests in agricultural sustainability and yet likely vary in the relative weight they place on different goals. As scientists, we cannot answer the question of what the goals and priorities should be; this is fundamentally a "social choice" that needs to be arrived at through negotiation and political process (NRC, 2010). In the case of Africa, this point is reflected in calls for African countries themselves to take the lead in development efforts to improve food production and sustainability (Interacademy Council, 2004; African Union Report, 2008; Flora, 2010) and for farmer voices to be heard in setting research agendas (African Farmers Organization, 2009).

What scientists can and must do is to inform this process of negotiation and decision making by providing answers to the "what is" and "what if" questions; that is, given what is known about the status quo, what are the likely outcomes and effects of choosing certain specific interventions over others? Investigations by agricultural researchers of the impacts of legumes, manure application, and fertilizers on crop yields, soil nutrients, and microbial communities address reasonable topics. However, to answer what-if types of questions most effectively, attention must be given to the socioeconomic and cultural context of systems being targeted and the needs and desires of the farm households in the region. These contextual factors will also determine sustainability outcomes and impacts of the different systems being tested. Further, consideration of context will increase the likelihood that any technologies or knowledge developed will be appropriate and meet the needs of farmers, be feasible given their circumstances, and thus be more likely to be adopted and adapted by farmers. Accomplishing this requires integration of different disciplinary approaches and a close connection with and participation of key stakeholders.

9.4.1 *The importance of the time dimension*

At a practical level, we suggest that the sustainability of existing or proposed agricultural systems is determined by their ability to meet immediate livelihood needs, provide reasonable stability in performance from year to year, and over the longer term to maintain or improve the natural resource base on which the system depends. There is often a trade-off between maximizing immediate livelihood benefits and conserving or enhancing the soil resource (Bezner Kerr et al., 2007), yet to be sustainable over the longer term, the cropping systems must at least maintain or ideally improve soil fertility and quality. Resource-poor farmers may be more likely to adopt cropping systems that benefit them both immediately and consistently (i.e., show improved yields, income, or food supply and involve less risk on a year-to-year basis), however, rather than systems that are most optimal for long-term fertility improvement.

Thus, we suggest that the most sustainable cropping systems are likely to represent a compromise between short- and medium-term benefits to farmers' livelihoods and long-term maintenance or improvement of the natural resource base, as Ashby et al. (1996) also observed. Stoorvogel et al. (2004) discussed the trade-offs between desirable traits associated with agricultural systems that operate over various time and spatial scales and have employed trade-off analysis models to investigate environmental and economic impacts. We propose that the trade-offs between goals operating at different time frames will be based on unique local socioeconomic, cultural, and biophysical parameters, which emphasizes the importance of performing locally relevant research.

9.4.2 *Socioeconomic variability, vulnerability, and distributional impacts*

It is important to recognize that specific trade-offs (and their extent) are likely to vary among households within communities based on factors such as differential economic endowments and landholding sizes, location of landholdings (biophysical context), and even within households based on factors such as gender, age, and health status. For example, a system that improves yields but requires more labor may be of value only to wealthier farmers. Researchers have placed increasing emphasis on understanding the role of *social* vulnerability in risk-based analyses (e.g., Kelly and Adger, 2000; Wisner et al., 2004). While much of the research has focused on the impacts on climate change, hazards may also result from new technologies that have a potential to cause social, infrastructural, or environmental change (Oliver-Smith and Hoffman, 2002).

The sustainable livelihoods framework (Chambers and Conway, 1992; Department for International Development [DFID], 2001) emphasizes the importance of addressing vulnerability, defined as both the exposure to shocks, risk, and stress and the inability to cope without experiencing hardship. Vulnerability thus encompasses but looks beyond income-poverty, a concept typically quantified by per-capita wealth generation in that it considers individuals' security and well-being based on locally relevant, complex, and multidimensional factors (Chambers, 1995). Important factors include biophysical vulnerability based on risks related to soil type, slope, and landscape position, for example, and social vulnerability that relates to poverty, access to and dependency on purchased resources, diversity of income sources, and the social status of individuals or households within a community (Adger, 1999). As the risk introduced by new agricultural systems will likely vary based on farmers' socioeconomic resources and degrees of vulnerability, vulnerability-based analyses may help researchers determine the distributional impacts of new cropping systems among different community members. In fact, distributional economic analyses (Von Braun, 2003) have frequently determined that new agricultural technologies are most profitable, and at times only profitable, for better-resourced farmers, while fewer benefits have been realized for women and poor farmers. Sustainability assessments would therefore be improved by distributional analyses of cropping system impacts.

In summary, we argue for a framework in which sustainability is assessed as the ability to meet a balance of immediate-term livelihood needs, acceptable food/income stability from year to year, and longer-term soil fertility improvement. Further, relative sustainability of different management strategies should be addressed in a distributional manner, that is, considering the variability in biophysical and socioeconomic resource endowments and vulnerability of different households. We illustrate this framework using data from an on-farm study in southern Malawi. Maize legume relay intercrop systems were being promoted as potential strategies for improving both short-term productivity and livelihoods while simultaneously helping to rebuild SOM in systems where land was limited (Chirwa et al., 2006; Snapp et al., 1998). The project compared the performance of different fertilizer and maize/legume relay intercrop systems established across a regionally representative range of smallholder farms.

9.5 Case study: legume maize relay cropping in Southern Malawi

Here, we use the conceptual framework developed in a case study from southern Malawi to illustrate a model for interdisciplinary and participatory research examining the relative sustainability of different maize/legume relay cropping systems. Sustainability is considered in terms of immediate-term livelihood benefits (crop yields, net income, and secondary food or income provision), stability (risk of low yields over time and space), evidence of potential and actual adoption of the system, and longer-term soil fertility impacts. Our analysis also addresses the implications of distributional impacts (among different socioeconomic groups, genders, and landscape positions) on the relative performance and desirability of the different cropping systems.

9.5.1 Study location and farmer selection

Malawi is a small, landlocked country in southern Africa with high levels of poverty and a history of chronic food insecurity (Chinsinga, 2005). Smallholder farmers comprise 85% of the population, and maize, the staple crop, is planted to roughly 85% of arable agricultural lands (Smale and Heisey, 1997). Southern Malawi is the most impoverished region in Malawi and has high population density, limited landholdings (National Economic Council [NEC], 2000), and few livestock.

The study was an on-farm, farmer-/researcher-designed and managed project initiated by researchers at the University of Malawi's Bunda College of Agriculture in 1994 and continued with our participation through 2004. Participating farmers were located in villages within the Songani watershed, located approximately 15–20 km north of Zomba in southern Malawi. This region is subject to a unimodal rainfall pattern, with the wet season occurring between October and May. Average annual rainfall in the study area is 1,150 mm (Kamanga et al., 1999). The soils are mainly classified as alfisols and ultisols (Eswaran et al., 1996). They are typically well-drained loamy sands, with N as the most limiting nutrient (Snapp, 1998). Since agriculture has increasingly spread onto hillsides and steep slopes in this region (Banda et al., 1994), our research included farmers with plots at three different landscapes: (1) dambo (less than 12% slope and poorly drained), (2) dambo margin (less than 12% slope and well drained), and (3) hillside (greater than 12% slope) (Kamanga et al., 1999).

Forty-eight farm families, or households, were recruited for the initial project, selected at random along six transects spaced 0.6 km apart. Eight households were no longer participating in the project by 2004, and six other participants were either too old or ill to consistently participate in all components of the socioeconomic data collection. The ethnicity of participating farmers was predominantly Yao, with a minority from Chewa and Ngoni ethnic groups. Forty-one percent of the households participating in the socioeconomic analyses were female-headed households (FHHs), defined as households where women were divorced, widowed, or separated from their husbands (Bezner Kerr, 2005).

9.5.2 Legume species

Prior to the project's initiation, University of Malawi, Bunda College of Agriculture researchers held community meetings to ascertain farmers' assessments of local agricultural constraints and opportunities. Soil fertility concerns predominated, and given high fertilizer prices at the time, limited landholding sizes in the region, and the lack of access to livestock manure, relay cropping of deep-rooted N-fixing legumes alongside maize was identified as a research priority. All three AF species in this study are short-lived deciduous shrubs of the family Leguminosae. *Sesbania sesban* generally grows between 4 and 8 m tall, while *Tephrosia vogelii* and pigeon pea (*Cajanus cajan*) are typically 1.3–3 m tall (Bunderson et al., 1995). Both *T. vogelii* and pigeon pea were cultivated in southern Malawi prior to the project's inception. Pigeon pea is also a perennial grain legume, and the seed is eaten to provide an important secondary protein source (Snapp, Blackie, et al., 2003). Pigeon pea is the most common intercrop with maize in southern Malawi (Chirwa et al., 2003). Farmers traditionally incorporated pigeon pea leafy biomass after the leaves had senesced and fallen, but researchers also incorporated any fresh leafy biomass that remained on the legume plants into the soil, leaving the woody portion to be used as fuel wood at farmers' request. *T. vogelii* contains a toxic compound called *tephrosin* and historically was used to poison fish for consumption but had not previously been used as a green manure. The farmers also had no prior experience cultivating *S. sesban*, but researchers chose to include it in the study because it produces large amounts of biomass, and the leaves can be used as green manure. The woody portions of all three legumes were used by farmers as fuel wood.

9.5.3 Experimental design and management

The project encompassed two distinct experimental designs: design 1 (D1) from 1995 to 2000 and design 2 (D2) from 2001 to 2004 (Figure 9.1). The designs primarily differed in the rate and timing of inorganic fertilizer application. In 1995, four rain-fed plots were established within each participating farmer's field; plots remained fixed in the same location and position for the remainder of the project. At the onset of rains (typically in late October or November), a maize hybrid was planted in each of the four plots and harvested the following year. As is traditional in Malawi, three maize seeds were sown together, in planting stations 90 cm apart on ridges also 90 cm apart. The legume species were planted directly between maize stations. *T. vogelii* and pigeon pea seeds were each relay cropped into one of the four plots within 2 weeks of maize planting, whereas due to a very low seed germination rate, *S. sesban* seedlings were grown in a nursery and transplanted into the third plot roughly 2 months after planting maize, one plant between each maize station. The fourth plot served as the maize-only control; however, farmers generally intercropped some plants in these plots.

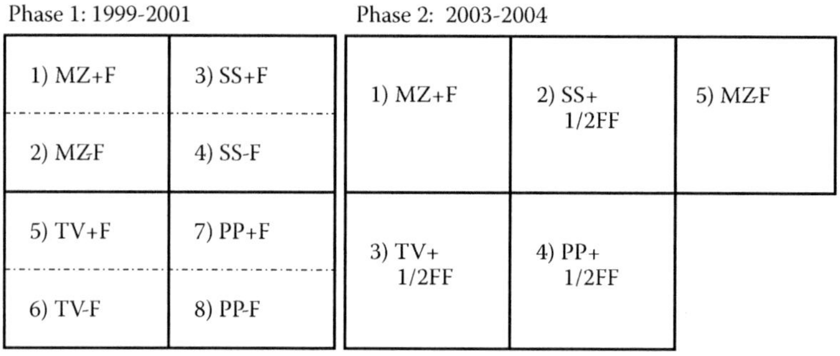

Figure 9.1 Project designs 1 (phase 1) and 2 (phase 2). Phase 1 treatments: (1) maize control plus full fertilizer (MZ+F); (2) maize control without fertilizer (MZ-F); (3) maize intercropped with *S. sesban* plus full fertilizer (SS+F); (4) maize intercropped with *S. sesban* without fertilizer (SS-F); (5) maize intercropped with *T. vogelii* plus full fertilizer (TV+F); (6) maize intercropped with *T. vogelii* without fertilizer (TV-F); (7) maize intercropped with pigeon pea plus full fertilizer (PP+F); and (8) maize intercropped with pigeon pea without fertilizer (PP-F). Phase 2 treatments: (1) maize plus full fertilizer (MZ+F); (2) maize intercropped with *S. sesban* plus the second fertilizer application (SS+1/2F); (3) maize intercropped with *T. vogelii* plus the second fertilizer application (TV+1/2F); (4) maize intercropped with pigeon pea plus the second fertilizer application (PP+1/2F); and (5) maize receiving no fertilizer (MZ-F). When sampling for phase 2, we always sampled from the center of the plot. In this way, we averaged the variation between previously fertilized and unfertilized plots.

During D1, one-half of each plot received an inorganic fertilizer treatment at a rate of 45 kg N ha^{-1} (see Figure 9.1 for cropping system abbreviations). As is typical for the region, fertilizer was applied twice during the cropping cycle: one-half at maize planting (formulated as 23N:21P:0K and 4S) and the other half as a side-dress of calcium ammonium nitrate when maize was roughly 60 cm in height. For D2, the experimental design was modified to determine if legumes could replace the first inorganic fertilizer application. At the time, few farmers were able to afford the full recommended rate of fertilizer as used in D1. Thus, in 2001 new treatments were superimposed over the same plots previously used; each legume treatment remained in the same location, but the plots were no longer split (Figure 9.1). The three legume systems did not receive the first inorganic fertilizer application, but each still received the second fertilizer application (22.5 kg N ha^{-1}). The entire maize control plot received the full rate of fertilizer (45 kg N ha^{-1}) at the two standard application timings. A fifth plot, MZ-F, was added at each farmer's field as the untreated control with maize alone. Based on this design, there was no true control to determine the impact of legumes versus fertilizer. Legume systems (SS+1/2F, TV+1/2F, and PP+1/2F) could only be compared to one another and again unfertilized or fertilized maize controls. Data from the 2001/2002 cropping cycle were not included in the analysis due to the recent change in the experimental design.

To fit within smallholder farming systems and minimize competition with maize, the perennial legume species were annually replanted. For both designs, nonwoody portions of the legumes were cut and incorporated into the soil, typically in late September or early October. Researchers applied fertilizer, provided much of the labor related to legumes (sowing, transplanting, and incorporation), and harvested maize from subplots. While this was done to ensure uniform management, it would have been preferable in retrospect to have farmers directly involved in these activities (see further discussion). Farmers performed land preparation, sowed maize, weeded, and harvested maize outside the experimental subplots. D1 data were collected by researchers at the University of Malawi, Bunda

College of Agriculture and are used here with permission from Dr. Kanyama-Phiri. D2 data were collected by us and local assistants employed by Bunda College of Agriculture for the complete duration of the project.

9.5.4 Maize yields

Mature maize was harvested from subplots at the center of each plot. For D1, grain yields were taken when plants had dried in the field, and 12% moisture content was assumed for the grain (Kamanga et al., 1999). For D2, we harvested maize slightly earlier as per farmers' requests and therefore calculated the actual dry weight based on a subsample of the fresh material. Although the study began in 1995, we only present phase 1 data from the 1999 and 2000 maize harvests and phase 2 data from the 2003 and 2004 maize harvests since these were years with the most complete data sets and greatest number of replicates available. Also, 2001–2002 was a transition period between the two designs. Maize yield data, as well as other forms of data, were not available for all farms in any given year for a variety of reasons. For instance, many farmers harvested maize early due to concerns of crop theft from their fields, and sometimes fields had been set ablaze by youth trying to capture mice as a protein source. Occasionally, some farmers applied small quantities of fertilizer in unknown quantities, precluding their inclusion in the analysis. These kinds of problems are to be expected in on-farm participatory work with resource-limited farmers, and we acknowledge that this poses challenges for data analysis. The resulting unbalanced data structure was considered throughout our statistical analyses in both our choice of methods and in evaluating the assumptions for each test. Data are presented for 29 farmers in 1999, 28 in 2000, 21 in 2003, and 14 in 2004. For each design, maize yields were evaluated using two-way analyses of variance (ANOVAs) ($*P < 0.05$), with cropping system and year as the main factors and maize yield as the response variable. For both D1 and D2, Tukey's post hoc tests ($*P < 0.05$) were employed to compare treatments.

9.5.5 Soil analysis

Soil was sampled from each treatment with the exception of D1 legume treatments receiving full fertilizer (the latter were not collected due to a variety of logistical constraints). A composite soil sample of six subsamples was collected from 0 to 10 cm depth in 1994 (prior to the establishment of treatments), 2000, and 2004. Soils were air dried, passed through a 2-mm sieve, and analyzed for pH, extractable Bray-P (Diamond, 1995), and percentage total C and percentage total N using a Carlo Erba 1108 elemental analyzer (Smith and Tabatabai, 2003). Soil data were analyzed from a subset of 12 farmers, balanced across the three landscapes. The same set of farmers was used for 1994, 2000, and 2004.

9.5.6 Maize foliar analysis

Foliar tissue testing was performed for D2 years 2003 and 2004. Samples were collected using a composite of 12 ear-leaf blades located immediately above the primary ear node (Jones and Eck, 1973) when maize was at peak tassel. They were then analyzed for percentage C, N, and S (Smith and Tabatabai, 2003) using high-temperature dry combustion. Foliar percentage P was determined using a nitric acid/hydrogen peroxide microwave digestion and analysis by inductively coupled plasma atomic emission spectrometry (Meyer and Keliher, 1992). Maize foliar data were collected from 21 farmers in 2003 and 12 farmers in 2004.

9.5.7 Legume biomass and tissue analysis

We did not collect D1 legume biomass data but determined D2 legume biomass by harvesting from a 5 × 5-mm subplot at the time of incorporation, which was October 2002 and ran from August to September in 2003 (the timing was changed due to farmers' requests). Foliar legume tissue samples, collected at time of incorporation, included a composite of green leafy materials from 12 plants. They were dried, ground, and analyzed for percentage N as described previously for maize foliar samples. Legume biomass data were collected from 13 farmers in 2002 and 17 farmers in 2003.

To determine the impact of legume treatments and landscapes, we performed separate two-way ANOVAs ($^*P < 0.05$) using the following dependent variables: pH, extractable P, soil percentage C, soil percentage N, legume biomass, maize foliar percentage N, maize foliar percentage S, and maize foliar percentage P. Legume treatment and landscape were the two independent factors. Total N input was calculated by combining organic legume-based N contribution (legume biomass × legume foliar percentage N), when present, and inorganic N quantities. We were unable to transform total N input to achieve normality or homogeneity of variance and therefore could not perform ANOVAs. Instead, we performed nonparametric median tests for the independent variable cropping system ($^*P < 0.05$) (Norušis, 2003).

9.5.8 Economic analysis

We performed a distributional cost-benefit analysis for the cropping systems in which we separately investigated costs and benefits for the wealthiest and poorest farmers because their marketing strategies varied substantially (see Sirrine, Shennan, Snapp, et al., 2010, for detailed methods). While wealthier and very impoverished farmers typically sell proportionally similar quantities of maize (10% of their yields), wealthier farmers often retain their maize to sell when prices are high, and highly impoverished farmers, in need of cash after the hungry season, generally sell when prices are low (Center for Regional Agricultural Trade Expansion Support [RATES], 2003; Peters, 2006). Due to substantial intra- and interannual fluctuations in costs and benefits (Sirrine, Shennan, Snapp, et al., 2010), we evaluated profitability separately for the two different design years. We were unable to present cost-benefit data for middle-income farmers because their marketing strategies were less well defined. Crop prices and input and labor costs used to estimate profitability can be found in the work of Sirrine, Shennan, Snapp, et al. (2010).

The methods for our participatory wealth-ranking exercise are described in detail in the work of Sirrine, Shennan, Snapp, et al. (2010). Briefly, farmers were placed into socioeconomic categories using a participatory wealth-ranking method described and validated by Adams et al. (1997), in which a few community members helped researchers place farmers into one of three categories: wealthiest, middle-income bracket, and poorest. Farmers were placed into these categories based on wealth and vulnerability indicators specific to the region, including selling their own labor (*ganyu*), hiring casual labor, ability to afford fertilizer, food availability throughout the year, and landholding size, among others. Indicators were chosen based on both locally based literature (e.g., Ellis, 1998) and key community members' perceptions. We later verified whether farmers had been placed in correct categories through household visits and interviews.

9.5.9 Socioeconomic data collection and analysis

Socioeconomic data were collected through farmer interviews and focus groups. For detailed methods, see the work of Sirrine, Shennan, and Sirrine (2010). Briefly, semiformal farmer interviews were carried out in 2001 to obtain demographic data and farmers' experience with the legumes and in 2003 to obtain farmers' assessments of the legume systems. For the latter, we then solicited information on farmers' preferred legume system(s) and farmers' perceptions of the legume systems' labor requirements, secondary benefits, impacts on food security, biophysical performance, and variability. Husbands and wives were interviewed separately for both interviews. From the 34 farm households still participating, 47 individual farmers were interviewed in 2001 and 51 in 2003, which represented 87% and 94% of the potential study population, respectively.

Adoption surveys were also performed in 2001 and 2003 to collect information on legume system adoption in farmers' fields. We carried out 34 surveys in 2001, interviewing husbands and wives together since they generally farm the same plots of land. In 2003, the adoption survey was combined with the preference survey; thus, we interviewed 51 farmers. Although heads of households were interviewed separately, data were only reported for each household. The first adoption survey also requested farmers to recall on-farm presence of any of the three legume species prior to the inception of the project's experimental phase in 1995. We analyzed legume system adoption and intensity of cropping chronologically using farmer recall for 1994 and actual adoption results for 2001 and 2003.

In 2004, we held four focus groups that included five to nine farmers each with the dual purpose of informing farmers of the research results and to obtain further insights into cropping system impacts on livelihoods. The composition of the focus groups was reflective of the overall population of farmers participating in the project and included men, women, FHHs, and a range of poor to wealthier farmers. Following the focus groups, we also held a training session focused on legume management.

9.6 Case study results and discussion

When averaged across all landscapes for D1, the highest yields were obtained in all treatments receiving inorganic fertilizer (F) (Table 9.1), with a consistent, but nonsignificant, trend toward higher average yields when legumes were present, especially *S. sesban*. In general, yields were lowest on the hillside and highest in the dambo margin, with similar patterns of response to fertilizer and legume treatments. Legume treatment differences were not significant, however, on the hillside, whereas they were in the dambo and dambo margin (Figure 9.2). In the last cases, the legumes with fertilizer treatment yields were all significantly higher than the unfertilized control, with the maize with fertilizer and legume-only treatment yields between the unfertilized control and the fertilized legumes. D1 and D2 fertilized maize control yields (Table 9.1) were substantially higher than those found by Harawa et al. (2006), Kamanaga, Waddington, et al. (2010), and Snapp et al. (2010) from other on-farm research trials in Malawi; but unfertilized maize yields from D2 were similar. Maize yields were found to increase by between 0.2 and 4 ton ha^{-1} when legumes were introduced in a similar study (Kamanga, Waddington, et al., 2010), which is in line with the increases observed in this study. Snapp et al. (2010) also observed improved fertilizer response with long-lived legumes, such as pigeon pea and *Tephrosia*, as was seen in D1 here.

Table 9.1 Mean Maize Yields and Profitability by Cropping System and Wealth Group

Cropping system	Mean maize yield (kg/ha)[1]	Wealth group[2]	Mean yield by wealth group: 1999 (kg/ha)[3]	Mean yield by wealth group: 2000 (kg/ha)[3]	Mean profitability by wealth group: 1999 (MKw/ha)[3]	Mean profitability by wealth group: 2000 (MKw/ha)[3]
D1/year			1999	2000	1999	2000
MZ+F	3,691[abc]	Wealthiest	4,476*	3,705*	28,098**	13,822*
		Poorest	1,470*	2,760*	5,532**	8,591*
MZ-F	2,407[ac]	Wealthiest	2,260	2,142	16,995	8,359
		Poorest	1,464	1,872	7,762	6,699
SS+F	4,422[bd]	Wealthiest	5,333***	4,837	33,182***	18,338
		Poorest	2,110***	4,230	8,830***	14,601
SS-F	2,795[ac]	Wealthiest	3,372***	3,180	21,006***	12,403
		Poorest	1,520***	2,785	6,928***	9,965
TV+F	3,811[ab]	Wealthiest	4,227*	4,577	25,982*	17,968
		Poorest	2,655*	3,820	13,637*	13,533
TV-F	2,279[c]	Wealthiest	2,452**	2,937	15,302**	12,238
		Poorest	1,570**	2,420	8,353**	9,239
PP+F	3,825[ad]	Wealthiest	4,451*	4,790*	34,233*	25,651*
		Poorest	1,593*	3,340*	12,725*	17,768*
PP-F	2,591[ac]	Wealthiest	2,624*	2,983	23,167*	19,074
		Poorest	1,363*	2,300	13,560*	15,247
D2/Year			2003	2004	2003	2004
MZ+F	4,201[a]	Wealthiest	4,457	4,183*	38,388	47,948*
		Poorest	3,256	2,283*	23,956	22,694*
MZ-F	1,233[c]	Wealthiest	1,786	885	14,685	7,868
		Poorest	1,104	1,055	6,941	10,225
SS+1/2F	3,150[ab]	Wealthiest	2,785	2,194	21,815	22,137
		Poorest	2,751	2,409	19,787	25,176
TV+1/2F	3,046[ab]	Wealthiest	3,029	2,776	25,361	30,929
		Poorest	2,874	2,028	22,000	21,057
PP+1/2F	1,756[bc]	Wealthiest	2,509	1,876	26,564	25,356
		Poorest	1,443	1,120	14,639	15,436

[1] Lowercase letters indicate statistical differences based on one-way ANOVAs ($P < 0.05$, Tukey's post hoc test).
[2] Profitability was not quantified for moderately wealthy farmers due to minimal information on their marketing strategies.
[3] The asterisks *, **, *** indicate significant *t*-test result comparing wealthiest and poorest farmers' yield or profitability for a single cropping system at $P < 0.05, 0.01, 0.001$, respectively. Data from Sirrine, Shennan, S. Snapp, et al. 2010.

When averaged across all landscapes in D2, the maize plus full fertilizer produced the highest yields, the legumes plus half fertilizer intermediate yields, and the unfertilized maize controls the lowest yields (Table 9.1). Once again, yields were lowest across all systems on the hillside and highest in the dambo margin (Figure 9.3). In a study in the same area, Harawa et al. (2006) also found that *S. sesban* and *T. vogelii* performed poorly on the hillside landscape, which had extremely rocky topography and shallow soils (Figure 9.4).

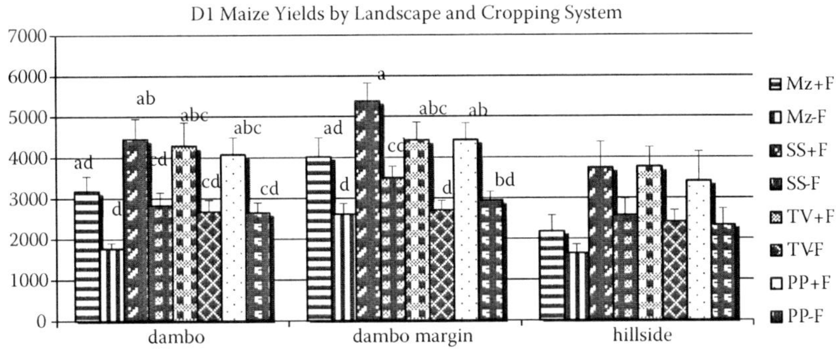

Figure 9.2 D1 (1999 and 2000 average) maize yields by treatment and landscape. Different letters indicate statistical significance at the $P < 0.05$ level.

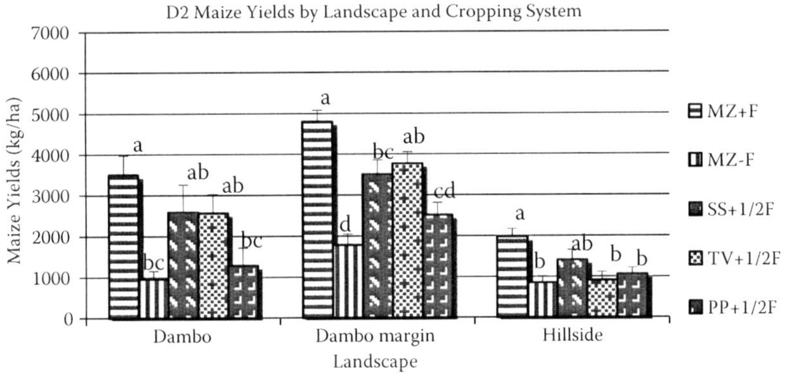

Figure 9.3 D2 (2003 and 2004 average) maize yields by treatment and landscape. Different letters indicate statistical significance at the $P < 0.05$ level.

Figure 9.4 Typical hillside field.

The greatest benefit of the legumes plus half fertilizer were apparent in the dambo margin and the dambo, where both *S. sesban* and *T. vogelii* with half fertilizer had higher yields than the unfertilized control, but less than the fully fertilized maize. When disaggregated in this way, pigeon pea plus half fertilizer did not yield significantly higher than unfertilized maize in any landscape. This indicates a more positive effect of *T. vogelli* and *S. sesban* on maize yields than pigeon pea since they received the equivalent quantities of inorganic fertilizer. In lab trials, Sakala et al. (2000) found that mixed pigeon pea and maize residues experienced prolonged net N immobilization greater than that predicted based on mineralization patterns of sole maize and sole pigeon pea. When there is no initial fertilizer application as in our D2 legume plus half fertilizer treatments, temporary N immobilization by the pigeon pea residue may have occurred. However, there was no evidence of immobilization by pigeon pea in the D1 period (Figure 9.2), when unfertilized pigeon pea yields were equal to or greater than the unfertilized no legume control across all landscapes. The reason for the different maize yield response to pigeon pea across the two designs is unclear but may have been related to poorer pigeon pea growth in the D2 years.

Both unfertilized and legume plus half fertilizer maize yields from D2 (Figure 9.3) were low relative to those of D1 (Figure 9.2) but were still in the range or higher than those reported by others for on-farm research in this region (Harawa et al., 2006; Kamanga et al., 2010; Snapp et al., 2010). It is interesting to note that maize yields in the legume treatments receiving half fertilizer were similar to those of D1 legumes receiving no fertilizer. The lack of initial fertilizer may have negatively affected early growth and nodulation of the legumes. Previous work has shown that small quantities of inorganic N fertilizer often have a stimulatory impact on legume growth, nodulation, and BNF (Giller and Cadisch, 1995), especially in low-N soils (Hardarson and Atkins, 2003). As Sirrine, Shennan, Snapp, et al. (2010) suggested, the legumes may perform best with an immediately available N input from inorganic fertilizer applied at maize planting to establish well and to avoid any nutrient immobilization following incorporation of the preceding legume residue. Further, small amounts of P fertilizer (20 kg P ha^{-1}) have also been found to improve pigeon pea growth (Kamanga, Whitbread, et al., 2010) and to improve nodulation in *S. sesban* and other legumes (Uddin et al., 2008). Even if the full recommended amount of fertilizer is unavailable, it may still be beneficial to apply at least some N and P fertilizer at maize planting rather than assuming that the previously incorporated legume biomass can replace the first application.

9.7 Legume yields and N input

Legume leafy biomass yields were low in D2 (mean = 472 kg ha^{-1} in 2002 and 204 kg ha^{-1} in 2003) but in line with those obtained by Snapp et al. (2010) for on-farm trials in this region (430 kg ha^{-1}). These numbers are lower, however, than others reported for the region by Harawa et al. (2006) and by Kamanga et al. (1999), particularly for *S. sesban* and *T. vogelii*. The difference may relate to the amount of woody biomass included in the measurement; here, only leafy biomass was incorporated according to farmers' requests. In this study, there was no difference in the quantity of leafy biomass produced by the different legume species (Table 9.2). Yet, despite the relatively low biomass amounts, subsequent maize yields were positively associated with legume biomass in both years (simple regression $P < 0.01$; Table 9.3), suggesting that legumes were still playing a role in improving maize productivity.

Nitrogen is limiting in this region (Snapp, 1998), and total N input (which combines inorganic fertilizer and legume N sources) was also strongly correlated with maize yields in both years (simple regression $P < 0.001$, 30% of yield variation in 2003 and 44% in 2004)

Table 9.2 Mean Legume Biomass and Total N Input[1] for 2002 and 2003

Cropping system	2002 Legume biomass (kg/ha)	2002 Total N input (kg/ha)	2003 Legume biomass (kg/ha)	2003 Total N input (kg/ha)
MZ+F	NA	45[a]	NA	45[a]
MZ-F	NA	0[b]	NA	0[b]
SS+1/2F	360	31[c]	208	27[c]
TV+1/2F	628	38[c]	197	27[c]
PP+1/2F	409	34[c]	208	28[c]

[1] Total N input combines organic and inorganic N sources. NA, not applicable. Lowercase letters indicate statistical differences based on nonparametric median tests ($P < 0.05$). Absence of letters indicates the factor was not significant.

Table 9.3 D2 Simple Regressions of Foliar Nutrient Content, Legume Biomass, and N Input with Maize Yields from 2003 and 2004

Variable	2003 Beta	2003 R^2	2003 *P* value	2003 N	2004 Beta	2004 R^2	2004 *P* value	2004 N
Foliar %N	0.489	0.239	0.000***	103	0.507	0.257	0.000***	60
Foliar %S	0.509	0.259	0.000***	103	0.513	0.263	0.000***	55
Foliar %P	NA	NA	NA	NA	0.233	0.054	NA	45
Legume biomass[1]	0.442	0.190	0.003**	45	0.531	0.282	0.003**	30
Total N input[2]	0.544	0.296	0.000***	79	0.662	0.438	0.000***	50

Note: Beta, slope of the regression line when predictor and independent variables are standardized (indicates the direction of the relationship); R^2, proportion of variability in the dependent variable attributable to the regression equation; *P* value, significance level, with $^*P < 0.05$, $^{**}P < 0.01$, and $^{***}P < 0.001$. N, total number of cases; NS, not significant; WS, whole soil; LF, light fraction.

[1] Legume biomass does not include MZ+F or MZ-F treatments.
[2] Total N includes both organic and inorganic fertilizer sources; also includes all five treatments.

(Table 9.3). Others have also found maize yields to be linearly related to the amount of N recycled from *Sesbania* and pigeon pea fallows (Ndufa et al., 2009) or to the total amount of inorganic and organic N sources applied (Mtambanengwe and Mapfumo, 2006). Kamanga, Waddington, et al. (2010) suggested that to have a positive impact on maize yields, legume dry matter biomass should be at least 2,000 kg ha^{-1}, but here we found that much smaller quantities of legume biomass combined with modest amounts (22.5 kg N ha^{-1}) of inorganic N provided benefits for farmers with access to limited quantities of fertilizer. It is also possible that there are cumulative benefits from annual legume relay cropping since 1996 on overall soil quality that contribute to enhanced maize yields, even though we were unable to demonstrate changes over time in major soil nutrients with legume use (see Section 9.7.1). Others have suggested that long-term soil quality benefits accrue from repeated legume use (Kamanga, Waddington, et al., 2010a; Franzel and Scherr, 2002).

A focus on improving legume biomass production in combination with the use of moderate amounts of fertilizer clearly has the potential to greatly improve maize yields in this region. Efforts are needed both to improve agronomic management of the legumes and to better understand the mechanisms by which legumes improve maize yields. Legume density and planting arrangement can have major effects on intercrop performance (Mucheru-Muna et al., 2010; Snapp and Silim, 2002), and timing of incorporation will affect residue quality, which will in turn affect N release dynamics. If residue quality is high, N will be released more quickly, whereas residue with a lower N content would decompose more slowly (Baijukya et al., 2006). If released too quickly, N will be vulnerable to leaching losses, whereas early maize growth could be reduced if it is released too slowly. Phiri et al. (1999) found that *S. sesban*, a moderate-quality residue, mineralized slowly in southern Malawi, with the strongest correlation between biomass N added and soil mineral N occurring 85 days after incorporation. Release dynamics from residues are also affected by fertilizer additions (Gentile et al., 2011; Kwabiah et al., 1999).

Increased N provision is obviously an important mechanism by which legumes improve maize yields. High rates of nodulation and nitrogen fixation are desirable, and one study found that more than 93% of the nitrogen in pigeon pea plants in farmers' fields in Malawi came from fixation (Adu-Gyamfi et al., 2007), and that annual rates of fixation ranged between 37 and 117 kg N ha^{-1}. This is greater than the total N content of pigeon pea harvested in D2 here, suggesting that growth and nodulation were relatively poor in most fields in these 2 years. Similarly, the biomass N from *T. vogelii* and *S. sesban* were well below amounts found in some other studies (Akinnifesi et al., 2010). As discussed previously, it may be beneficial to apply even small amounts of N and P fertilizer, if available, at maize planting to help stimulate early legume root growth and nodulation (Hardarson and Atkins, 2007; Kamanga, Waddington, et al., 2010; Kamanga, Whitbread, et al., 2010). It would also be useful to measure levels of viable rhizobia in the soil and degree of nodulation to determine if inoculant additions would be beneficial.

In addition to N input through BNF, there is evidence that deep-rooted legumes, such as those used here, can recycle nitrogen that has leached deeper in the soil profile back into surface soils on incorporation (Akinnifesi et al., 2010; Snapp et al., 1998). Interestingly, pigeon pea has also been found to improve water availability for associated maize plants by accessing deep water and subsequently releasing water back into the soil through its shallower roots (Sekiya and Yano, 2004). Enhanced soil moisture could also benefit the soil microbial community and hence nutrient mineralization. Further, Makumba et al. (2009) found that decomposing pigeon pea roots contributed substantially to mineral N in the soil profile; *T. vogelii* and *S. sesban* roots could possibly have similar effects.

Maize tissue percentage N levels were highest in both 2003 and 2004 in the full-fertilizer treatment, followed by similar but lower levels in the various legume plus half fertilizer treatments, and were lowest in unfertilized control (Table 9.4). Maize yields were positively correlated with both tissue percentage N and percentage S (Table 9.3). While nitrogen is the most commonly limiting nutrient in the region (Snapp, 1998), Weil and Mughogho (2000) found that maize yields in Malawi responded to sulfur applications in scenarios in which adequate N was available. Further, extensive S deficiency has been observed in farmer fields in West Africa (Nziguheba et al., 2009) and elsewhere (Scherer, 2009).

Table 9.4 Maize Foliar %N and %S by Treatment and Landscape for 2003 and 2004

	2003		2004		
Cropping system	%N	%S	%N	%S	%P
MZ+F	2.473[a]	0.188[a]	2.038[a]	0.210[a]	0.311
MZ-F	1.342[b]	0.121[b]	1.262[b]	0.133[b]	0.250
SS+1/2F	1.828[c]	0.145[bc]	1.715[c]	0.156[b]	0.248
TV+1/2F	1.852[c]	0.150[c]	1.723[c]	0.149[b]	0.264
PP+1/2F	1.769[c]	0.139[bc]	1.711[c]	0.153[b]	0.277

Note: Lowercase letters indicate statistical differences based on one-way ANOVAs ($P < 0.05$, Tukey's post hoc test). Absence of letters indicates the factor was not significant.

Interestingly, we observed higher tissue S levels in the legume plus half fertilizer treatments as compared to unfertilized maize, suggesting that the legumes enhanced S availability in the soil since the fertilizer applied with the legumes in D2 did not contain any S (whereas that applied to the full fertilizer MZ+F at planting did). If microbial activity were increased in the legume treatments, this could have resulted in greater mineralization of organic S pools in the soil. Studies have shown that crops acquire the majority of their S from organic forms, even in the presence of sulfate fertilizer (Boye et al., 2010). Alternatively, it is possible that sulfate leached deeper in the soil profile was accessed by the legumes and recycled into the upper soil layers. Interactions between legume use and S dynamics warrant further investigation. There is also some evidence that inclusion of legumes in rotations can benefit crop P nutrition through the stimulation of AM infection (Bagayoko et al., 2000; Alvey et al., 2001) or by increasing soil P availability (Randhawa et al., 2005; Mweta et al., 2007; Alvey et al., 2001). However, we found no evidence for legume-mediated effects on maize tissue percentage P here (Table 9.4) or soil-available P (see next section).

9.7.1 *Soil fertility*

Neither treatment nor landscape position resulted in measurable changes in pH and total soil percentage N in 2000 or 2004 (Table 9.5), which is perhaps not surprising given the high variability across sites and the low replicate numbers for soil analysis (12 farmers). Interestingly, soil percentage C was significantly lower in the dambo margin than on the hillside in 2000 (Table 9.6), although there were no percentage C treatment effects on maize yields in 2000 or 2004. The trend toward higher percentage C on the hillside may be related to the small number of replicates or the more recent transition of agriculture to the hillside (Banda et al., 1994), whereas the dambo margin has historically been the prime agricultural land in highly populated southern Malawi. Furthermore, the hillside landscape was often highly rocky, and farmers pushed together boulders to collect pockets of soil in which to crop maize (see Figure 9.4), which may have resulted in higher soil C levels in these pockets than in the surrounding soil.

To address the variability issue, we also looked at the percentage change over time of each parameter for a given plot, and even then there were no discernible changes over time in either soil percentage C or percentage N or effects of treatment or landscape. Furthermore, despite the range of soil C and N contents present across the sites, there was no correlation between maize yields and either percentage C or percentage N (Table 9.7). Snapp et al. (2010) also did not detect changes in soil C in Malawi trials of long-lived legumes that included pigeon pea, noting the long time frame necessary to detect changes

Table 9.5 Two-Way ANOVA Results for Long-Term Soil Indicators for D1 and D2

	D1-2000				D2-2004			
	Legume treatment		Landscape		Legume treatment		Landscape	
Variable	F	*P* value	F	*P* value	F	*P* value	F	*P* value
pH	1.15	NS	1.6	NS	1.48	NS	3.15	NS
Extractable P (ppm)	0.07	NS	20.2	0.000***	0.2	NS	9.65	0.001**
%C-WS	1.13	NS	3.39	0.042*	0.43	NS	2.064	NS
%N-WS	1.41	NS	1.95	NS	0.356	NS	1.47	NS

Note: The asterisks *, **, *** indicate significance at $P < 0.05, 0.01, 0.001$, respectively. NS, not significant.

Table 9.6 Whole-Soil %C and %N, pH, and Extractable P (ppm) for Baseline, 2000, and 2004 Soils

	pH	Whole-soil %C	Whole-soil %N	Extractable P (ppm)
		Baseline		
Average	5.76	1.20	0.091	13.90
		By landscape		
Dambo	5.68	1.60	0.122	2.60
Dambo margin	5.83	0.90	0.067	12.37
Hillside	5.78	0.97	0.076	26.73
		D1		
		Cropping system		
MZ+F	5.83	1.13	0.079	13.21
MZ-F	5.71	1.20	0.089	13.70
SS-F	5.74	1.18	0.087	12.00
TV-F	5.61	0.87	0.063	14.03
PP-F	5.70	1.22	0.086	19.68
		By landscape		
Dambo	5.73	1.29[ab]	0.095	1.89[a]
Dambo margin	5.64	0.91[a]	0.068	13.31[b]
Hillside	5.78	1.23[b]	0.083	28.38[b]
		D2		
MZ+F	5.60	1.01	0.076	26.71
MZ-F	5.80	1.00	0.071	19.85
SS+1/2F	5.52	1.11	0.083	27.64
TV+1/2F	5.49	1.14	0.084	21.46
PP+1/2F	5.64	0.98	0.073	23.73
		By landscape		
Dambo	5.49	1.29	0.085	5.16[a]
Dambo margin	5.58	0.90	0.068	15.64[a]
Hillside	5.76	1.13	0.082	50.55[b]

Note: Lowercase letters indicate significant differences based on one-way ANOVAs ($P < 0.05$ Tukey's post hoc test).

in soil percentage C. Even without changes in total percentage C or percentage N, it is possible that changes in more labile fractions of SOM may have occurred as discussed previously. We attempted to isolate the more labile light fraction of the SOM using a size density fractionation method (Meijboom et al., 1995; Barrios et al., 1996) but were unable to isolate sufficient quantities of soil from the initial size fractionation to complete the density fractionation, despite using the recommended 500 g of soil.

Some studies have shown benefits of legume incorporation on soil P availability due to increased mineralization of organic P (Randhawa et al., 2005) or solubilization of inorganic P fixed by Fe and Al in the soil (Mweta et al., 2007). Here, no legume effects on available

Table 9.7 Simple Regressions of Long-Term Soil Quality Indicator Variables and Maize Yields for D1 and D2

	D1				D2			
Soil indicators	Beta	R^2	*P* value	*N*	Beta	R^2	*P* value	*N*
pH	0.150	0.000	NS	23	–0.261	0.068	NS	24
Extract P (ppm)	–0.449	0.164	0.032*	23	0.097	0.009	NS	24
%C-WS	0.930	0.090	NS	38	–0.237	0.056	NS	34
%N-WS	0.085	0.007	NS	38	–0.180	0.032	NS	34

Note: Beta, slope of the regression line when predictor and independent variables are standardized (indicates the direction of the relationship); R^2, proportion of variability in the dependent variable attributable to the regression equation; *P* value, significance level, with $^*P < 0.05$, $^{**}P < 0.01$, and $^{***}P < 0.001$. *N*, total number of cases; NS, not significant; WS, whole soil; LF, light fraction.

Bray-extractable soil P were observed; however, soil P was highly variable across sites and time, with averages across treatment and year ranging between 2 and 51 ppm (Table 9.6). The trend was for the dambo to have the lowest levels and the hillside the highest, with the dambo margin intermediate (Tables 9.5 and 9.6). Soil P was significantly lower in the dambo than the other landscapes in 2000 and lower in the dambo and dambo margin than the hillside in 2004 (Table 9.6). Low extractable P levels in the dambo (Table 9.3) may be consistent with P fixation by iron and aluminum complexes, which occurs as seasonally flooded soils dry (PPI, 2005). Soil P may have built up on the hillside due to low maize productivity (Figures 9.2 and 9.3), the practice of creating pockets of soil between rocks for planting maize, or related to landscape geology, but more extensive sampling is needed to determine how robust and widespread this pattern is. Harawa et al. (2006) performed research in a nearby location and found P levels to be lowest in the dambo and hillside and highest in the dambo margin. Snapp et al. (1998), however, found a very high level of variability in soil P among smallholder farms in Malawi.

9.7.2 *Distributional economic analysis*

Yields varied depending on the wealth status of the farmers (Table 9.1), with a strong trend for wealthier farmers to have the highest yields in most cropping systems, and significantly higher yields in 1999, and in the fully fertilized treatments in 2003 (*t* tests $P < 0.05$). This difference likely reflected the preponderance of poor farmers cultivating on the marginal hillside soils, the frequency of poorer farmers selling their labor during critical periods in the maize cropping cycle (Sirrine, Shennan, Snapp, et al., 2010; Alwang and Siegel, 1999), and differences in prior field management practices, including history of fertilizer use. Kamanga, Waddington, et al. (2010) also found that better-resourced farmers in central Malawi experimenting with maize-legume intercrops had higher yields than poorly resourced farmers, attributing differences to disparate field management practices prior to the project's inception.

Given that the wealthier farmers could afford to wait for higher prices when selling maize (see Sirrine, Shennan, Snapp, et al., 2010b), these differences in yield resulted in even greater disparities in profitability for each system for wealthy versus poor farmers (Table 9.1). Design 1 legume-based systems were generally more profitable than the sole maize systems receiving equivalent fertilizer quantities, with pigeon pea systems typically the most profitable. Under D2, the legume effect could not be isolated, but the different legume plus half fertilizer treatments were more profitable than the unfertilized control and less than the full fertilizer treatment (Table 9.1). In a study conducted in the area, Snapp et al. (2010) ran an economic analysis that varied fertilizer input prices and found that as input prices increased, diversified legume-based systems maintained their value, while the profitability of fertilized maize monocultures decreased. Interestingly, the profitability of D1 fertilizer-based systems was diminished for the poorer farmers compared to wealthier ones (Table 9.1), and a similar pattern was seen for fertilized maize in 2004. This raises concerns about the distributional impacts of fertilizer subsidies for farmers from different resource groups.

9.7.3 Legume system adoption and preference

A detailed discussion of adoption and preference for the different legume systems can be found in the work of Sirrine, Shennan, and Sirrine (2010). Here, we present a brief synopsis of the findings to illustrate the importance of addressing the socioeconomic context when developing and recommending alternative production strategies. In 2003, farmers were asked which of the three legume species they preferred to intercrop with maize. The majority of farmers (55%) stated a preference for pigeon pea, primarily due to its versatility. Farmers valued it as a secondary food source, as a source of firewood, and for soil improvement. The farmers who preferred *T. vogelii* (26%) referred to its ability to improve soil quality, its secondary use as a fish poison, and its low labor demand. The remaining 19% preferred *S. sesban* for its capacity to enhance soil quality, increase maize yields, and perceived larger growth compared to the other two legumes.

We also asked farmers what they actually planted on their own land to determine if stated preferences were reflected in actual adoption. At the onset of the project, essentially all farmers recalled growing some pigeon pea, and 16% grew some *T. vogelii* (none grew *S. sesban*). Farmer recall data on pigeon pea presence and *S. sesban* absence coincide with that reported in the regional literature. Regional cropping history of *T. vogelii* is less well known in terms of quantity, although interviews consistently clarified it had been used in the region as a fish poison for quite some time. In 2003, pigeon pea was still grown on 97% of farms, and *T. vogelii* planting had risen to 20% (Figure 9.5). Only 6% of farmers had adopted *S. sesban* by 2003, with the additional labor demands (including growing it in a nursery, transplanting it during a time of peak agricultural labor needs, and cutting the larger trunks it tended to produce) and poorly understood germination requirements being stated as impediments to its adoption.

Patterns of both preference and adoption varied among farmers depending on wealth ranking, gender, landscape position, and other factors. We hypothesized that the poorest farmers would have the strongest preference for pigeon pea due to the immediate livelihood benefits it had as a food source, but while the poorest farmers did have a strong preference for pigeon pea, farmers in the middle-income bracket had an even stronger preference (Figure 9.5). Interestingly, 40% of the poorest farmers preferred *T. vogelii* due to the immediate livelihood benefit it offers by selling or using the biomass as a fish poison. Clearly, short-term considerations were critical to the poorest farmers who only planted pigeon pea or *T. vogelii* on their own land (Figure 9.5). Bezner Kerr et al. (2007) also found that immediate food security concerns were more influential than soil quality in dictating northern Malawian farmers' cropping system preference. Here, the wealthiest farmers were most likely to prefer, and the only ones to adopt, *S. sesban*, indicating they were best positioned to accommodate the higher labor requirements necessary to benefit from yield gains and the longer-term soil quality improvement likely to accrue over a number of years (Franzel and Scherr, 2002).

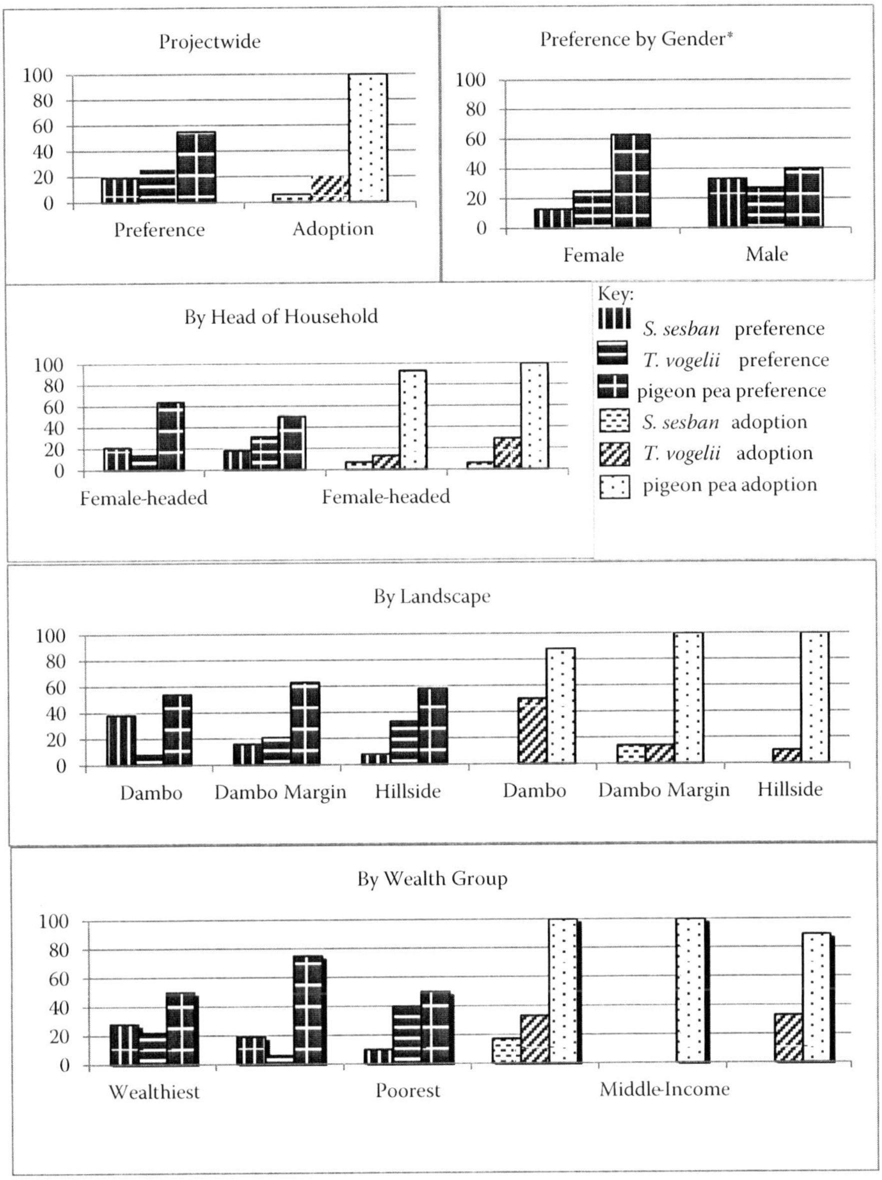

Figure 9.5 Percentage farmer adoption and preference of the legume species in 2003, both projectwide and disaggregated by gender, head of household, wealth rank, and landscape. (Data from Sirrine, D., C. Shennan, and J.R. Sirrine. 2010. Comparing agroforestry systems' ex ante adoption potential and ex post adoption: on-farm participatory research from southern Malawi. *Agroforest. Syst.* 79(2):253–266.)

We found that pigeon pea was often stereotyped as women's and children's food in this region, so it was not surprising that females were more likely to prefer pigeon pea than males (Figure 9.5). Women spoke of its role as an additional food source, and a few mentioned the ability to sell excess seed and retain the money themselves for household purchases. Conversely, many men said they did not like to consume pigeon pea but rather preferred to consume meat and fish. FHHs were also more likely to prefer pigeon pea than were male-headed households (MHHs) (Figure 9.5). Indeed, FHHs are disproportionately poor (NEC, 2000; Simtowe, 2010; Takane, 2009) and likely in greater need of short-term livelihood support in the form of food. Nonetheless, although many men voiced their dislike of pigeon pea's taste, all MHHs planted pigeon pea (Figure 9.5). Some FHHs had a difficult time saving seeds for subsequent planting due to food insecurity, which may explain why a few FHHs did not plant pigeon pea in 2003. Likewise, Ferguson (1994) found that Malawian land-limited, resource-poor females struggled to save seed from season to season as they either consumed it or sold it shortly after harvest.

We had expected pigeon pea preference to be highest on the hillside, the landscape with the highest concentration of the poorest farmers. In actuality, pigeon pea preference was highest at the dambo margin and slightly lower for farmers on the hillside and dambo (Figure 9.5). Hillside farmers explained that baboons from the adjacent forest stole pigeon pea from their fields, and as a result they had low yields, making *T. vogelii* more attractive. Nonetheless, pigeon pea adoption was still much higher than *T. vogelii* and *S. sesban* across all three landscapes (Figure 9.5), suggesting that it continues to play an important role for hillside farmers despite theft by baboons.

9.8 Assessing sustainability

Using the framework outlined, we can examine sustainability as the ability to meet a combination of goals, including immediate-term livelihood needs, food/income stability over time, and long-term improvement of soil quality. In addition, the distributional dimension to sustainability must be considered since the sustainability of each system clearly depends on the socioeconomic status of the farmers, their gender, and their position across the landscape.

Short-term livelihood concerns relate not only to maize yields but also to immediate benefits such as provision of a secondary food or income source, as is the case with pigeon pea and *T. vogelii*, and untimely labor requirements associated with managing *S. sesban*, which limited its use to a few wealthier farmers. In terms of short-term livelihoods, for the most vulnerable farmers the unfertilized pigeon pea system provided the best option among the D1 treatments in the period 1999–2000 based on its high returns and provision of a secondary food or income source for women. The fertilized *T. vogelii* system gave similar returns, but it is unlikely that the poorest farmers would be able to purchase the necessary amount of fertilizer. For the wealthier farmers, fertilized pigeon pea gave the highest returns in both years, but all fertilized systems performed well.

Among the D2 treatments, the highest returns even for the poorest farmers came from the fully fertilized system, which appears to represent the best option but may be unrealistic for most farmers, especially poorer farmers, for two reasons. First, until the government's recent fertilizer subsidy programs, initiated in 2005, only roughly 25% of smallholders in southern Malawi could afford to purchase inorganic fertilizers (Minot et al., 2000), and many applied quantities far smaller than the recommended rate (Dorothy Sirrine, personal observation, 2002–2004). Second, southern Malawian smallholders, even wealthier smallholders, rarely monocrop maize due to limited landholdings. Thus, for the poorest farmers either *T. vogelii* plus half fertilizer or pigeon pea plus half fertilizer was a promising option, providing they could afford a smaller amount of fertilizer. While *S. sesban* plus half fertilizer gave good returns, poorer farmers would be unlikely to have the necessary additional labor available for growing *S. sesban* since poorer households in general sell their own labor, especially during the peak agricultural season, as a livelihood strategy (Alwang and Siegel, 1999). This is especially true for many FHHs, further reducing the time they can dedicate to their own plots of land (Simtowe, 2010; Takane, 2009).

Interestingly, there is evidence that in addition to providing short-term livelihood benefits, legumes help reduce the risk of very low yields. Sirrine, Shennan, Snapp, et al. (2010) measured risk of low yields as 75% lower confidence limits (LCLs), which is measured as a one-in-four chance of yields falling below this level. Systems receiving fertilizer had higher LCLs than those with no fertilizer, but notably, they also found that the maize/legume intercrops had consistently higher LCLs than the equivalent nonlegume controls. That is, in both the presence and absence of fertilizer, legumes reduced the risk of low yields. Similar patterns of relative risk have been found in two other studies in Malawi. Kamanga et al. (2009) also looked at risk using LCLs and found that addition of fertilizer reduced risk, but that maize intercropped with pigeon pea had the lowest risk of any system. In another study, spatial variability in yields was consistently higher in unfertilized than in fertilized maize (Snapp et al., 2010), and yet superior yield stability as measured by coefficients of variation was observed in maize/shrubby grain legume rotations. Greater yield stability with legumes also translates into more stability of returns since using legumes either avoids (if no fertilizer is used) or reduces the vulnerability of farmers to fluctuations in fertilizer pricing (Snapp et al., 2010).

Overall, from livelihood, maize yield, and risk perspectives, we found that relay intercropping pigeon pea with maize offered the most sustainable and low-risk, low-cost option for the poorest farmers to improve production and food supply. If fertilizer is available at low cost, even limited amounts could improve the pigeon pea system for these farmers. In contrast, the wealthier farmers have many more options. They are well positioned to benefit from the fertilizer and legumes due to their flexible marketing strategies, higher-quality landholdings, access to labor, and the ability to afford inputs. Based on net returns, under D1 the full fertilizer, followed by *S. sesban* plus full fertilizer, were the most promising systems, whereas under D2 the fully fertilized maize gave the highest returns. When the reduced risk associated with legumes is taken into account, as well as their soil improvement potential, any of the legumes together with fertilizer are good options for wealthier farmers. Again, pigeon pea with full fertilizer may be the best option given the value placed on it by women and the potential for women to gain control over any money generated from pigeon pea sales—a point that was raised in our focus groups, where women spoke of the importance of having income that they could use to purchase household necessities.

In terms of long-term soil quality improvement, we were unable to detect significant changes in soil percentage C or percentage N over time under any of the systems tested, so we cannot use these indicators to differentiate among the systems. More nuanced indicators of nutrient cycling and SOM dynamics were needed to better assess legume impacts on soil fertility. One challenge is coming up with indicators, especially biological indicators that are low cost and feasible, to measure at multiple field sites distant from laboratory or cooling/freezing facilities, as in this case. Yet, it is critical to collect this kind of data under realistic field conditions and not only at experiment stations, where legume productivity tends to be much higher (Mafongoya et al., 2006). Moebius-Clune et al. (2011) argued for a suite of relatively low-cost soil quality indicators that reflect a combination of soil physical, chemical, and biological properties, each linked to important ecosystem processes. These include a range of macro- and micronutrient levels, water-stable aggregates, available water capacity, penetrometer resistance, and biologically active soil carbon measured with a very dilute potassium permanganate method and a handheld colorimeter. Penetrometer resistance was measured in the field, and the remaining measurements were made using sieved dried soil, making them appropriate for sampling large numbers of fields in distant sites. It would be interesting to see if this suite of indicators effectively captures management impacts on soil quality across a range of locations and if the permanganate measure proves to be a good indicator of soil biological changes.

It is interesting to note that there is close correspondence between the systems recommended and what farmers actually adopted during the course of the study. Essentially all farmers planted some pigeon pea with their maize, about 20% planted *Tephrosia,* and only 6%, all wealthier farmers, planted *Sesbania.* Overall, our results concur with others (Snapp et al., 2010; Kamanga, Waddington, et al., 2010; Kanyama-Phiri et al., 1998), which shows that legumes could play an important role in increasing sustainability of maize production systems in Malawi, but that the benefits of using legumes were greatly enhanced by their use in combination with moderate amounts of inorganic fertilizer (Table 9.1, Figure 9.2). The higher productivity under D1 as compared to D2 suggests that farmers should ideally use as much fertilizer as they can afford up to the recommended amount and, if possible, apply it in two applications as done in D1. However, it is unclear at this point when to recommend application of more limited amounts of fertilizer or indeed at what stage it is best to incorporate the legume biomass. Our understanding of the dynamics of N release from different age legume residues is currently limited, as is our understanding of effects of amounts and timing of fertilizer application on growth and nitrogen fixation by different legume species.

Legume productivity was less than optimal in these on-farm trials given the low biomass obtained (Table 9.2) relative to other values in the literature, suggesting there is considerable potential to further improve system productivity. There is a real opportunity for researchers or extension agents to work with farmers in an iterative and adaptive way to optimize legume management for different landscape positions and resource levels (Shennan, 2008). Research is needed on basic agronomic questions (optimizing seeding rates, planting arrangements, fertilizer amounts and timing, timing of biomass incorporation, etc.), assessment of rhizobium levels and BNF, and the use of improved legume and maize cultivars as suggested by Snapp, Blackie, et al. (2003). Work is also needed to better understand legume and fertilizer effects on labile organic matter pools, microbial activity, and patterns of mineralization/immobilization from different age residues to optimize legume management.

Focus groups held at the end of the project showed that farmers had ideas for improving legume management and suggestions to make the system work better in terms of timing of labor demands. Farmers frequently suggested earlier incorporation of the legumes so that the biomass was further along in decomposition and would not interfere with land preparation (creating ridges) for the upcoming cropping season. Sirrine, Shennan, and Sirrine (2010) noted that this would require farmers to forgo a small second harvest of dry pigeon pea but would also incorporate residue earlier before the leaves begin to senesce, perhaps minimizing N loss from decomposing leaves. Farmers also suggested that less woody material be incorporated because they felt it decomposed too slowly, which was primarily an issue for *S. sesban,* which has smaller leaves and stems that were more challenging to thresh. As discussed, changing the timing and composition of residue incorporation will impact N release dynamics and perhaps increase vulnerability to leaching losses. Reducing total biomass inputs by not incorporating woody material may also have an impact on long-term SOM changes. These trade-offs would need to be evaluated along with the economics of the different incorporation strategies.

In many on-farm projects, farmer involvement is limited, as in this one, due to the trade-off between including farmers in all facets of cropping system management and maintaining standard management practices across farmers' treatments. With hindsight, we think that greater farmer involvement in project management could have provided results that reflected realistic farmer management strategies and provided more opportunities for social learning to occur among researchers and farmers during the project. Other work has demonstrated the benefits of facilitating farmer innovation in adapting legume (Snapp, Jones, et al., 2003), alley cropping (Kanmegne and Degrande, 2002), and tree mulch (Stoate and Jarju, 2008) systems to better meet their needs and constraints.

In terms of social learning, it would also have been beneficial for farmers to be apprised of the research results at regular intervals throughout the duration of our on-farm projects. We held focus groups at the end of the project to present findings to farmers; these findings were understood well and resulted in many additional suggestions and invaluable feedback. A subsequent training session on legume management clearly demonstrated missed opportunities to both educate and collaborate with farmers on improved legume management.

Nonetheless, farmers provided us with a vivid picture of the limitations and constraints they face when trying to adopt and continue the legume systems being tested. This feedback was extremely valuable in helping us understand how to improve the sustainability of the current cropping systems and underscored the socioeconomic and cultural heterogeneity that will determine whether the next green revolution attempts will be met with success or failure. The aim of tripling cereal grain yields to achieve a green revolution (Sanchez, 2010) and improving soil fertility appears to make sense. However, in the case of FHHs and poorer households in southern Malawi, a strong focus on improving maize yields without an understanding of how diversified cropping systems with pigeon pea can contribute to food security and reduce risk might overlook a critical goal: that of alleviating hunger for the most vulnerable populations. That is, it is essential that a new green revolution be comprehensive and inclusive, encompassing considerations of food security, vulnerability, nutrition, and gender dimensions (Negin et al., 2009).

We cannot emphasize enough the importance of careful investigation of the distributional impacts of proposed changes to production systems, coupled with an understanding of the need to balance short- and long-term sustainability goals. This is particularly true for the neediest farmers, often FHHs. In this regard, the frequently diminished effectiveness of fertilizer on maize yields and profitability for the poorest farmers (Table 9.1) raises concerns about the impacts of fertilizer subsidies for Malawi's neediest farmers. Moreover, inorganic fertilizer prices are both volatile and increasing in the world market (Woods et al., 2010). Malawi's current input subsidy program represents a substantial part of the national budget (Snapp et al., 2010) and may not be sustainable even into the immediate future. While legume/maize relay crops benefited from the addition of inorganic fertilizers (Table 9.1), given the important role legumes played in production risk reduction, household food security, and maintaining maize yields, increased emphasis should also be placed on improving their use and effectiveness through agronomic and soil microbiological research, extension, and potentially even subsidies for acquiring legume seeds.

The conceptual framework we present for evaluating cropping system sustainability emphasizes the importance of identifying trade-offs across time frames and distributional impacts of modifying cropping systems on those potentially impacted. It can accommodate assessment of spatial impacts and could be applied at larger spatial scales than used here. The fundamental framework could equally well be applied in developed nations, using appropriate measures of differential social and biophysical vulnerability within and among communities and incorporating additional indicators of environmental impacts as sustainability measures. As in the developing world, a nuanced understanding of landscape, social, and cultural heterogeneity will also influence the ability to transition to more sustainable cropping systems.

9.9 Conclusion

There is ample evidence that the use of legumes, ideally in conjunction with some fertilizer, provides significant benefits in terms of improved crop productivity in a variety of sub-Saharan Africa cropping systems. Less widely documented are measures of changes in different aspects of soil quality, especially in terms of impacts on soil C and N pools and microbial ecology, and how these relate to crop growth and nutrient cycling. It is also apparent that there is great variability in the performance of legumes in different systems and locations, leaving significant opportunities for improved management to increase BNF and biomass production. Improved understanding of residue quality and decomposition dynamics would also help to synchronize N release dynamics better with crop demand.

The question of how to assess the relative sustainability of different management options was discussed, framing sustainability as the ability to meet a combination of goals, including immediate-term livelihood needs, food/income stability over time, and long-term improvement of soil quality. This framing makes it clear that in addition to understanding system impacts on the biophysical components of an agroecosytem, it is critical to consider the socioeconomic and cultural contexts where a system is targeted. These arguments were illustrated through a case study of legume relay cropping in Malawi.

The case study provided a realistic assessment of the performance and sustainability of different fertilizer and legume relay intercropped systems by testing them across a representative range of smallholder farms in this region of Malawi. We used a framework that examined sustainability as a combination of short-term livelihood benefits measured as crop yields and net returns, farmer preference, adoptability, risk, and longer-term soil fertility impacts, although we were limited in our ability to differentiate among the treatments in terms of changes in soil fertility. While presenting some logistical challenges, the project design allowed us to consider the effects of different landscape positions, resource endowments, and gender on the desirability and sustainability of each system tested. By considering variability across time and space, we obtained information on the risks of low yields associated with each system. The incorporation of surveys, interviews, participant observation, and focus groups provided insights on the socioeconomic and cultural realities of the participating farmers and their opinions of the systems being tested and enabled a distributional assessment of performance to be made.

We found that relay intercropping maize with pigeon pea offered the most sustainable low-cost, low-risk option of the systems tested for improving food production and net income for the poorest farmers. However, further improvements are needed to move these farmers to greater food security. In contrast, any of the legumes plus a moderate amount of fertilizer offered higher returns and benefits for wealthier farmers. Pigeon pea, however, has the advantage of being a secondary food and a potentially valuable source of income specifically for women.

For this, and other legume-based systems, a focus on improving legume management to increase growth and BNF together with a better understanding of the dynamics of residue decomposition, nutrient cycling, and SOM changes could greatly enhance system productivity and sustainability. More effective use of legumes can help reduce risk and vulnerability to fertilizer price fluctuations by reducing fertilizer requirements and improving soil quality over the long term. The choice of systems to recommend needs to reflect socioeconomic and cultural considerations, as well as biophysical performance, and take into account food security and livelihood needs of farmers with different resource levels.

References

Adams, A.M., T.G. Evans, R. Mohammed, and J. Farnsworth. 1997. Socioeconomic stratification by wealth ranking: is it valid? *World Dev.* 25(7):1165–1172.

Adger, N. 1999. Social vulnerability to climate change and extremes in coastal Vietnam. *World Dev.* 27:249–269.

Adu-Gyamfi, J., F.A. Myaka, W.D. Sakala, R. Odgaard, J.M. Vesterager, and H. Hogh-Jensen. 2007. Biological nitrogen fixation and nitrogen and phosphorus budgets in farmer-managed intercrops of maize-pigeonpea in semi-arid southern and eastern Africa. *Plant and Soil* 295:127–136.

African Farmers Organization. 2009. African Farmers Organization declaration on "The GCARD regional face to face consultation" held in Accra, Ghana. http:/gcardblog.wordpress.com/2009/10/13/ssa17/.

African Green Revolution. 2009. AGR conferences. http://www.africangreenrevolution.com/en/conferences/index.html.

African Union Report. 2008. *Progress Report on Implementing the Comprehensive Africa Agriculture Development Programme: Agricultural Growth, Poverty Reduction and Food Security in Africa.* http://www.africa-union.org/root/ua/Conferences/2008/avril/REA/01avr/Experts_CAADP_Progress.pdf

Akinnifesi, F.K., O.C. Ajayi, G. Sileshi, P.W. Chirwa, and J. Chianu. 2010. Fertilizer trees for sustainable food security in the maize-based production systems of East and Southern Africa. A review. *Agron. Sustain. Dev.* 30:615–629.

Akinnifesi, F.K., W. Makumba, G. Sileshi, O. Ajayi, and D. Mweta. 2007. Synergistic effects of inorganic N and P fertilizers and organic inputs from *Gliricidia sepium* on productivity of intercropped maize in southern Malawi. *Plant Soil* 294:203–217.

Alvey, S., M. Bagayoko, G. Neumann, and A. Buerkert. 2001. Cereal/legume rotations affect chemical properties and biological activities in two West African soils. *Plant Soil* 231:45–54.

Alwang, J., and P.B. Siegel. 1999. Labor shortages on small landholdings in Malawi: implications for policy reforms. *World Dev.* 27(8):1461–1475.

Ashby, J.A., J.A. Beltran, M.D.P. Guerrero, and H.F. Ramos. 1996. Improving the acceptability to farmers of soil conservation approaches. *J. Water Soil Conserv.* 51(4):309–314.

Bagayoko, M., A. Buerkert, G. Lung, A. Bationo, and V. Romheld. 2000. Cereal/legume rotation effects on cereal growth in Sudano-Sahelian West Africa: soil mineral nitrogen, mycorrhizae and nematodes. *Plant Soil* 218:103–116.

Baijukya, F.P., N. de Ridder, and K. Giller. 2006. Nitrogen release from decomposing residues of leguminous cover crops and their effect on maize yield on depleted soils of Bukoda District, Tanzania. *Plant Soil* 279:77–93.

Banda, A.Z., J.A. Maghembe, D.N. Ngugi, and V.A. Chome. 1994. Effect of intercropping maize and closely spaced leucaena hedgerows on soil conservation and maize yield on a steep slope at Ntcheu, Malawi. *Agroforest. Syst.* 27:17–22.

Barrios, E., R.J. Buresh, and J.I. Sprent. 1996. Organic matter in soil particle size and density fractions from maize and legume cropping systems. *Soil Biol. Biochem.* 28(2):185–193.

Barthes, B., A. Azontonde, E. Blanchart, C. Girardin, C. Villenave, S. Lesaint, R. Oliver, and C. Feller. 2004. Effect of a legume cover crop (*Mucuna puriens* var *utilis*) on soil carbon in an ultisol under maize cultivation in southern Benin. *Soil Use Manage.* 20:231–239.

Bationo, A., and A. Buerkert. 2001. Soil organic carbon management for sustainable land use in Sudano-Sahelian West African. *Nutr. Cycl. Agroecosyst.* 61:131–142.

Bationo, A., J. Kihara, B. Vanlauwe, B. Waswa, and J. Kimetu. 2007. Soil organic carbon dynamics, functions and management in West African agroecosystems. *Agric. Syst.* 94:13–25.

Beedy, T.L., S.S Snapp, F.K. Akinnefesi, and G.W. Sileshi. 2010. Impact of *Gliricidia sepium* intercropping on soil organic matter fractions in a maize-based cropping system. *Agric. Ecosyst. Environ.* 138:139–146.

Bekunda, M., N. Sangina, and P.L. Woomer. 2010. Restoring soil fertility in sub-Saharan Africa. *Adv. Agron.* 108:183–236.
Bezner Kerr, R. 2005. Informal labor and social relations in northern Malawi: the theoretical challenges and implications of ganyu labor for food security. *Rural Sociol.* 70:167–187.
Bezner Kerr, R., S. Snapp, M. Chirwa, L. Shumba, and R. Msachi. 2007. Participatory research on legume diversification with Malawian smallholder farmers for improved human nutrition and soil fertility. *Expl. Agric.* 43:37–453.
Bossio, D.A., M.S. Girvan, L. Verchot, J. Bullimore, T. Borelli, A. Albrecht, K.M. Scow, A.S. Ball, J.N. Pretty, and A.M Osborn. 2005. Soil microbial community response to land use change in an agricultural landscape of Western Kenya. *Microb. Ecol.* 49:50–62.
Boye, K., J. Eriksen, S.I. Nilsson, and L. Mattsson. 2010. Sulfur flow in a soil-plant system—effects of long-term treatment history and soil properties. *Plant Soil* 334:323–334.
Bunderson, W.T., F. Bodnar, W.A. Bromley, and S.J. Nanthambwe. 1995. *A Field Manual for Agroforestry Practices in Malawi.* Publication No. 6. Malawi Agroforestry Extension Project, Lilongwe, Malawi.
Center for Regional Agricultural Trade Expansion Support (RATES). 2003. *Maize Market Assessment and Baseline Study for Malawi.* RATES Center, Nairobi, Kenya.
Chambers, R. 1995. Poverty and livelihoods: whose reality counts? *Environ. Urbanization* 7:173–204.
Chambers, R., and G.R. Conway. 1992. Sustainable rural livelihoods: practical concepts for the 21st century. IDS Discussion Paper 296. Institute of Development Studies, Brighton, UK.
Chinsinga, B. 2005. The clash of voices: community-based targeting of safety-net interventions in Malawi. *Social Policy and Administration* 39(3):284–301.
Chirwa, P.W., C.R. Black, C.K. Ong, and J.A. Maghembe. 2003. Tree and crop productivity in *gliricidia*/maize/pigeonpea cropping systems in southern Malawi. *Agroforest. Syst.* 59:265–277.
Chirwa, P.W., C.R. Black, C.K. Ong, and J.A. Maghembe. 2006. Nitrogen dynamics in cropping systems in southern Malawi containing *Gliricidia sepium*, pigeonpea and maize. *Agroforest. Syst.* 67:93–106.
Chivenge, P., B. Vanlauwe, R. Gentile, and J. Six. 2011b. Comparison of organic versus mineral resource effects on short-term aggregate carbon and nitrogen dynamics in a sandy soil versus a fine textured soil. *Agric. Ecosyst. Environ.* 140:361–371.
Chivenge, P., B. Vanlauwe, and J. Six. 2011a. Does the combined application of organic and mineral nutrient sources influence maize productivity? A meta analysis. *Plant Soil* 342:1–30.
Denning, D., P. Kabambe, P. Sanchez, A. Malik, R. Flor, R. Harawa, P. Nkhoma, C. Zamba, C. Banda, C. Magombo, M. Keating, J. Wangila, and J. Sachs. 2009. Input subsidies to improve smallholder maize productivity in Malawi: toward an African green revolution. *Plos. Biol.* 7:e1000023.
Department for International Development (DFID). 2001. *Sustainable Livelihoods Guidance Sheets.* DFID, London.
Diamond, D. 1995. *Phosphorus in Soil Extracts.* Quickchem Method 10-115-01-1-A. Lachat Instruments, Milwaukee, WI.
Diao, X., P. Hazell, D. Resnick, and J. Thurlow. 2007. The role of agriculture in development: implications for sub-Saharan Africa. Research Report 153. IFPRI, Washington, DC.
Ellis, F. 1998. Household strategies and rural livelihood diversification, *J. Dev. Stud.* 35(1):1–38.
Eswaran, H., R. Almaraz, E. Van den Berg, and P. Reich. 1996. *An Assessment of the Soil Resources of Africa in Relation to Productivity.* USDA NRCS, Washington, DC.
European Technology Assessment Group. 2009. *Agricultural Technologies for Developing Countries.* Brussels, Belgium. http://www.itas.kit.edu/downloads/etag_meye09a.pdf
Ferguson, A. 1994. Gendered science: a critique of agricultural development. *Am. Anthropol.* 96:540–552.
Flora, C. 2010. Food security in the context of energy and resource depletion: sustainable agriculture in developing countries. *Renew. Agric. Food Syst.* 25:118–128.
Food and Agricultural Organization (FAO). 2006. *The State of Food Insecurity in the World 2006.* FAO, Rome.
Franke, A.C., G. Laberge, B.D. Oyewole, and S. Schulz. 2008. A comparison between legume technologies and fallow, and their effects on maize yields and soil traits, in two distinct environments of the West African Savannah. *Nutr. Cycl. Agroecosyst.* 82:117–135.
Franzel, S., and S.J. Scherr. 2002. Introduction. In *Trees on the Farm: Assessing the Adoption Potential of Agroforestry Practices in Africa*, eds. S. Franzel and S.J. Scherr. CABI, Wallingford, UK, 1–10.

Gentile, R., B. Vanlauwe, P. Chivenge, and J. Six. 2011. Trade-offs between the short- and long-term effects of residue quality on soil C and N dynamics. *Plant Soil* 338:159–169.

Gentile, R., B. Vanlauwe, A. Kavoo, P. Chivenge, and J. Six. 2010. Residue quality and N fertilizer do not influence aggregate stabilization of C and N in two tropical soils with contrasting texture. *Nutr. Cycl. Agroecosyst.* 88:121–131.

Giller, K.E., and G. Cadisch. 1995. Future benefits from biological nitrogen fixation: an ecological approach to agriculture. *Plant Soil* 174:255–277.

Giller, K.E., P. Titonnell, M.C. Rufino, et al. 2010. Communicating complexity: integrating assessment of trade-offs concerning soil fertility management within African farming systems to support innovation and development. *Agric. Syst.* 104:191–203.

Giller, K.E., E. Witter, M. Corteels, and P. Tittonell. 2009. Conservation agriculture and smallholder farming in Africa: the heretics' view. *Field Crops Res.* 114:23–34.

Gowing, J.W., and M. Palmer. 2008. Sustainable agricultural development in sub-Saharan Africa: the case for a paradigm shift in land husbandry. *Soil Use Manage.* 24:92–99.

Harawa, R., J. Lehmann, F. Akkinifesi, E. Fernandes, and G. Kanyma-Phiri. 2006. Nitrogen dynamics in maize-based agroforestry systems as affected by landscape position in southern Malawi. *Nutr. Cycl. Agroecosyst.* 75:271–284.

Hardarson, G., and C. Atkins. 2003 Optimizing biological N_2 fixation by legumes in farming systems. *Plant Soil* 252:41–54.

Haynes, R.J. 2008. Soil organic matter quality and the size and activity of the microbial biomass: their significance to agricultural soil quality. In *Soil Mineral-Microbe-Organic Interactions: Theories and Applications*, eds. Q. Huang, P.K. Huang, and A. Violante, 201–231. Springer, New York.

InterAcademy Council. 2004. *Realizing the Promise and Potential of Africa Agriculture*. Amsterdam, the Netherlands. http://www.interacademycouncil.net/24026/AfricanAgriculture.aspx

International Assessment of Agricultural Knowledge, Science and Technology for Development (IAASTD). 2009. *Summary for Decision Makers of the Global Report*. IAASTD, Washington, DC.

Joergensen, R.G. 2010. Organic matter and microorganisms in tropical soils. In *Soil Biology and Agriculture in the Tropics*, Soil Biology Series, Vol. 21. ed. P. Dion. Springer, New York, 17–44.

Jones, J.B., and H.V. Eck. 1973. Plant analysis as an aid in fertilizing corn and grain sorghum. In *Soil Testing and Plant Analysis*, eds. L.M. Walsh and J.B. Beaton, 349–364. SSSA, Madison, WI.

Kamanga, B.C.G., G.Y. Kanyama-Phiri, and S. Minae. 1999. Intercropping perennial legumes for green manure additions to maize in southern Malawi. *African Crop Sci. J.* 7:355–363.

Kamanga, B.C.G., S.R. Waddington, M.J. Robertson, and K.E. Giller. 2010. Risk analysis of maize-legume crop combinations with smallholder farmers varying in resource endowment in central Malawi. *Expl. Agric.* 46:1–21.

Kamanga, B.C.G., A. Whitbread, P. Wall, S.R. Waddington, and C. Almekinders. 2010. Farmer evaluation of phosphorus fertilizer application to annual legumes in Chisepo, Central Malawi. *Afr. J. Agric. Res.* 5(8):668–680.

Kanmegne, J., and A. Degrande. 2002. From alley cropping to rotational fallow: farmers' involvement in the development of fallow management techniques in the humid forest zone of Cameroon. *Agroforest. Syst.* 54:115–120.

Kanyama-Phiri, G.Y., S.S. Snapp, and S. Minae. 1998. Partnership with Malawian farmers to develop organic matter technologies. *Outlook Agr.* 27:167–175.

Kelly, P.M., and W.N. Adger. 2000. Theory and practice in assessing vulnerability to climate change and facilitating adaptation. *Climatic Change* 47(4):325–352.

Kone, A.W., J.E. Tondoh, P.K.T. Angui, F. Bernhard-Reversat, G. Loranger-Merciris, D. Brunet, and S.T.K. Bredoumi. 2008b. Is soil quality improvement by legume cover crops a function of initial soil chemical characteristics? *Nutr. Cycl. Agroecosyst.* 82:89–105.

Kone, A.W., J.E. Tondoh, F. Bernard-Reversat, G. Loranger-Merciris, D. Brunet, and Y. Tano. 2008. Changes in soil biological quality under legume-and maize-based farming systems in a humid savanna zone of Cote d'Ivoire. *Biotech. Agron. Soc. Environ.* 12:147–155.

Kwabiah, A.B., R.P. Voroney, C.A. Palm, and N.C. Stoskopf. 1999. Inorganic fertilizer enrichment of soil: effect on decomposition of plant litter under subhumid tropical conditions. *Biol. Fertil. Soils* 30:224–231.

Lal, R. 2006. Enhancing crop yields in the developing countries through restoration of the soil organic carbon pool in agricultural lands. *Land Degrad. Dev.* 17:197–209.

Lal, R. 2009. Enhancing eco-efficiency in agroecosystems through soil carbon sequestration. *Crop Sci.* 50:S120–S131.

Mafongoya, P.L., A. Bationo, J. Kihara, and B.S. Waswa. 2006. Appropriate technologies to replenish soil fertility in southern Africa. *Nutr. Cycl. Agroecosyst.* 76:137–151.

Makatiani, E.T., and D.W. Odee. 2007. Response of *Sesbania sesban* (L.) Merr. to rhizobial inoculation in an N-deficient soil containing low numbers of indigenous rhizobia. *Agroforest. Syst.* 70:211–216.

Makumba, W., F.K. Akkinifesi, and B.H. Janssen. 2009. Spatial rooting patterns of gliricidia, pigeon pea and maize intercrops and effect on profile soil N and P distribution in southern Malawi. *Afr. J. Agric. Res.* 4(4):278–288.

Meijboom, F.W., J. Hassink, and M. van Noordwijk. 1995. Density fractionation of soil macroorganic matter using silica suspensions. *Soil Biol. Biochem.* 27:1109–1111.

Meyer, G.A., and P.N. Keliher. 1992. An overview of analysis by inductively coupled plasma-atomic emission spectrometry. In *Inductively Coupled Plasmas in Analytical Atomic Spectrometry*, eds. A. Montaser and D.W. Golightly, 473–516. VCH, New York.

Meyer, R. 2009. Agricultural technologies for developing countries. Final report for European Technology Assessment Group. http://www.ernaehrungsdenkwerkstatt.de/fileadmin/user_upload/EDWText/TextElemente/Soziologie/TAB-ITAS_OTA/Agriculture_Technology_Developing_LDC_STOA_Report_2009.pdf.

Minot, N., M. Kherallah, and P. Berry. 2000. Fertilizer market reform and the determinants of fertilizer use in Benin and Malawi. Markets and Structural Studies [MSSD] Discussion Paper No. 40. International Food Policy Research Institute (IFPRI), Washington, DC.

Mishra, S., S. Sharma, and P. Vasudevan. 2008. Role of bioinoculants and organic fertilizers in fodder production and quality of leguminous tree species. *J. Environ. Biol.* 32:57–64.

Moebius-Clune, B.N., H.M. van-Es, O.J. Idowu, R.R. Schindelbeck, J.M. Kimetu, S. Ngoze, J. Lehmann, and J.M. Kinyangi. 2011. Long-term soil quality degradation along a cultivation chronosequence in western Kenya. *Agric. Ecosyst. Environ.* 141:86–99.

Morris, M., V.A. Kelly, R.J. Kopicki, and D. Byerlee. 2007. *Fertilizer Use in African Agriculture: Lessons Learned and Good Practice Guidelines*. World Bank, Washington, DC.

Mtambanengwe, F., and P. Mapfumo. 2006. Effects of organic resource quality on soil profile N dynamics and maize yields on sandy soils in Zimbabwe. *Plant Soil* 281:173–191.

Mtambanengwe, F., and P. Mapfumo. 2008. Smallholder farmer management impacts on particulate and labile carbon fractions of granitic sandy soils in Zimbabwe. *Nutr. Cycl. Agroecosyst.* 81:1–15.

Mucheru-Muna, M., P. Pypers, D. Mugendi, J. Kung'u, J. Mugwe, R. Merckx, and B. Vanlauwe. 2010. A staggered maize-legume intercrop arrangement robustly increases crop yields and economic returns in the highlands of Central Kenya. *Field Crops Res.* 115:132–139.

Mugwe, J., D. Mugendi, M. Mucheru-Muna, D. Odee, and F. Mairura. 2009. Effects of selected organic materials and inorganic fertilizer on the soil fertility of a humic nitisol in the central highlands of Kenya. *Soil Use Manage.* 25:434–440.

Murage, E.W., N.K. Karanji, P.E. Smithson, and P.L. Woomer. 2000. Diagnostic indicators of soil quality in productive and non-productive smallholders' fields of Kenya's Central Highlands. *Agric. Ecosyst. Environ.* 79:1–8.

Mweta, D.E., F.K. Akinnifesi, J.D.K. Saka, W. Makumba, and N. Chokotho. 2007. Use of pruning and mineral fertilizer affects soil phosphorus availability and fractionation in a gliricidia/maize intercropping system. *Afr. J. Agric. Res.* 2:521–527.

National Economic Council (NEC). 2000. *Profile of Poverty in Malawi, 1998—Poverty Analysis of the Malawi Integrated Household Survey, 1997–98*. Revised version. Poverty Monitoring System, Government of Malawi, Lilongwe.

National Research Council (NRC). 2008. *Emerging Technologies to Benefit Farmers in Sub-Saharan Africa and South Asia*. National Academies Press, Washington, DC.

National Research Council (NRC). 2010. *Sustainable Agricultural Systems in the 21st Century*. National Academies Press, Washington, DC.

Naudin, K., E. Goze, O. Balarabe, K.E. Giller, and E. Scopel. 2010. Impact of no tillage and mulching practices on cotton production in North Cameroon: a multi-locational on-farm assessment. *Soil Tillage Res.*108:68–76.

Ndufa, J.K., S.M. Gathumbi, H.W. Kamiri, K. Giller, and G. Cadisch. 2009. Do mixed-species legume fallows provide long-term maize yield benefit compared with monoculture legume fallows? *Agron. J.* 101:1352–1362.

Negin, J., R. Remans, S. Karuti, and J.C. Fanzo. 2009. Integrating a broader notion of food security and gender empowerment into the African green revolution. *Food Security* 1:351–360.

Norušis, M.J. 2003. *SPSS 12.0 Statistical Procedures Companion.* Prentice Hall, Englewood Cliffs, NJ.

Nyamadzawo, G., J. Nyamangara, P. Nyamugafata, and A. Muzulu. 2009. Soil microbial biomass and mineralization of aggregate protected carbon in fallow-maize systems under conventional and no-tillage in Central Zimbabwe. *Soil Tillage Res.* 102:151–157.

Nyamadzawo, G., P. Nyamugafata, R. Chikowo, and K. Giller. 2007. Residual effects of fallows on selected soil hydraulic properties in a kaolinitic soil subjected to conventional tillage and no tillage. *Agroforest. Syst.* 47:161–168.

Nziguheba, G., B. Tossah, J. Diels, A. Franke, K. Aihou, E. Iwuafor, C. Nwoke, and R. Merckx. 2009. Assessment of nutrient deficiencies in maize in nutrient omission trials and long-term field experiments in the West African savanna. *Plant Soil* 314:143–157.

Okalebo, J.R., C.O. Othieno, P.L. Woomer, et al. 2006. Available technologies to replenish soil fertility in East Africa. *Nutr. Cycl. Agroecosyst.* 76:153–170.

Oliver-Smith, A., and S.M. Hoffman. 2002. Introduction: why anthropologists should study disasters. In *Catastrophe and Culture,* eds. S.A. Hoffman and A. Oliver-Smith, 3–22. School of American Research Press, Santa Fe, NM.

Peters, P. 2006 Rural income and poverty in a time of radical change in Malawi. *J. Dev. Stud.* 42(2):322–345.

Phiri, A.D.K., G.Y.K. Kanyama-Phiri, and S. Snapp. 1999. Maize and sesbania production in relay cropping at three different landscape positions in Malawi. *Agroforest. Syst.* 47:153–162.

PPI (Potash and Phosphorous Institute). 2005. Dealing with seasonal variability in soil test results. Agri-Brief #5, Fall 2005. Norcross, GA. http://www.ipni.net/ppiweb/agbrief.nsf/5a4b8be72a35cd46852568d9001a18da/39518d8891eaa74f8525707000071d89!OpenDocument

Pretty, J. 2008. Agricultural sustainability: concepts, principles and evidence. *Phil. Trans. Roy. Soc. B* 363:447–465.

Pypers, P., M. Huybrighs, J. Diels, R. Abaidoo, E. Smolders, and R. Merckx. 2007. Does the enhanced P acquisition by maize following legumes in a rotation result from improved soil P availability? *Soil Biol. Biochem.* 39:2555–2566.

Rajendran, G., F. Sing, A.J. Desai, and G. Archana. 2008. Enhanced growth and nodulation of pigeon pea by co-inoculation of *Bacillus* strains with *Rhizobium* spp. *Bioresource Technol.* 99:4544–4550.

Randhawa, P.S., L.M. Condron, H.J. Di, S. Sinaj, and R.D. McLenaghen. 2005. Effect of green manure addition on soil organic phosphorus mineralisation. *Nutr. Cycl. Agroecosyst.* 73:181–189.

Sakala, W.D., G. Cadisch, and K.E. Giller. 2000. Interactions between residues of maize and pigeon-pea and mineral N fertilizers during decomposition and N mineralization. *Soil Biol. Biochem.* 32:679–688.

Sanchez, P.A. 2010. Tripling crop yields in tropical Africa. *Nature Geosci.* 3:299–300.

Sanginga, N. 2010. Innovations as key to the green revolution in Africa: exploring the scientific facts. *Nutr. Cycl. Agroecosyst.* 88:1–2.

Scherer, H.W. 2009. Sulfur in soils. *J. Plant Nutr. Soil Sci.* 172:326–335.

Sekhon, G.K., R.P. Gupta, M.S. Pandher, and J.K. Arora. 1992. Symbiotic effectiveness of HUP + rhizobium, VAM fungi and plant growth of *Cajanus cajan*. *Folia Microbiol.* 37:210–214.

Sekiya, N., and K. Yano. 2004. Do pigeon pea and sesbania supply groundwater to intercropped maize through hydraulic lift? Hydrogen stable isotope investigation of xylem waters. *Field Crop Res.* 86:167–173.

Shennan, C. 2008. Biotic interactions, ecological knowledge, and agriculture. *Phil. Trans. Roy. Soc. B* 363:717–739.

Sileshi, G., F.K. Akinnifesi, O.C. Ajayi, and F. Place. 2008. Meta-analysis of maize yield response to woody and herbaceous legumes in sub-Saharan Africa. *Plant Soil* 307:1–19.

Sileshi, G., and P.L. Mafongoya. 2006. Long term effect of legume-improved fallows on soil invertebrates and maize yields in eastern Zambia. *Agric. Ecosyst. Environ.* 115:69–78.

Simtowe, F.P. 2010. Livelihoods diversification and gender in Malawi. *Afr. J. Ag. Res.* 5:204–216.

Singh, C.S. 1996. Arbuscular mycorrhiza (AM) in association with *Rhizobium* sp. improves nodulation, N-2 fixation, and N utilization of pigeon pea (*Cajanus cajan*), as assessed with a N-15 technique, in pots. *Microbiol. Res.* 151:87–92.

Sirrine, D., C. Shennan, and J.R. Sirrine. 2010. Comparing agroforestry systems' ex ante adoption potential and ex post adoption: on-farm participatory research from southern Malawi. *Agroforest. Syst.* 79(2):253–266.

Sirrine, D., C. Shennan, S. Snapp, G. Kanyama-Phiri, B. Kamanga, and J.R. Sirrine. 2010. Improving recommendations resulting from on-farm research: agroforestry, risk, profitability, and vulnerability in southern Malawi. *Int. J. Agric. Sustain.* 8(4):290–304.

Smale, M., and P.W. Heisey. 1997. Maize technology and productivity in Malawi. In *Africa's Emerging Maize Revolution*, eds. D. Byerlee and C.K. Eicher, 63–79. Lynne Reiner, Boulder, CO.

Smith, K., and M. Ali Tabatabai. 2003. Automated instruments for the determination of total carbon, hydrogen, nitrogen, sulfur, and oxygen. In *Soil and Environmental Analysis; Modern Instrumental Techniques*, eds. K.A. Smith and M.S. Cresser. Dekker, New York, 202–246.

Snapp, S.S. 1998. Soil nutrient status of smallholder farms in Malawi. *Commun. Soil Sci. Plant Anal.* 29:2571–2588.

Snapp, S.S., M.J. Blackie, and C. Donovan. 2003. Realigning research and extension to focus on farmers' constraints and opportunities. *Food Policy* 28:349–363.

Snapp, S.S., M.J. Blackie, R.A. Gilbert, R. Bezner-Kerr, and G.Y. Kanyama-Phiri. 2010. Biodiversity can support a greener revolution in Africa. *Proc. Natl. Acad. Sci. USA* 107(48):20840–20845.

Snapp, S.S., R.B. Jones, E.M. Minja, J. Rusike, and S.N. Silim. 2003. Pigeon pea for Africa: a versatile vegetable—and more. *HortScience* 38:1073–1078.

Snapp, S.S., P.L. Mafongoya, and S. Waddington. 1998. Organic matter technologies for integrated nutrient management in smallholder cropping systems of southern Africa. *Agr. Ecosyst. Environ.* 71:185–200.

Snapp, S.S., and B. Pound, eds. 2008. *Agricultural Systems: Agroecology and Rural Innovation for Development*. Academic Press, Burlington, MA, 1–386.

Snapp, S.S., and S.N. Silim. 2002. Farmer preferences and legume intensification for low nutrient environments. *Plant Sci.* 245:181–192.

Stoate, C., and A.K. Jarju. 2008. A participatory investigation into multifunctional benefits of indigenous trees in West African savanna farmland. *Int. J. Agric. Sustain.* 6:122–132.

Stoorvogel, J.J., J.M. Antle, C.C. Crissman, and W. Bowen. 2004. The tradeoff analysis model: integrated bio-physical and economic modeling of agricultural production systems. *Agric. Syst.* 80:43–66.

Sugihara, S., S. Funakawa, M. Kilasara, and T. Kosaki. 2010. Dynamics of microbial biomass nitrogen in relation to plant nitrogen uptake during the crop growth period in a dry tropical cropland in Tanzania. *Soil Sci. Plant Nutr.* 56(1):105–114.

Takane, T. 2009. Disparities and diversities among female headed households in rural Malawi after 20 years of economic liberalization. *Singapore J. Trop. Geog.* 30:358–372.

Toenniessen, G., A. Adesina, and J. Devries. 2008. Building an alliance for a green revolution in Africa. *Ann. N.Y. Acad Sci.* 1136:233–242.

Uddin, M.B., M.A.S. Khan, S.A. Mukul, and M.K. Hussain. 2008. Effects of inorganic fertilizers on biological nitrogen fixation and seedling growth of some agroforestry trees in Bangladesh. *J. Forestry Res.* 19:303–306.

Uribe, D., J. Sanchez-Nieves, and J. Vanegas. 2010. Role of microbial biofertilizers in the development of a sustainable agriculture in the tropics. In *Soil Biology and Agriculture in the Tropics*, Soil Biology series Vol. 21. ed. P. Dion, Chap. 11. Springer, New York.

Vanlauwe, B., A. Bationo, J. Chianu, et al. 2010. Integrated soil fertility management: operational definition and consequences for implementation and dissemination. *Outlook Agric.* 39:17–24.

Verchot, L.V. 2010. Impacts of forest conversion to agriculture on microbial communities and microbial functions. In *Soil Biology and Agriculture in the Tropics*, Soil Biology series Vol. 21, ed. P. Dion. Springer, New York, 45–64.

Von Braun, J. 2003. Agricultural economics and distributional effects. Presidential address at the Proceedings of the 25th International Conference of Agricultural Economists. 16–22 August, Durban, South Africa. http://www.iet.ru/afe/conferences/Durban/braun.pdf (accessed 28 May 2007).

Weil, R.R., and S.K. Mughogho. 2000. Sulfur nutrition of maize in four regions of Malawi. *Agron. J.* 92:649–656.

Wisner, B., P. Blaikie, T. Cannon, and I. Davis. 2004. *At Risk: Natural Hazards, People's Vulnerability and Disaster*, 2nd ed. Routledge, New York.

Woods, J., A. Williams, J.K. Hughes, M. Black, and R. Murphy. 2010. Energy and the food system. *Phil. Trans. Roy. Soc. B* 365:2991–3006.

Yarzabal, L.A. 2010. Agricultural development in tropical acid soils: potential and limitations of phosphate-solubilizing bacteria. In *Soil Biology and Agriculture in the Tropics*, Soil Biology series Vol. 21, ed. P. Dion. Springer, New York, 209–234.

Yusuf, A.A., E.N.O. Iwuafor, R.C. Abaidoo, O. Olufajo, and N. Sangina. 2009. Grain legume rotation benefits to maize in the northern Guinea savanna of Nigeria: fixed-nitrogen versus other rotation effects. *Nutr. Cycl. Agroecosyt.* 84:129–139.

chapter 10

Making soil biodiversity matter for agriculture

Ecosystem services and challenges

Diana H. Wall
Colorado State University

Contents

10.1 Introduction

Throughout the 1960s and 1970s, during modernization and implementation of the "green revolution" in agriculture, soil and its biodiversity were largely invisible and considered almost external to the system. The continued focus in agriculture worldwide on the use of insecticides, nematicides, and herbicides for control of soil pathogens and pests rather than on biocontrol of pathogens, herbivore-resistant crop varieties, or other management strategies has furthered the impression that soil biodiversity is of little relevance to agricultural production. However, a suite of sustainability issues over the past half century combined with new scientific knowledge and economic factors has steadily altered this perception. Soil biodiversity, here defined as the myriad soil invertebrates and microbes, is increasingly known by ecologists as the basis for provision of ecosystem services such as improving soil fertility, regulating disease, remediating pollution, storing carbon, improving water quality, and reducing erosion and nutrient leaching (Millennium Ecosystem

Assessment [MA], 2005). Yet, the living components of soils and their multiple interactions above- and belowground still receive relatively little attention globally in agriculture.

Here, I enlarge the discussion of biodiversity in agricultural ecosystems from the current focus on soil parasites and pathogens to a broader overview of soil biodiversity; this is particularly needed when planning for long-term sustainability of ecosystems worldwide. Sustainability, a concept of treating the environment carefully so that the needs of future generations may be met, is not new and has been employed from the time of the earliest hunting and farming societies as resource limitation of one form or another affected the ability to catch prey or produce food. I begin with several examples of global environmental agreements and policies that have already focused attention on components of soil biodiversity, discuss the role of soil biodiversity in the provision of ecosystem services, and conclude with a discussion of biocontrol or disease regulation as a service. Ecosystem services here are defined according to the global analysis of the MA (2005) as benefits people obtain from ecosystems. These benefits were termed provisioning services (e.g., food, fiber, and clean water), regulating services (e.g., climate regulation, disease regulation), cultural services (e.g., aesthetic, recreation, spiritual values), and supporting services (e.g., primary production, decomposition, and soil formation).

10.2 Global challenges and international agreements

Globally, agriculture is faced with dynamic and complex challenges, with the most urgent being the growing demand for food production, particularly in the developing world, as populations and consumption increase. This, compounded with interactions of multiple issues of climate change, increased demands for energy, water scarcity, land-use change and biodiversity loss, and the necessity to develop solutions that include knowledge of ecosystem services, presents formidable obstacles for all (Zimmerer, 2010). Facing the simple statistics, it is estimated that by 2050 nearly 30–40% more food will need to be produced with about half the inputs and on a decreased area of agriculturally productive lands (Fedoroff et al., 2010; Godfray et al., 2010; Lambin and Meyfroidt, 2011). Past efforts that modernized agriculture and vastly increased food production were achieved by creating homogeneous large-scale solutions using irrigation and chemical inputs across heterogeneous landscapes, often unaware of the growing need for using principles of sustainability. The concepts developed for these intensive large-scale agricultural systems have been reexamined for more innovative small-scale and site-specific management options, including new farming systems, agricultural biotechnology, and consideration of organisms from multiple trophic levels to improve food production and global sustainability (Fedoroff et al., 2010). Crop production systems are no longer regarded as isolated ecosystems. Instead, they are included in broader implications of sustainability science through consideration of the benefits of interactions with nearby urban and wild lands, by judicious use of water and chemical inputs, conserving present crop diversity, hosting nonmarketable wild diversity, and altering the distribution of crops and their interactions with pests, pathogens, and weeds.

Whether these changes will constrain or contribute to food production markets will be balanced against the effects of multiple drivers (land-use change, climate change, nitrogen use, water availability) affecting agricultural dynamics. Human actions directly affect agriculture through small-scale alterations of biological interactions or, at a larger scale, through land use change. Pests and diseases, for example, were estimated as responsible for up to 40% of worldwide crop losses. Yet, there is little comprehension of how this estimate might be affected with climate change (Garrett et al., 2006; Gregory et al., 2009).

Bioengineering of food crops, large-scale international land acquisitions, and shifts in crops to biofuel production are just a few examples of management that will need to be reassessed under interacting global changes. Reduced levels of applied chemicals such as herbicides, fertilizers, and pesticides that were often nonrenewable and toxic and produced externally can now be combined with the use of biocontrol (e.g., parasites, predator-pest interactions) in trials to manipulate belowground diversity for enhanced plant production. Increasing knowledge of these and other new options for managing agriculture and connected ecosystems sustainably must include belowground diversity as a component if we are to discover new longer-term, larger-scale strategies and avoid ecosystem surprises and further degradation of ecosystems (MA, 2005; Clark, 2007).

The earliest farmers and hunters practiced sustainable land management, although they may not have been aware of the multiple organisms interacting in soils. Management of soil was known to be an important link to production of food, although there is also a history of civilizations declining because of inattention or abuse of soil. This long history of human involvement in agriculture was primarily at small scales (compared to current large-scale land-use change). Since the advent of cheaper chemical inputs, many have not considered management of soil biodiversity as linked to sustainability of human populations and to the overall global picture of growing populations and food demand. However, in the past 30 to 40 years, numerous international agreements, policies, and scientific consensus reports have documented results that together show that for environmental sustainability to occur, soil biodiversity must be considered as part of soil productivity, not ignored or left out of discussions on the flows, properties, and connections of global cycles. A few examples of environmental issues, global assessments, and policy agreements that directly involve soils as critical components show that inclusion of more information on soils and soil biodiversity can be valuable for environmental management and long-term provision of ecosystem services.

10.2.1 The Montreal Protocol

The Montreal Protocol is one of the most famous agreements known for its success in galvanizing nations toward a common solution. It is, however, an example of misunderstanding the magnitude of connections between the atmosphere and belowground-aboveground conditions. The 1995 Montreal Protocol on Substances that Deplete the Ozone Layer required a phaseout of certain halogenated hydrocarbons, including the soil fumigant methyl bromide. This compound is used globally in agricultural forestry and produce to suppress soil pest sand pathogens (such as species of fungi, nematodes, insects, mites, rodents, weeds, and some bacteria) and to enhance root health, plant growth, and yield. But, it also has broad soil biocidal activity affecting nontarget organisms, including beneficial organisms (Nyczepir and Thomas, 2009). Methyl bromide emissions account for about 5–10% of global stratospheric ozone depletion, and although being phased out since 2005 in the United States, exemptions exist for many crops and countries (Schneider et al., 2003). Today, as a result of the Montreal Protocol, and a new awareness of the impact of pesticides, many alternatives to methyl bromide are being tested, including solarization of soils, pathogen-resistant crop varieties, methyl iodide, and other less-ozone-depleting pesticides. While these alternatives meet the criterion of not harming the ozone layer, they are evaluated for additional effects, such as toxicity problems for water and effects on soil food webs, nontarget organisms, and humans (Ingham and Coleman, 1984; Duniway, 2002; Sanchez-Moreno et al., 2009, 2010). It is doubtful that the food production needed for tomorrow's populations can be reached in larger-scale agriculture without fertilizers and other chemical inputs, such as the alternatives to methyl bromide, especially in tropical countries, where pests, weeds, and pathogens can cause devastating loss (Fedoroff et al., 2010; Godfray et al., 2010). Thus, manipulation of soil biodiversity in small-scale farming as a means to increase biological control options becomes a more feasible option.

10.2.2 The U.N. Convention to Combat Desertification

Another example of an environmental issue in which soils play a primary role is the soil-plant-climate connection. On large scales, disruption of this connection can have devastating effects. For instance, the Dust Bowl phenomenon of the United States in the 1930s caused large-scale human migrations and socioeconomic and environmental disruption. Currently, the desertification of one-third of Earth's land area (arid and semiarid lands as well as dry subhumid tropics) greatly affects the world's poorest and most marginalized people. The U.N. Convention to Combat Desertification (UNCCD) since its beginnings in 1996 has had failures and changed focus to a more regional approach in 2005, with the growing recognition that desertification is expected to worsen in the next 50 years. The causes of cultivated and rangeland degradation are many: Droughts and human activities (e.g., deforestation, inappropriate irrigation practices leading to salinization, invasive species introduction, and overgrazing) all alter biodiversity and plant productivity and result in fewer available soil nutrients and massive erosion. Goals of the Convention are to improve ecosystems and living conditions at regional scales. Management and restoration of soils are linked to types of plant cover and stabilization of soils (e.g., use of nitrogen-fixing plants, locally produced manures, securing soils with trees instead of fences), but new knowledge is needed for sustaining and maintaining these dryland systems (Dregne, 1996; UNCCD, 2011).

10.2.3 The Convention on Biodiversity

A third example in which soils are recognized within a global environmental issue is the Convention on Biodiversity (CBD), an international treaty that in 1993 called for action on the rapid loss of biodiversity. It includes a focus on soil biodiversity through the objectives of conservation, sustainable use of biodiversity, and the fair and equitable sharing of benefits arising from genetic resources. The CBD is striking in that it is not focused solely on protection of species or ecosystems but instead notes that biodiversity will be used for the benefit of humans and promotes sustainable use to avoid increased rates of decline in biological diversity. The definition of biodiversity is broad and encompasses the biological variation across all scales from genes to whole ecosystems; as a result, soil biodiversity and its role in above- and belowground ecosystems is now considered in discussions on climate change, farming, water and biogeochemical cycles, species and habitat exploitation, and desertification (Wall et al., 2010). The endorsement of the International Initiative for the Conservation and Sustainable Use of Soil Biodiversity by the CBD Conference of the Parties in 2006 acknowledged the importance of belowground diversity and called for further research.

The Convention has had major effects in many countries, particularly in Europe, where the Joint Research Center (JRC) of the European Commission provides scientific support and technical support for E.U. policies, including the E.U. commitment to make significant progress toward CBD targets to halt biodiversity loss. Besides publishing the *European Atlas of Soil Biodiversity* (2010), the Commission has linked business and natural capital (living and nonliving resources) by showing how soil biodiversity can be integrated into business plans and market opportunities as part of a U.N. Environmental Program and European Commission and a multicountry initiative called The Economics of Ecosystems and Biodiversity (TEEB) (ten Brink, 2011). The JRC had a working group on soil biodiversity that helped inform science and policy in Europe by contributing through the CBD on areas such as protecting soil biodiversity, examining the taxonomy and roles soil organisms play in ecosystem function, and building databases on soil biodiversity. This acknowledgment of soil biodiversity as a critical component of all terrestrial ecosystems is a major step in addressing knowledge gaps and promoting syntheses, inventories, monitoring, and management (Decaëns et al., 2006).

10.2.4 The Millennium Ecosystem Assessment

The international assessment and resulting synthesis report of the Millennium Ecosystem Assessment (MA, 2005) on the status of the world's ecosystems concluded that Earth's biodiversity was being significantly degraded by human activities at a pace that affected the ability of ecosystems to continue to provide essential services for life support. In the analysis, soils and their biodiversity were considered as a component of ecosystems that provide essential services to society. This report, requested by governments through four international environmental conventions (including the CBD and UNCCD), involved over 1,000 scientists and took 5 years to complete. The framework went beyond more commonly used terms in environmental economics (ecosystem goods and services) to include all the ways that people benefit from ecosystems: Twenty-four ecosystem services were analyzed to measure the life support systems or natural capital of ecosystems. The MA framework stimulated use of the concepts of ecosystem services by teachers, conventions, nongovernmental organizations (NGOs), businesses, and policy makers, for example, for meeting the U.N. Millennium Development Goals (Wall et al., 2005), and resulted in new research and studies, including a detailed analysis of how services provided by soil biodiversity differ across managed and unmanaged ecosystems (van der Putten et al., 2004; Wardle, Brown, et al., 2004). The MA conclusions on the status of the world's ecosystems and the wider realization that direct (e.g., the application of fertilizers) and indirect (population, technology, lifestyle) changes in biodiversity affect the ability of ecosystems to provide services, which in turn modify food, water, air, overall human well-being, and national economies, were a wake-up call. That ecosystems are degrading rapidly and urgently need protecting for future generations continues to accelerate new interdisciplinary knowledge and analyses of sustainability science (Carpenter et al., 2009; Raudsepp-Hearne et al., 2010).

10.2.5 The U.N. Intergovernmental Science-Policy Platform on Biodiversity and Ecosystem Services

The newest body, the U.N. Intergovernmental Science-Policy Platform on Biodiversity and Ecosystem Services (IPBES), aims to ensure that policy decisions are made on the best available scientific information for conservation and sustainable use of biodiversity and ecosystem services. The need for information that is accurate and objective has grown partly as a result of the MA (Perrings et al., 2011). Soil biodiversity and ecosystem services are key to managing ecosystem sustainability. Syntheses to contribute to the IPBES and other policy frameworks for land management could draw on measures such as those used by van der Putten et al. (2004) and Wardle, Brown, et al. (2004) to quantify the relative importance of soil organisms for each ecosystem service across a number of ecosystems.

10.2.6 The Intergovernmental Panel on Climate Change

The Intergovernmental Panel on Climate Change (IPCC), which began in 1988, is a scientific body that focuses on developing syntheses from data to report a clear scientific view on current information regarding the status of climate change and its potential consequences for society and the environment. Soil organic matter, soil greenhouse gas flux, and nutrient availability are assessed for different land/environmental management types over time, but inclusion of soil biodiversity is limited. The IPCC has identified gaps in our knowledge of how carbon is fragmented within different pools in the soil and how temperature and moisture in different ecosystems influence soil carbon sequestration (Paustian, 2009). Soil biota, key regulators of decomposition and thus soil carbon turnover, are influenced by climatic conditions at regional scales. Including them in predictions of decomposition at the regional or biome scale could help reduce uncertainties in regional decomposition models (Wall et al., 2008).

These examples of internationally acknowledged environmental issues that relate soil or soil biodiversity to global scales indicate their relevance to society. More importantly, these few examples illustrate that research findings on soil and biodiversity science cannot be considered as separate disciplinary silos if we are to address interacting global changes and feed the growing population sustainably.

10.3 Incorporating soil biodiversity and ecosystem services

Soil biota provide many ecosystem services, including contribution to landscape heterogeneity and stability; contribution to production of food, fuel, and fiber; provision of habitat and food for above- and belowground wildlife; provision of habitats for recreation and natural history; regulation of air and water quality; sequestration of carbon; recycling of nutrients and energy from waste and dead organisms; translocation of organic matter, nutrients, and organisms within and from soil; erosion control; and regulation of pests and pathogens and soil fertility (Wall, 2004; Lavelle et al., 2006; Singh et al., 2010). Over the past few decades, research to generate this knowledge has followed two paths, largely driven by recognition of accelerated biodiversity loss due to human activities. In one path, scientists explored how species loss would affect the operation of different ecosystems. This was scientific recognition that biodiversity plays a role in ecosystem processes (Daily et al., 1997; Wall, 2004; Nielsen et al., 2010). Because environmental changes are affecting society, a separate research path developed by economists, ecologists, social scientists, NGOs, and businesses focused on how the products of ecosystem processes resulted in the goods and services used by society (Karieva et al., 2011). This latter question was relevant to agricultural forests, crops, and rangeland as well as to fisheries, but it also addressed how biodiversity that is central to ecosystem functioning would provide services in the face of continued human alterations to climate, water, land, and biodiversity. Critically, this latter question was of interest to policy makers internationally because all countries and peoples are dependent on ecosystem services for their economies and societal well-being (President's Council of Advisors on Science and Technology, 2011; U.K. National Ecosystem Assessment, 2011).

Evidence for the effect of soil biodiversity on ecosystem processes has been variable. In most soils, particularly in wild and less-managed lands, research on the role of *individual species* in ecosystem functioning has been more difficult to assess than in actively managed agricultural systems. This is partly because the larger diversity and abundance of species and their interactions on processes generally hinder causative analyses (Nielsen et al., 2010; Wu et al., 2011). Nevertheless, there is sufficient evidence from microcosm and field experiments to show that interactions between individual species can have major consequences for ecosystem processes (Heemsbergen et al., 2004; Barrett et al., 2008). There are also many reports showing that soil biodiversity is altered due to land-use change (including intensification), climate change, and chemical inputs to soils (Wardle, Bardgett, et al., 2004; Wardle, Brown, et al., 2004; Bardgett, 2005); that loss of symbionts (nitrogen-fixing bacteria, mycorrhizae) in ecosystems has dramatic effects for nutrient cycling and plant communities (van der Heijden and Horton, 2009); and that reduction of species numbers of mycorrhizal fungi in grasslands reduces primary production (van der Heijden et al., 1998). Thus, loss of biodiversity in some natural soil systems can have a positive, negative, or no effect on ecosystem functioning, depending on the particular ecosystem and type of disturbance (Nielsen et al., 2010).

In contrast, agriculture has a long history of evidence for the role of individual soil-dwelling species in an ecosystem process (e.g., plant production). Pests, pathogens, or herbivores of roots have long been known at the species level for their individual and characteristic effect on plant physiology, metabolism, productivity, and crop yield. Also known from years of greenhouse and field experiments is evidence of when control should be initiated for species of root parasitic nematodes. This evidence includes an understanding of the nematode biology (the plant host range, soil temperature and moisture range, the feeding site preference [plant root tip, meristem, zone of elongation], life history) and, for some species, the density of nematodes at which control should be initiated to prevent increasing populations from causing plant economic damage (Bridge and Starr, 2007). Agricultural evidence demonstrates that root parasitic nematode biodiversity at the individual species or multiple species level can decrease an ecosystem process (e.g., net plant productivity) and affect an ecosystem service (e.g., provision of food). Most agronomic field research on target pests and pathogens, however, has not included multiple trophic levels (van der Putten et al., 2010). There has been successful research based on including an understanding of microbes, predators, and parasites into crop production as biocontrol measures (e.g., Borneman and Becker, 2007; Gao et al., 2008).

The provision of ecosystem services provided by soils has been the subject of many articles (Daily et al., 1997; Swift et al., 2004; van der Putten et al., 2004; Wardle, Brown, et al., 2004; Lavelle et al., 2006; Barrios, 2007; Velasquez et al., 2007; Weber, 2007; Zhang et al., 2007; Kibblewhite et al., 2008; Dominati et al., 2010; Nielsen et al., 2010; Sylvain and Wall, 2011; Barrios et al., 2012). Among these studies is a new economic framework for soil ecosystem services based on nonliving physical soil properties as the natural capital and soil biological activity as part of supporting services (Dominati et al., 2010). The Dominati et al. article is a substantial departure from the MA (2005) that is based on biodiversity and nonliving resources—the ecosystem—that through functioning provide ecosystem services for human well-being. Others noted a research need is quantification of ecosystem services provided by soil biodiversity (Swift et al., 2004; Wall, 2004; Barrios et al., 2012). Swift et al. (2004) questioned whether biodiversity and ecosystem services in agricultural landscapes could be measured at field scales in industrialized agriculture, particularly since an objective of agriculture was to reduce diversity and to substitute fertilizer for the functions of soil biodiversity. Their article suggested that biodiversity and ecosystem services could be maintained at landscape and farm scales rather than the field scale, but this would require economic change and policies that encouraged land-use diversity. Velasquez et al. (2007) tested a general indicator of soil quality (GISQ), with macrofaunal biodiversity representing an integrating component that could be used to evaluate ecosystem services across national boundaries and as a basis for quantifying soil quality (Lavelle et al., 2009).

The research initiatives on soil biodiversity and ecosystem functioning and their resulting benefits to society have developed rapidly and are providing new interdisciplinary challenges. However, integrating soil science and soil biodiversity into the larger issues of biodiversity loss and conservation requires evidence of what we know now and our collective synthetic expertise for predictions of the future. Questions that need to be resolved for sustainability of ecosystems and for society are simply as follows: Does it matter if species or certain trophic groups are lost when soils are heavily perturbed (such as with intensive agriculture)? If so, does this loss alter ecosystem functioning and the ability to manage land sustainability to enhance soil ecosystem services? Certain groups, such as nitrogen-fixing bacteria or larger soil animals, including enchytraeids, earthworms, termites, or ants, are all known to be vulnerable to disturbance (Wall et al., 2010), and their loss would significantly affect an ecosystem function and the provision of a service. But are other groups of soil biota and ecosystem services equally vulnerable? Hunt and Wall (2002) modeled the loss of soil biodiversity for an arid grassland and showed that ecosystems could sustain loss of some functional groups. For example, only a 10% change in function was noted with deletion of bacteria, saprophytic fungi, and root-feeding nematodes. However, this study did not include effects of climate change drivers.

To address these questions of vulnerability of soil species under multiple global change drivers, there was a thorough assessment of soil biodiversity and provision of ecosystem services at various management regimes and across spatial and temporal scales (Wall, 2004). First, van der Putten et al. (2004) assessed the contribution of each group of soil biotic and abiotic factors (scaled from unimportant to highly important) to the provision of a range of ecosystem services (or disservices) in several ecosystems, including tilled agricultural systems. Then, using this synthesis (i.e., van der Putten et al., 2004) and other reports showing that soil taxonomic groups contributing to an ecosystem service vary in their response to global change drivers (e.g., land-use change, climate change, pollution, agricultural intensification, and invasive species) across various spatial scales, Wardle, Brown, et al. (2004) evaluated the vulnerability of each biotic group to global change drivers and the consequence for the provision of an ecosystem service. This assessment emphasized not only the impacts of global change drivers on the three ecosystems studied but also on the linkages above- and belowground. Direct effects of the driver on organisms on one side of the aboveground-belowground interface will likely stimulate feedbacks from organisms on the other side of the interface. The analysis showed that food and fiber production of arable lands was highly vulnerable to global change drivers such as drought, invasive species, and land-use change because soil biota were altered and supporting services such as plant production were reduced. This type of assessment analysis used current knowledge about soil biodiversity and ecosystem functioning to predict detailed scenarios of the vulnerability of soil biota, ecosystems, and ecosystem services. Lavelle et al. (2004) explored the scales at which processes that sustain a service operate and the scales of the organisms responsible and how they could be used for managing provision of services (protection, limited use, rotations, etc.). This analysis expanded from soil to sediment because the scale of disturbances affecting organisms in one habitat can affect organisms and services farther away and at a larger scale (such as excess fertilizer in a field affecting nitrogen retention, resulting in runoff and algal blooms in rivers) (Lavelle et al., 2004).

These and other studies (Wall et al., 2008; Ostle et al., 2009; Woodward et al., 2009) showed that soil biodiversity, ecosystem processes, and ecosystem services are vulnerable to threats, depending on the type of ecosystems, the spatial and temporal scale, and the organisms present. This reflects the fact that the climatic and edaphic factors of each soil type define the organisms in it and is valuable knowledge for planning future land management for food and fiber.

10.4 *Management of soil biodiversity for disease regulation and crop yield*

An ecosystem service provided by soil biodiversity is the control or suppression of organisms that cause damage to or kill a plant, animal, or human. There have been substantial gains in our understanding of belowground organisms as biocontrol agents of root parasites and herbivores because of their effect on other ecosystem services (plant production and crop yield) (Brown and Gange, 1989; Kerry, 1998; Blackshaw and Kerry, 2008; Seastedt and Murray, 2008; Johnson, 2010). This gain has been due to the search for alternative pesticides, the growing awareness of harm to humans and the environment resulting from pesticide use, and curiosity-driven ecological research. The history of research to control plant parasitic nematodes in agriculture has included, but is not limited to, use of parasitoids, predaceous microbes, and invertebrates, planting cover crops such as marigold (*Tagetes* spp.), which produces allelopathic compounds such as alpha-terthienyl (Riga et al., 2005), resistant plants, and designed landscapes for enhanced growth of biological control organisms (see review by van der Putten et al., 2006).

Disturbance to soils with increasing agricultural intensification can reduce species diversity and affect the biological regulation of parasites and predators, which in turn could affect the delivery of ecosystem services (Figure 10.1; Tian et al., 1992; Bloemers et al., 1997; Eggleton et al., 2002; Heemsbergen et al., 2004; Hol et al., 2010). However, effects of soil disturbances on species diversity can vary considerably, depending on the biology of the organism (Wall et al., 2010). Knowing the plant host and the biology, the spatial and temporal patterns, and the food preferences of the soil pathogen/pest, as well as the types of associated organisms that constrain growth of the pathogen, pest, or herbivore, is key

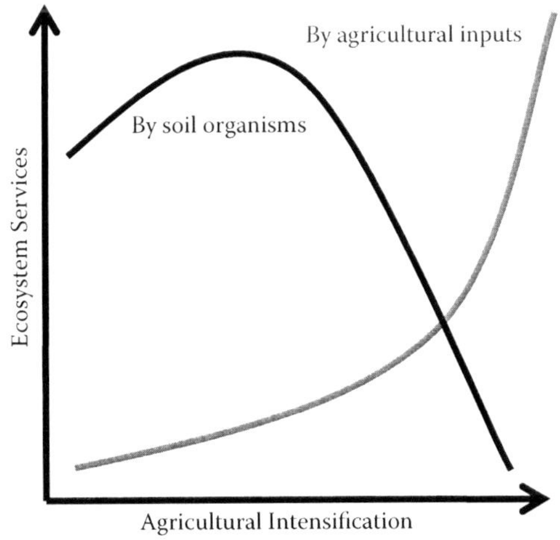

Figure 10.1 Increasing intensification of agriculture, including increased chemical (e.g., pesticides, fertilizers) inputs, adversely affects the benefits provided by soil microbes and invertebrates such as those that contribute to the biocontrol of root pathogens and regulate soil fertility.

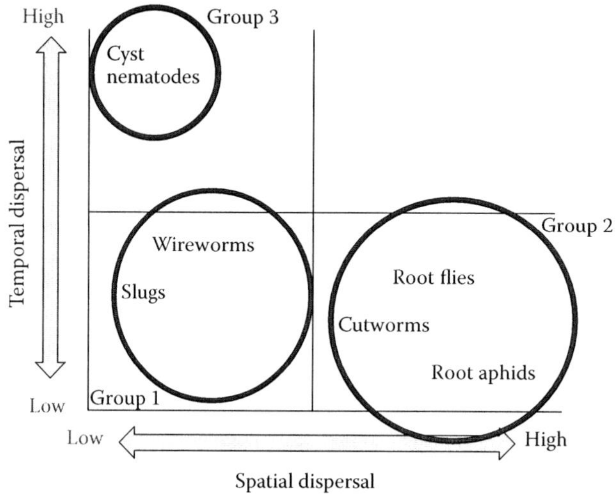

Figure 10.2 Knowledge of species biology is key to control of root herbivores and pathogens. The diagram shows how various species of soil pests in the United Kingdom vary in their dispersal across fields: Group 1 pests increase numbers within fields over several seasons and build to damaging populations when a crop is sown; group 2 pest females seek specific crops each year from outside the immediate field at a larger spatial distance; group 3 plant parasites maintain populations within a field between suitable host crops but have poor spatial dispersal characteristics. (Modified from Blackshaw, R., and B. Kerry. 2008. Root herbivory in agricultural ecosystems. In *Root Feeders: An Ecosystem Perspective*, eds. S. Johnson and P. Murray, 35–53. CAB International, Wallingford, UK. With permission.)

to control and understand how environmental change may affect soil biotic-plant relationships (Figure 10.2; Blackshaw and Kerry, 2008; Gao et al., 2008). Root parasitic nematodes cause substantial economic damage and are formidable pests. Sedentary obligate root parasites are highly specialized for living on plant hosts, and in their esophageal gland cells they produce numerous secretory proteins with several functions in parasitism: They can manipulate the plant cell development, modify cell walls with cellulases and other enzymes, alter cell metabolism, modulate protein degradation, and can remarkably elude plant defenses.

Applying knowledge of the biocontrol organisms in natural systems to agricultural systems can suggest alternative management strategies. Seastedt and Murray (2008) suggested that in wild systems, damage to plants by root herbivores was dependent on the root herbivores' size, and that low levels of feeding or parasitism by smaller microanimals, such as root parasitic nematodes, had fewer plant costs than one large animal (Figure 10.3), but the authors noted that this differed considerably between wild and agricultural systems. This suggests that plant-feeding and plant-pathogenic animals are better controlled in natural systems to levels that can benefit plant growth. In wild systems, damage by root parasitic nematodes ranges from significant to nonsignificant. While there is a high diversity of potential organisms (e.g., bacteria, fungi, tardigrades, nematodes, mites) to control parasitic populations, the mechanisms (predation, competition, etc.) are less known compared to monocultural or perennial agroecosystems. In such systems, natural enemies such as a fungal or bacterial species may build up over time (4–5 years) to suppress a specific nematode (Kerry, 1998; De Deyn et al., 2004). Indigenous organisms can successfully suppress a specific nematode root parasite in agricultural systems more so than can

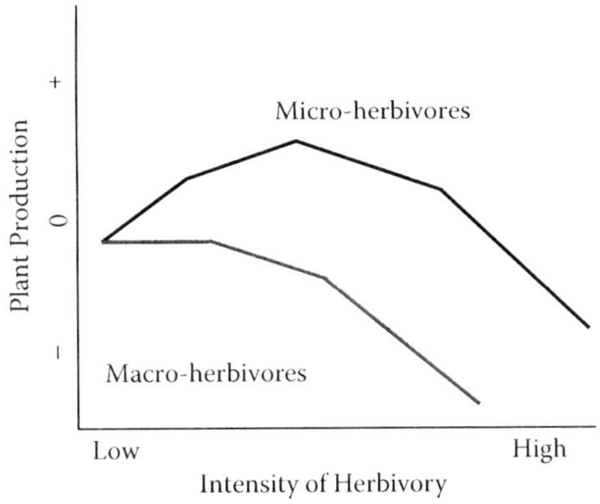

Figure 10.3 Low levels of micro-root herbivores such as nematodes and other small organisms can result in compensatory growth or overcompensation by the plant in contrast to that caused by macroherbivores. Larger soil invertebrates feeding on roots can cause significant damage or death of roots. (Modified from Seastedt, T., and P. Murray. 2008. Root herbivory in grassland ecosystems. In *Root Feeders: An Ecosystem Perspective*, ed. S. Johnson and P. Murray, 54–67. CAB International, Wallingford, UK. With permission.)

exotic biocontrol organisms (Kerry and Hominick, 2002; van der Putten et al., 2006; Gao et al., 2008). And, there can be positive benefits. van der Putten et al. (2006) noted that in wild systems, root parasitic nematodes, other root feeders, and pathogens drive community processes such as plant species diversity and succession because they have coevolved with the plants and their enemies. These examples illustrate that knowledge of the biology of the pathogen and plant host and pests and predators from wild land and agricultural systems can be successfully transferred into some managed agriculture strategies (van der Putten et al., 2006; Crute and Muir, 2011; Pritchard, 2011).

Managing soil biodiversity for small-scale agriculture and promotion of ecosystem services was the focus of a United Nations Environment Program (UNEP)-Global Environmental Fund project in several tropical countries. The Conservation and Sustainable Management of Below-ground Biodiversity (BDBD) Project of the Tropical Soil Biology and Fertility (TSBF) Institute of the International Center for Tropical Agriculture (CIAT) based in Nairobi, Kenya, spent more than a decade working with expert scientists from seven countries to develop international standards for the inventory of belowground biota under a range of land uses in the humid tropics. Although the early work was on the use of organic matter and how carbon and nitrogen are supplied to plants, the result of the program was a focus on the organisms that mediate the processes, functions, and services of soil.

The *Handbook of Tropical Soil Biology* (Moreira et al., 2008) is a primary example of recognition by experts who work in the poorest regions of the world's most rapidly deforested areas that management of soil biodiversity can be employed to enhance soil ecosystem services such as soil fertility and biological control of soil pathogens in agroecosystem management. Methods in the book were tested in the seven participating tropical countries on three continents by local experts, biologists, and managers who had varying levels of expertise in identification. The program included building capacity of the scientists in aspects of soil biology, soil fertility, crops, and ecosystem services and experiments that could be compared on measures to manipulate soil biodiversity to increase biological control and plant health. The framework that was developed for manipulating soil biodiversity to reduce pests and pathogens can be useful and should be considered for the maintenance and sustainability of agriculture worldwide.

Progress on biocontrol can and is being made, but to achieve sustainability, other disciplines must recognize the crucial importance of the belowground-aboveground connection. The development of plants with genes to withstand drought or other targeted characteristics is not useful unless the interactions of the belowground organisms common to the soil type are included. The Global Soil Map (http://www.globalsoilmap.net) (see Barrios et al., 2012), a digital soil map of the world that is freely accessible online, is providing new chemical and physical soil data on Africa's degraded and fertile soils to encourage better land use and soil conservation. Sachs et al. (2010) have called for this effort to be part of an agricultural monitoring network. However, unless an understanding of soil biodiversity underpins this monitoring, control of pests and diseases will likely continue to be through synthetic chemical application. A pilot project is examining soil biodiversity in a few countries to coincide with the abiotic data of the Global Soil Map Project (D.H. Wall and N. Fierer, unpublished data, 2012). This will provide information on soil habitat, plant and animal pests and pathogens, as well as multitrophic levels of organisms that are responsible for long-term, sustainable soil fertility. The value of linking projects such as the Global Soil Map with soil biodiversity is the ability to design new ways for agriculture to reach the potential to feed people and to cope with climate change in the future (Ostle et al., 2009; Woodward et al., 2009; van der Putten et al., 2010). In the absence of international climate change policy, Woodward et al. (2009) urged scientists to move forward with research addressing biological uncertainties, one of which, carbon sequestration, involves the soil food web in both natural and agricultural ecosystems.

10.5 Concluding comments

Many of the most pressing and globally recognized challenges to human societies are tied to soil and soil organisms. Human activities are irreversibly impacting Earth's ecosystems at a rapid pace, and these activities are having major global impacts on soils. Sustaining soils, their biodiversity, and their ability to provide ecosystem services is now one of society's major global challenges, as evidenced by numerous environmental governance policies that include soil. Hans Jenny, a consummate early soil scientist and pedologist, included soil biota as a soil-forming factor key to the study of soil science. However, in recent years, the use of fertilizers and the need for strategies to reduce greenhouse gases have somewhat marginalized the study of soil invertebrate-microbial interactions from much of soil science. What is yet to be resolved by the scientific community as a whole, and politicians and funding agencies, is that addressing these environmental challenges will require increased understanding and integration of the physical aspects of soil science with soil biology and ecosystem services (Woodward et al., 2009; van der Putten et al., 2010). While soil biodiversity alone will not provide solutions, including understanding of its interactions with biocontrol, climate mitigation, and water quality may encourage new scenarios of landscape management. We need to make better use of what we know, generate key new knowledge, and expand our use of this knowledge and technology. Beyond this, there is a need to agree on measures and standards for quantifying ecosystem services provided by soil organisms and to implement these into sustainable management strategies and ecosystem monitoring assessments at landscape scales. Gathering the evidence and deciding on the essential new knowledge to reduce uncertainties in response to multiple global change drivers of climate change and loss of species on ecosystem processes require that we move quickly to include the soil biological processes and their services.

Acknowledgments

I gratefully acknowledge the reviews of U. N. Nielsen, two anonymous reviewers, and particularly the editing by A. Shaw and figures and editing by Z. Sylvain.

References

Bardgett, R. D. 2005. *The Biology of Soil: A Community and Ecosystem Approach*. Oxford University Press, Oxford, UK.

Barrett, J. E., R. A. Virginia, D. H. Wall, and B. J. Adams. 2008. Decline in a dominant invertebrate species contributes to altered carbon cycling in a low-diversity soil ecosystem. *Global Change Biology* 14:1734–1744.

Barrios, E. 2007. Soil biota, ecosystem services and land productivity. *Ecological Economics* 64:269–285.

Barrios, E., G. Sileshi, K. Shepherd, and F. Sinclair. 2012. Agroforestry and soil health: linking trees, soil biota and ecosystem services. In *Soil Ecology and Ecosystem Services*, eds. D. H. Wall, R. D. Bardgett, V. Behan-Pelletier, J. E. Herrick, H. Jones, K. Ritz, J. Six, D. R. Strong, and W. H. van der Putten. Oxford University Press, Oxford, UK.

Blackshaw, R., and B. Kerry. 2008. Root herbivory in agricultural ecosystems. In *Root Feeders: An Ecosystem Perspective*, eds. S. Johnson and P. Murray, 35–53. CAB International, Wallingford, UK.

Bloemers, G., M. Hodda, P. Lambshead, J. Lawton, and F. Wanless. 1997. The effects of forest disturbance on diversity of tropical soil nematodes. *Oecologia* 111:575–582.

Borneman, J., and J. O. Becker. 2007. Identifying microorganisms involved in specific pathogen suppression in soil. *Annual Review of Phytopathology* 45:153–172.

Bridge, J., and J. Starr. 2007. *Plant Nematodes of Agricultural Importance: A Color Handbook*. Academic Press, Boston.

Brown, V. K., and A. C. Gange. 1989. Herbivory by soil-dwelling insects depresses plant species richness. *Ecology* 3:667–671.

Carpenter, S., H. Mooney, J. Agard, D. Capistrano, R. DeFries, S. Diaz, T. Dietz, A. Duraiappah, A. Oteng-Yeboah, H. Pereira, C. Perrings, W. Reid, J. Sarukhan, R. Scholes, and A. Whyte. 2009. Science for managing ecosystem services: beyond the Millennium Ecosystem Assessment. *Proceedings of the National Academy of Sciences* 106:1305–1312.

Clark, W. 2007. Sustainability science: a room of its own. *Proceedings of the National Academy of Sciences of the United States of America* 104:1737–1738.

Crute, I., and J. Muir. 2011. Improving the productivity and sustainability of terrestrial and aquatic food production systems: future perspectives. *Journal of Agricultural Science* 149:1–7.

Daily, G., P. Matson, and P. Vitousek. 1997. Ecosystem services supplied by soils. In *Nature's Services: Societal Dependence on Natural Ecosystems*, ed. G. Daily, 113–132. Island Press, Washington, DC.

Decaëns, T., J. Jimenez, C. Gioia, G. Measey, and P. Lavelle. 2006. The values of soil animals for conservation biology. *European Journal of Soil Biology* 42:S23-S38.

De Deyn, G., C. Raaijmakers, J. van Ruijven, F. Berendse, and W. H. van der Putten. 2004. Plant species identity and diversity effects on divergent trophic levels of nematodes in the soil food web. *Oikos* 106:576–586.

Dominati, E., M. Patterson, and A. Mackay. 2010. A framework for classifying and quantifying the natural capital and ecosystem services of soils. *Ecological Economics* 69:1858–1868.

Dregne, H. E. 1996. Desertification: challenges ahead. *Annals of Arid Zone* 35:305–311.

Duniway, J. 2002. Status of chemical alternatives to methyl bromide for pre-plant fumigation of soil. *Phytopathology* 92:1337–1343.

Eggleton, P., R. Davies, S. Connetable, D. Bignell, and C. Rouland. 2002. The termites of the Mayombe Forest Reserve, Congo (Brazzaville): transect sampling reveals an extremely high diversity of ground-nesting soil feeders. *Journal of Natural History* 36:1239–1246.

Fedoroff, N. V., D. S. Battisti, R. N. Beachy, P. J. M. Cooper, D. A. Fischhoff, C. N. Hodges, V. C. Knauf, D. Lobell, B. J. Mazur, D. Molden, M. P. Reynolds, P. C. Ronald, M. W. Rosegrant, P. A. Sanchez, A. Vonshak, and J. K. Zhu. 2010. Radically rethinking agriculture for the 21st century. *Science* 327:833–834.

Gao, X. B., B. Yin, J. Borneman, and J. O. Becker. 2008. Assessment of parasitic activity of *Fusarium* strains obtained from a *Heterodera schactii*-suppressive soil. *Journal of Nematology* 40:1–6.

Garrett, K. A., S. P. Dendy, E. E. Frank, M. N. Rouse, and S. E. Travers. 2006. Climate change effects on plant disease: genomes to ecosystems. *Annual Review of Phytopathology* 44:489–509.

Godfray, H. C. J., J. R. Beddington, I. R. Crute, L. Haddad, D. Lawrence, J. F. Muir, J. Pretty, S. Robinson, S. M. Thomas, and C. Toulmin. 2010. Food security: the challenge of feeding 9 billion people. *Science* 327:812–818.

Gregory, P. J., S. N. Johnson, A. C. Newton, and J. S. I. Ingram. 2009. Integrating pests and pathogens into the climate change/food security debate. *Journal of Experimental Botany* 60:2827–2838.

Heemsbergen, D. A., M. P. Berg, M. Loreau, J. R. van Haj, J. H. Faber, and H. A. Verhoef. 2004. Biodiversity effects on soil processes explained by interspecific functional dissimilarity. *Science* 306:1019–1020.

Hol, W., W. de Boer, A. Termorshuizen, K. Meyer, J. Schneider, N. van Dam, J. van Veen, and W. H. van der Putten. 2010. Reduction of rare soil microbes modifies plant-herbivore interactions. *Ecology Letters* 13:292–301.

Hunt, H. W., and D. H. Wall. 2002. Modelling the effects of loss of soil biodiversity on ecosystem function. *Global Change Biology* 8:33–50.

Ingham, E., and D. Coleman. 1984. Effects of streptomycin, cycloheximide, fungizone, captan, carbofuran, cygon and PCNB on soil microorganisms. *Microbial Ecology* 10:345–358.

Jefferey, S., C. Gardi, A. Jones, L. Montanarella, L. Marmo, L. Miko, K. Ritz, G. Peres, J. Rombke, and W. H. van der Puttem (eds.). 2010. *European Atlas of Soil Biodiversity.* European Commission, Publications Office of the European Union, Luxembourg.

Jenny, H. 1941. Factors of soil formation. *A System of Quantitative Pedology.* McGravitz, Hill Book Company, New York, NY.

Johnson, K. 2010. Pathogen refuge: a key to understanding biological control. *Annual Review of Phytopathology* 48:141–160.

Karieva, P., H. Tallis, T. H. Ricketts, G. C. Daily, and S. Polasky, eds. 2011. *Natural Capital: Theory and Practice of Mapping Ecosystem Service.* Oxford University Press, Oxford, UK.

Kerry, B. 1998. The dynamics of the decline of the cereal cyst nematode, *Heterodera avenae,* in four soils under intensive cereal production. *Fundamental and Applied Nematology* 21:617–625.

Kerry, B., and W. Hominick. 2002. Biological control. In *Biology of Nematodes,* ed. D. Lee, 483–509. Taylor & Francis, London.

Kibblewhite, M., K. Ritz, and M. Swift. 2008. Soil health in agricultural systems. *Philosophical Transactions of the Royal Society B* 363:685–701.

Lambin, E., and P. Meyfroidt. 2011. Global land use change, economic globalization, and the looming land scarcity. *Proceedings of the National Academy of Sciences of the United States of America* 108:3465–3472.

Lavelle, P., D. Bignell, M. Austen, V. Brown, V. Behan-Pelletier, J. Garey, P. Giller, S. Hawkins, G. Brown, M. St. John, H. W. Hunt, and E. A. Paul. 2004. Connecting soil and sediment biodiversity: the role of scale and implications for management. In *Sustaining Biodiversity and Ecosystem Services in Soils and Sediments,* ed. D. H. Wall, 193–224. Island Press, Washington, DC.

Lavelle, P., T. Decaëns, M. Aubert, S. Barot, M. Blouin, F. Bureau, P. Margerie, P. Mora, and J.-P. Rossi. 2006. Soil invertebrates and ecosystem services. *European Journal of Soil Biology* 42:S3–S15.

Lavelle, P., E. Velasquez, and M. Andrade. 2009. "Indicating soil quality and the GISQ": reply to the comments by Rossi et al. *Soil Biology and Biochemistry* 41:446–447.

Millennium Ecosystem Assessment (MA). 2005. *Ecosystems and Human Well-Being: Synthesis.* Island Press, Washington, DC.

Moreira, F., E. Huising, and D. E. Bignell. 2008. *A Handbook of Tropical Soil Biology: Sampling and Characterization of Below-Ground Biodiversity.* Earthscan, London.

Nielsen, U. N., E. Ayres, D. H. Wall, and R. D. Bardgett. 2010. Soil biodiversity and carbon cycling: a synthesis of studies examining diversity-function relationships. *European Journal of Soil Science* 62:105–116.

Nyczepir, A., and S. Thomas. 2009. Current and future management strategies in intensive crop production systems. In *Root-Knot Nematodes,* ed. R. Perry, M. Moens, and J. Starr, 412–435. CAB International, Cambridge, MA.

Ostle, N. J., P. Smith, R. Fisher, F. I. Woodward, J. B. Fisher, J. U. Smith, D. Galbraith, P. Levy, P. Meir, N. P. McNamara, and R. D. Bardgett. 2009. Integrating plant-soil interactions into global carbon cycle models. *Journal of Ecology* 97:851–863.

Paustian, K., J. Brenner, M. Easter, K. Killian, S. Ogle, C. Olson, J. Schuler, R. Vining, and S. Williams. 2009. Counting carbon on the farm: reaping the benefits of carbon offset programs. *Journal of Soil and Water Conservation* 64:36–40.

Perrings, C., A. Duraiappah, A. Larigauderie, and H. Mooney. 2011. The biodiversity and ecosystem services science-policy interface. *Science* 331:1139–1140.

President's Council of Advisors on Science and Technology (PCAST). 2011. *Sustaining Environmental Capital: Protecting Society and the Economy*. Executive Office of the President, Washington, DC.

Pritchard, S. 2011. Soil organisms and global climate change. *Plant Pathology* 60:82–99.

Raudsepp-Hearne, C., G. Peterson, M. Tengo, E. Bennett, T. Holland, K. Benessaiah, G. MacDonald, and L. Pfeifer. 2010. Untangling the environmentalist's paradox: why is human well-being increasing as ecosystem services degrade? *Bioscience* 60:576–589.

Riga, E., C. Hooper, and J. Potter. 2005. In vitro effect of marigold seed exudates on plant parasitic nematodes. *Phytoprotection* 86:31–35.

Sachs, J., R. Remans, S. Smukler, L. Winowiecki, S. Andelman, K. Kassman, D. Castle, R. DeFries, G. Denning, J. Fanzo, L. Jackson, R. Leemans, J. Lehmann, S. Naeem, G. Nziguheba, C. Palm, P. Pingali, J. Reganold, D. Richter, S. Scherr, J. Sircely, C. Sullivan, T. Tomich, and P. Sanchez. 2010. Monitoring the world's agriculture. *Nature* 466:558–560.

Sanchez-Moreno, S., E. Alonso-Prados, J. Alonso-Prados, and J. Garcia-Baudin. 2009. Multivariate analysis of toxicological and environmental properties of soil nematicides. *Pest Management Science* 65:82–92.

Sanchez-Moreno, S., L. Jimenez, J. Alonso-Prados, and J. Garcia-Baudin. 2010. Nematodes as indicators of fumigant effects on soil food webs in strawberry crops in southern Spain. *Ecological Indicators* 10:148–156.

Schneider, S., E. Rosskopf, J. Leesch, D. Chellemi, C. Bull, and M. Mazzola. 2003. Research on alternatives to methyl bromide: pre-plant and post-harvest. *Pest Management Science* 59:814–826.

Seastedt, T., and P. Murray. 2008. Root herbivory in grassland ecosystems. In *Root Feeders: An Ecosystem Perspective*, eds. S. Johnson and P. Murray, 54–67. CAB International, Wallingford, UK.

Singh, B., R. Bardgett, P. Smith, and D. Reay. 2010. Microorganisms and climate change: terrestrial feedbacks and mitigation options. *Nature Reviews: Microbiology* 8:779–790.

Swift, M., A. Izac, and M. van Noordwijk. 2004. Biodiversity and ecosystem services in agricultural landscapes—are we asking the right questions? *Agriculture, Ecosystems and Environment* 104:113–134.

Sylvain, Z., and D. H. Wall. 2011. Linking soil biodiversity and vegetation: implications for a changing planet. *American Journal of Botany* 98:517–527.

ten Brink, P., ed. 2011. *The Economics of Ecosystems and Biodiversity (TEEB) in National and International Policy Making*. Earthscan Press, London.

Tian, G., B. Kang, and L. Brussaard. 1992. Biological effects of plant residues with contrasting chemical-compositions under humid tropical conditions—decomposition and nutrient release. *Soil Biology and Biochemistry* 24:1051–1060.

U.K. National Ecosystem Assessment. 2011. *The UK National Ecosystem Assessment Technical Report*. UNEP-WCMC, Cambridge.

United Nations Convention to Combat Desertification (UNCCD). 2011. *Measuring the Value of Land. The Economics of Desertification, Land Degradation and Drought*. UNCCD, Bonn, Germany.

van der Heijden, M., and T. Horton. 2009. Socialism in soil? The importance of mycorrhizal fungal networks for facilitation in natural ecosystems. *Journal of Ecology* 97:1139–1150.

van der Heijden, M. G. A., J. N. Klironomos, M. Ursic, P. Moutoglis, R. Streitwolf-Engel, T. Boller, A. Wiemken, and I. R. Sanders. 1998. Mycorrhizal fungal diversity determines plant biodiversity, ecosystem variability and productivity. *Nature* 396:69–72.

van der Putten, W. H., J. M. Anderson, R. D. Bardgett, V. Behan-Pelletier, D. Bignell, G. S. Brown, V. K. Brown, L. Brussaard, H. W. Hunt, P. Ineson, T. H. Jones, P. Lavelle, E. A. Paul, M. St. John, D. A. Wardle, T. Wojtowicz, and D. H. Wall. 2004. The sustainable delivery of goods and services provided by soil biota. In *Sustaining Biodiversity and Ecosystem Services in Soil and Sediments*, ed. D. H. Wall, 15–44. Island Press, Washington, DC.

van der Putten, W. H., R. Cook, S. Costa, K. Davies, M. Fargette, H. Freitas, W. Hol, B. Kerry, N. Maher, T. Mateille, M. Moens, E. de la Pena, A. Piskiewicz, A. Raeymaekers, S. Rodriguez-Echeverria, and A. van der Wurff. 2006. Nematode interactions in nature: models for sustainable control of nematode pests of crop plants? *Advances in Agronomy* 89:227–260.

van der Putten, W. H., M. Mace, and M. Visser. 2010. Predicting species distribution and abundance responses to climate change: why it is essential to include biotic interactions across trophic levels. *Philosophical Transactions of the Royal Society* B 365:2025–2034.

Velasquez, E., P. Lavelle, and M. Andrade. 2007. GISQ, a multifunctional indicator of soil quality. *Soil Biology and Biochemistry* 39:3066–3080.

Wall, D. H., ed. 2004. *Sustaining Biodiversity and Ecosystem Services in Soils and Sediments*. Island Press, Washington, DC.

Wall, D. H., R. D. Bardgett, and E. F. Kelly. 2010. Biodiversity in the dark. *Nature Geoscience* 3:297–298.

Wall, D. H., M. Bradford, M. St. John, J. Trofymow, V. Behan-Pelletier, D. Bignell, J. Dangerfield, W. Parton, J. Rusek, W. Voigt, V. Wolters, H. Zadeh Gardel, F. Ayuke, R. Bashford, O. Beljakova, P. Bohlen, A. Brauman, S. Flemming, J. Henschel, D. Johnson, T. Jones, M. Kovarova, J. Kranabetter, L. Kutny, K. Lin, M. Maryati, D. Masse, A. Pokarzhevskii, H. Rahman, M. Sabara, J. Salamon, M. Swift, A. Varela, H. Vasconcelos, D. White, and X. Zou. 2008. Global decomposition experiment shows soil animal impacts on decomposition are climate-dependent. *Global Change Biology* 14:2661–2677.

Wall, D., R. Rabbinge, G. Gallopin, K. Khoday, N. Lewis, J. Lubchenco, J. Melillo, G. Schmidt-Traub, M. Sombilla, and L. Cimarrusti. 2005. Implications for achieving the millenium development goals. In *Millenium Ecosystem Assessment. Policy Responses: Findings of the Responses Working Group*, eds. K. Chopra, R. Leemans, P. Kumar, and H. Simons, 549–584. Island Press, Washington, DC.

Wardle, D. A., R. D. Bardgett, J. N. Klironomos, H. Setälä, W. H. van der Putten, and D. H. Wall. 2004. Ecological linkages between aboveground and belowground biota. *Science* 304:1629–1633.

Wardle, D. A., V. K. Brown, V. Behan-Pelletier, M. S. John, T. Wojtowicz, R. D. Bardgett, G. G. Brown, P. Ineson, P. Lavelle, W. H. van der Putten, J. M. Anderson, L. Brussaard, H. W. Hunt, E. A. Paul, and D. H. Wall. 2004. Vulnerability to global change of ecosystem goods and services driven by soil biota. In *Sustaining Biodiversity and Ecosystem Services in Soils and Sediments*, ed. D. H. Wall, 101–135. Island Press, Washington, DC.

Weber, J. 2007. *Accounting for Soil in the SEEA*. European Environment Agency, Rome.

Woodward, F. I., R. D. Bardgett, J. A. Raven, and A. M. Hetherington. 2009. Biological approaches to global environment change mitigation and remediation. *Current Biology* 19:R615–R623.

Wu, T., E. Ayres, R. D. Bardgett, D. H. Wall, and J. R. Garey. 2011. Molecular study of worldwide distribution and diversity of soil animals. *Proceedings of the National Academy of Sciences of the United States of America* 108:17720–17725 doi:10.1073/pnas.1103824108.

Zhang, W., T. Ricketts, C. Kremen, K. Carney, and S. Swinton. 2007. Ecosystem services and dis-services to agriculture. *Ecological Economics* 64:253–260.

Zimmerer, K. 2010. Biological diversity in agriculture and global change. *Annual Review of Environment and Resources* 35:137–166.

Index

B

C

T

U

V

W

X

Y